Studies in Logic
Volume 28

Passed over in Silence
On Wittgenstein's *Tractatus*
and its System

Studies in Logic Series Editor
Dov Gabbay   dov.gabbay@kcl.ac.uk

Passed over in Silence

On Wittgenstein's *Tractatus* and its System

Jaap van der Does

© Individual author and College Publications 2011.
All rights reserved.

ISBN 978-1-84890-008-0

College Publications
Scientific Director: Dov Gabbay
Managing Director. Jane Spurr
Department of Computer Science
King's College London, Strand, London WC2R 2LS, UK

http://www.collegepublications.co.uk

Illustration for cover by Ellis van der Does
Cover produced by Laraine Welch
Printed by Lightning Source, Milton Keynes, UK

I like to have time for the things I do. I think that we're rushing too much nowadays. That's why people are nervous and unhappy – with their lives and with themselves. How can you do anything perfect under such conditions? Perfection takes time.

Marilyn Monroe

Ring the bells that still can ring
Forget your perfect offering
There is a crack in everything
That's how the light gets in.

Leonard Cohen

For Marjon, Ellis, Thomas to whom I belong in this or any other order...

Contents

Introduction

In philosophy, seen from a distance, there are authors and analysts. An author holds that true insight requires the beauty of everyday language. By contrast an analysts is quite indifferent to the force of a well-chosen phrase. In his opinion understanding only results from charting the logical structure of a problem in minute detail. Authors think such technicalities distracting and irrelevant. On this view: Sartre is an author, Carnap an analyst.

It is comforting at times to invent labels, but 'labels are for the things men make, not for men'. So if you label anyway, 'don't glue them on, and have replacements handy'. (Rex Stout, motto of Themerson (1974).) Early Wittgenstein, in particular, was an author who started his philosophical activity with a passion of the analyst: the nature of propositions and logic.

Due to the unusual combination of art and analysis, Wittgenstein's *Tractatus Logico-Philosophicus* was likely to attract a wide variety of readers. Those labeled at one extreme take the text as a logical poem with some Russellian formulas interspersed as ready-mades. Such readers leave the suggestion as is that major parts of the book are about a logical system, based on genuine logical insights, and feel justified in doing so due to the author's remarks that in the end all is nonsense. The readers labeled at the other extreme are impressed by the text's stark formulation, but still think there is no interest in contemplating the booklet for long. Some of its basic ideas on logic and semantics are also found by others, such as truth tables by Frege, Pierce, or Post, who present them in a less vague, more profound way. And obviously the sum of literature and logic is nonsense.[1]

I find myself keeping maximal distance from the readers as labeled at both extremes. There is no royal road to nonsense of the tractarian variety, because it is based on the logical insight that the system of descriptive language is not reflexive. This insight can only be grasped after a full study of the tractarian system and how it relates to Russell's paradox. Indeed, I hold that despite its artistic, somewhat paradoxical presentation, the system allows for a detailed interpretation, and that such an interpretation is helpful for readers with less firm logical intuitions as young Ludwig happened to have. Spelling out the system's particulars is gratifying as it helps making sense

---

[1] As to the one extreme, *The Times* obituary of Wittgenstein described the *Tractatus* as a logical poem (Edmonds and Eidinow, 2001, 228). As to the other extreme, quite a few logicians and mathematicians would be examples; e.g. Menger (1994), 89.

of statements that at first were fascinating but hard to follow. From a social point of view, the task may well be ungrateful – the result is likely to be too philosophical for the logician, and too logical for the philosopher, – but I cannot resist the temptation of trying to understand the *Tractatus* in this way.

Aim

This book sets out to show that in spite of its condensed literary presentation the *Tractatus* has a coherent reading, both philosophically and technically. It takes the *Tractatus* as an ethical deed, and a primary aim of the book is to show how Wittgenstein's ethics is related to his highly original 'symbolic turn'.[2] It is without doubt that human life poses a problem, and that a way of living has to be found to resolve it. The *Tractatus* strongly suggests that insight into the world's contingency via its close ties with meaningful language would pave the way toward the proper way of living.

In line with the *Tractatus*, the book mainly offers a detailed overview of the symbolic turn. The symbolic nature of contingent propositions in logical space is charted in detail, while logical propositions are characterized as empty forms about nothing. To this end, logical space, object, object-form, identity of object-form, state of things, picture, projection, proposition, sign, symbol, situation, sense, truth, logical consequences and their formal relationships are made as explicit as possible; sometimes even in the form of proofs. Yet, the main purpose is to attain clarity, not so much logical depth.

A prime advantage of the current approach is that it shows early Wittgenstein to have been sufficiently precise concerning his so-called perfect notation; be it, as always, without specifying it in much technical detail. The idea of a perfect notation allows for some alternatives, but minimally it captures the symbol that a proposition expresses, which is the contingent nucleus of its sense disregarding logical parts. In a perfect notation the representation of $(p \wedge q) \vee (p \wedge \neg q)$ is the same as that of p. Thus, a perfect notation ensures that equisignificant propositions have the same form and content. Insofar as the formalities of this core aspect of the *Tractatus* have received attention, it is assumed that Wittgenstein abandoned his quest for perfection because it is logically impossible. See e.g. Potter (2009), section 24.4. By contrast, I think one should distinguish between aiming for a perfect notation of sense, and aiming for a transparent notation to show logic's triviality. The latter is indeed impossible due to the undecidability of the logic. If Wittgenstein had paid sufficient attention to his perfect notation of sense for the infinite case, he would have noticed immediately that logic, although still *about* nothing, is not trivial at all.

Apart from treating of the relationship between ethics and the symbolic

---

[2]In 1967, Paul Engelmann's *Letters from Ludwig Wittgenstein* has published with an editor's appendix of Brian McGuinness (Engelmann, 1967). Engelmann was the architect of Margaret Stonborough's house at the *Kundmangasse* until Wittgenstein took over. He was also one of the first to publish on the ethical impact of the *Tractatus*. McGuinness' appendix has the famous quotation from Wittgenstein's letter to Von Ficker that made Wittgenstein's ethical intent certain; see page 2. Janik and Toulmin (1973) present this view on the *Tractatus* in a wider cultural-historic setting. 'Symbolic turn' is Michael Potter's apt term. See his history of ideas *Wittgenstein's Notes on Logic*.

turn, the current approach also gives insight into how objects compare with typed-entities; into the nature of signs in logical space; into the different ways in which the notion of projection can be interpreted; into the nature of truth-operations and how they compare with truth-functions; into the notion of sense for contingent and logical propositions; into the niceties of logical consequence; into substitution in a representation of sense; into how the tractarian system solves Russell's paradox. Not only the finite system is covered but also a natural infinite generalization that Wittgenstein's text just hints at. It is shown that Wittgenstein's use of truth-table and graphical signs anticipates the elegant tableau methods developed much later. Despite the bad press Wittgenstein's treatment of quantification received round about 1980, it is here argued to be correct.

Philosophical engineering

Understanding Wittgenstein's early philosophy seems to require a minute study of the tractarian system – often called 'the system' here, – but as far as I know a precise model that does justice to the entire text has never been defined. Why? Wittgenstein presents the system as a philosopher and writer, who would have been distracted by too much attention for detail. Primarily concerned with general philosophical ideas, he sketches the outlines of the system in a lucid style as its features come to his attention. One would presume that this gives enough information to enable philosophical engineers to fill out the details, in the non-literary, dry fashion, but few felt the urge to take up this rôle. (When working on his *hausgewordene Logik* – logic turned into a house, a term from the *Familienerinnerungen* of his sister Hermine Wittgenstein, – Wittgenstein assumed a similar rôle himself. Cf. McGuinness (2001), 21.) Roundabout 1930 Waismann seemed to have suggested writing an introductory book on the *Tractatus* enriched with some newer insights, but by then Wittgenstein saw too many shortcomings in his system to find it useful. See part IV of McGuinness' introduction to Wittgenstein (1984).

In this book I will mainly act as a philosophical engineer, working on the system's design as it is specified in the text. I will try to keep the philosophical debate concise, and only engage in it if I have to argue for a particular reading. The secondary literature of the *Tractatus* has grown into a *mer à boire*, which offers some beautiful vistas, but at the same time evaporates any ambition to concern oneself with more than a specific aspect of the philosophy and its the system. By contrast, to detail the system almost no thesis can be left out of consideration. To keep this manageable I should try to find an optimum between being clear and being brief.

An ethical deed

Philosophical engineering may leave the impression that the *Tractatus* is chiefly concerned with logico-linguistics, like on its reception by the logical positivists in the early twentieth century. To prevent this from happening, I stress its ethical concern from the outset. In a way this book is a technical sequel to Martin Stokhof's 'World and Life as One' (Stokhof (2002)). Even when his book was still in manuscript, I was particularly impressed by his view on the relation between ethics and ontology in Wittgenstein's early thought. Still, I felt that a more detailed presentation of the system would be helpful.

Advantages

Formalizing the system is not an aim in itself but has a few advantages.

An improved insight into the system of the *Tractatus* will lead to a better understanding of how ethics and the symbolic turn are related to each other. In this regard, a single, coherent model of the system amounts to giving a strong interpretation of the text that can be criticized more effectively.

A rather formal interpretation of the system allows one to gauge accurately how Wittgenstein's approach to logic and sense compares with and deviates from that of Frege and Russell. It makes clear how innovative Wittgenstein's ideas on the perfect notation are, in which the ideal representations of sense are logical structures that mirror possibilities in an intensional domain. It is also a little therapeutic to arrive at such insights: one has to unlearn modern logic to a certain extent to see the specifics of Wittgenstein's approach.

Studying the tractarian system makes one fully aware that logic and semantics are not philosophically neutral: a proper understanding of their basic concepts is at the heart of any philosophy of logic and language. E.g., there are crucial differences between Frege's and Russell's 'platonic' views on the one hand – with a third realm providing the objects and structure that logic is about, – and early Wittgenstein's 'aristotelean' view on the other – where language is a worldly matter concerning the structure and content of reality, pictured within logic's empty frame.

Insight in the early system gives a strong foothold when studying Wittgenstein's critique in his later works, namely: that not all elementary propositions are independent of each other and so that not all logic is truth-functional; that quantifiers should not be reduced to truth-functions; that the idea of absolutely simple names is based on confusing the meaning of a name with its bearer; that there are no absolute notions of simplicity and complexity that analysis should unveil; that language lacks a unique system of representation capturing the essentials of sense. A perspicuous view on the early system may allow one to trace what can be retained as one of the more modern outskirts of language, which should no longer be mistaken dogmatically for *the* system showing its semantic essence.

A full interpretation of the system should contribute to improving the quality of philosophical discussion. It naturally leads to the idea of ostensive philosophy, which allows for a strong alternative to so-called resolute readings of the *Tractatus*.[3]

---

[3]In this book I have little to say about resolute readings. In a way resolute readings of the *Tractatus* are prime examples of weak interpretation. They postulate a few theses to be 'frame remarks', which are the only theses that should be taken seriously. The frame remarks are then used to argue that the *Tractatus* and its early sources offer a deflationary philosophy of mere nonsense. To come to such a reading much evidence to the contrary must be resolutely ignored. See Hacker (2000) and Proops (2001b) for details.

Cloud and drop

For a text that aims to concentrate on the system, some paragraphs may still appear rather verbose. But besides presenting a more formal framework, I want to make transparent how the details given relate to the text. Also, there are passages where the brevity of the *Tractatus* requires me to argue for a particular reading on which the formalization is based. Freely paraphrasing later Wittgenstein: I cannot avoid showing how some philosophical or philological clouds are condensed into a drop of logic. That I present both cloud and drop is based on a principle of ethics: help your opponents to find a proper attack.

Attempting the impossible?

Some readers will observe that every now and then I attempt to say what according to early Wittgenstein cannot be said. This observation is not always correct. For instance, specifying a formal concept as an abstraction over logical space, e.g., the form of an object or the identity of such forms, is part of the system (4.122). Yet there are other areas where the observation must be granted. But here I find myself in good company, for of course early Wittgenstein accused himself of doing exactly the same.

If required, there are two ways to argue in favor of my approach. Firstly, before criticizing a philosophy one should make a serious attempt to grasp it. Without literary aspiration, my reconstruction of the tractarian system mainly uses the methods of the analyst. This approach may help some to get a better view on early Wittgenstein's ostensive philosophy – to show the limits of what can be said, – than his more artistic presentation. Secondly, when studying a system the methods used must leave it intact, but they do not have to comply with all of its principles. Whenever required I do step back and employ methods that from a tractarian perspective are way out of line. E.g. they use non-tractarian notions, or compare the system with other systems. Without being too finicky about it, I have labeled such parts as 'reflections'.

Overview

The structure of this book is concentric. Chapter 1 starts with a general overview, which indicates how the *Tractatus* brings us from the world as the basis for logic and language to the world as the totality with which one has to reconcile in order to live the good life. The chapter indicates how the analysis of descriptive language makes the general form of propositions manifest, and with that ontology. It also introduces basic themes, like ostensive philosophy and descriptive essentialism, which are developed step by step in the chapters to follow.

In chapter 2, it is argued that Wittgenstein held typing to be unnecessary, and so not given prior to analysis. The insight is developed into a holism of logical space that can only be captured semi-formally. The fine-structure of the system is given, introducing basic notions such as state of things, object, and identity of object-form in a detailed way.

Chapter 3 is about projection, which is crucial to understand propositions as logical pictures (models). After considering the pros and cons of a few local variants we arrive at a holistic notion of projection that does justice to the intensional nature of objects. Projections are allowed to vary in logical use, but per use they partition logical space in a pictorial and a non-pictorial area.

With logical space and projection in place, chapter 4 develops the notion of elementary proposition: its sign, sense and truth. Once these atoms of meaning are understood, proper expressions and material functions are introduced as abstractions over them.

The next four chapters are about propositions of finite logical complexity.

Chapter 5 begins with the tractarian view on truth-functions, and goes on to compare its three notations: truth-table signs, graphical signs and truth-operations. It is shown that truth-operations are truth-functionally complete.

Chapter 6 is about the ontological status of logical complex signs. It is a basic thought of the *Tractatus* that logical constants do not refer, but there is a clear tension between this thought and the idea that propositions are facts, and so part of ontology. To reconcile both aspects of the system, I will hold that only propositional signs show complex logical structure overtly, and that this structure is essentially a matter of form. The non-pictorial part of ontology consist of independent states of things, which allow the sense of a proposition to project logically incompatible possibilities onto them.

Chapter 7 concerns the sense of logically complex propositions against the backdrop of Frege's philosophy of language. Situations are introduced as the intensional interpretation of truth-table signs. Next it is observed that to get proper notions of contingent and logical proposition, a clear distinction must be made between the propositional sign and its symbol. By contrast to elementary sense, logically complex sense is shown to be atomistic and compositional.

The last step in developing the finite notion of proposition consists of clarifying truth. Chapter 8 combines the study of truth with that of logical consequence. This results in a formal characterization of logical propositions; i.e., a reflection on the soundness and completeness of the finite system, which in line with the philosophy is presented in a non-axiomatic way.

With the finite notion of proposition in place, chapter 9 shows how Wittgenstein has realized a perfect notation of sense. This notation vindicates speaking about *the* symbol of a proposition. It requires to discuss how the representation of sense fares with regard to different forms of substitution. The ideas are summarized in a reflection on symbol and Lindenbaum algebra's.

One way of looking at the tractarian system is that it gives Wittgenstein's solution to Russell's paradox. Chapter 10 recalls how Russell found his paradox and how he went about trying to solve it. Against this background, I discuss what Wittgenstein solution to Russell's paradox consists of. As a result, we see the system is non-reflexive; it is unable to describe its own sense conditions. Also, modern techniques to achieve reflexivity, like coding, are argued to be unavailable. Instead, Wittgenstein takes resort in an ostensive philosophy, which highlights the main features of the system to an understanding reader. The main philosophical activities are charted in a 'choreography for a swansong'.

Chapter 11 starts probing to what extent the system can be generalized into the infinite. This aspect of the system was left until after the introduction of its finite part, because trying to incorporate infinity from the start would introduce a chasm between the text and my interpretation of it. First, the reasons why the system should be infinite are given. Next, an analogy between truth-table signs and systematic analytical tableaux is used to suggest a countable system of representation. The chapter ends with a discussion on decidability, perspicuity and the independent nature of logic, to see whether the metaphysical and the human aspects of an infinitary system can be kept in balance.

Chapter 12 concerns the technical details of the infinitary system. After presenting the basic notion of a systematic sign, it is shown to what degree the infinite notation can still be regarded perfect. Next, the sense of a proposition is proved to be determinate, independent of knowing its truth value. A notion of truth is given that allows us to characterize the logical propositions from among the contingent ones. The chapter finishes with reflections on logical consequence, descriptive completeness, compactness and interpolation.

Chapter 13 is about the somewhat thorny issue of tractarian quantification. Quantifiers are abbreviations of truth-functional structure that should be considered in the context of analysis. I argue that in this context Wittgenstein's approach is correct. When considered in isolation the representations used may be ambiguous, but even then there are 'remedies' that hardly extend the system. The chapter concludes with showing how Wittgenstein's injective treatment to names and quantifiers can be had via a simple extension of the infinitary system presented in chapter 12.

Chapter 14 recapitulates the comparisons between the work of Frege, Russell, and Wittgenstein that were interspersed in the previous chapters. It concludes with giving an overview of the system as it has unfolded pleat after pleat in the remainder of the book.

Most chapters have 'interludes'. They offer historical or other points of detail without interrupting the flow of interpretation too much. Finally, there are four indices: a general one, one for persons, one for theses, and one for notation introduced.

Favorable heuristics

In hindsight it is not so hard to see that quite a few tractarian claims should be toned down. To begin with, its universality cannot be sustained. If there were a logical essence of everyday language, it would require a much richer system than an infinitary truth-functional logic based on logically independent elementary propositions. Elementary propositions can be logically dependent, as the color-exclusion problem strongly suggests, and more seems to be required than a system with truth-functional signs interpreted in an intensional space of states of things to handle opaque contexts such as 'I know that...', 'I believe that...', etc. It is also unclear whether the system is rich enough to capture all forms of extensional quantification. For example, non-first-order definable quantifiers such as 'most' seem problematic (as in: 'most books on the *Tractatus* are hard to read', meaning: there are more books on the *Tractatus* that are hard to read than there are books on the *Tractatus* that are not hard to read). But of course I will not argue that the tractarian philosophy can be upheld in all detail. My aim is rather to show that despite its limitations, the system that the book presents has a strong, coherent interpretation. Such an interpretation should highlight what may still be of value today; e.g. it's highly original route toward ethics, or its philosophy of logic.

Early Wittgenstein was a philosopher who combined valuable logical intuitions with a disinterest in technical detail. For some the lack of logical detail may be sufficient to disregard the logical value of the *Tractatus* entirely. In the spirit of fairly recent studies – e.g., Frascolla (1994), Marion (1998), Von Kibéd (2001), Stokhof (2002), and Potter (2009) – I prefer a favorable heuristics. Instead of observing that Wittgenstein's technical remarks do not fit the mold of modern logic, it is more rewarding to ask: Is the system coherent at all, and does it give insights that have been lost in current logics? Apart from the perfect notation of sense, Wittgenstein's now obscure techniques contain gems that do not seem to have been noted before. E.g., the combination of truth-tables and truth-operations are in the spirit of tableaux; cf. Beth (1955), Smullyan (1968).

Context

To the best of my knowledge, there is no formal approach to the system in the *Tractatus* that tries to do justice to the full text. Stegmüller (1966) is an interesting model-theoretic supplement to some of the ideas in Stenius (1960). Hintikka (1986) argues that the system is basically the same as Tarksi's semantics of predicate logic. Although it would surely be interesting to compare the two approaches, to do so requires developing the tractarian system in its own right. This will show that besides similarities, which depend on a quite specific interpretation, several differences between them remain; e.g., the holistic nature of elementary propositions and material functions; the use of truth-operations transforming signs; the treatment of sign and depicted as part of one logical space; the intensional interpretation of truth-table signs; the rôle of sense in the hyper-intensional system; the unique approach to quantification based on an uncommon notion of variable... Lokhorst (1988) is a first attempt of full coverage, but his axiomatic formalization makes heavy use of modern techniques, is hence only partly based on the text, and does not treat such key notions as operation, variable, form, or the distinction between saying and showing. Landini (2007) covers quite a few aspects of the system, but his aim is rather to rebalance the appreciation of early Wittgenstein philosophy in favor of the philosophies of Russell that had a strong influence on the *Tractatus*. Notions like operations and variables have been studied, in Sundholm (1992), Frascolla (1994), Marion (1998), Ule (2001), Potter (2000), and Von Kibéd (2001), but mostly in an informal manner or without covering other aspects of the system.

Sources

I much enjoyed using the *Kritische Edition* of McGuinness and Schulte

(Wittgenstein, 1922a). More in general, McGuinness' and Schulte's contributions to making sources and historic details available is of inestimable value.

Since I felt that the elegant translation of Pears and McGuinness (Wittgenstein, 1922b) does not always capture the sterner beauty of Wittgenstein's writing, the English translation is often my own. I did use the translations of Ramsey and Ogden and of Pears and McGuinness to check my translation for correctness. The title of the present book is from the translation of Pears and McGuinness.

I have developed the website www.tractatus.nl to facilitate research of the *Tractatus*. Based on modern web-techniques, the site enables one to view the text from different perspectives and to search it using regular expressions.

Manuscripts and typescripts are from Wittgenstein's *Nachlass* (Wittgenstein, 2000).

As to the secondary literature, apart from the now standard introductions Anscombe (1959), Stenius (1960) and Black (1964), my main influences are: Stokhof (2002), Hacker (1984), McGuinness (2001), Ishiguro (1969, 1981), Pears (1987), Von Kibéd (2001), Marion (1998), Janik and Toulmin (1973), Frascolla (1994), and Potter (2009). One may safely assume they do not agree with all that is presented here.

Acknowledgements

Round about 1975, the author W.F. Hermans brought the *Tractatus* to my attention. The first translation into Dutch is his. Although I no longer share his interpretation, I am thankful for his influence.

I am very grateful to my former teacher and colleague Martin Stokhof. Despite the demanding context of a Dutch university, he gave me some of his free time to discuss the topics addressed here. He also commented in detail on drafts of the manuscript. This was very helpful and a great delight. Without doubt it has improved the quality of the book.

I have also enjoyed and learned much from discussions with Michiel van Lambalgen, Göran Sundholm, Rob van der Sandt, Frank Veltman, Albert Visser, and Tine Wilde. Presenting my ideas to audiences at seminars in Amsterdam, Leyden and Nijmegen helped me to improve them. Brian McGuinness suggested a few amendments that I happily adopted. I enjoyed talking Wittgenstein, among other things, with Joachim Schulte while walking along the Züricher See on a sunny afternoon. It was a pleasure, finally, to discuss some aspects of my interpretation with P.M.S. Hacker in his beautiful room at St. John's College, Oxford.

The book is dedicated to my wife Marjon and our children Ellis and Thomas. Without their loving presence and help I would not have found the patience to finish this ambitious project.

My research is independent and financed entirely by myself. It is my hope that these lecture notes for imaginary students will also be of interest to some real ones.

1

World, life, language

Es handelt sich, ganz eigentlich, um die Darstellung eines Systems. Und zwar ist die Darstellung *äußerst* gedrängt, da ich nur das darin festgehalt habe, was mir – und wie es mir – wirklich eingefallen ist. (...) Die Arbeit ist streng Philosophisch und zugleich literarisch, es wird aber doch nicht darin geschwefelt.[4]

Briefe an Ludwig von Ficker, undated

In the last quarter of 1919 Wittgenstein wrote to Ludwig von Ficker, publisher of the journal *Der Brenner*, that he had written a philosophical work, and asked him to consider publishing it. To forestall Von Ficker expecting the usual philosophical tome – thick, erudite and vague – he gave an indication of its nature. It was a sixty-page brochure about philosophy, which presented a system in literary fashion. The key to the work was that it demarcated ethics from within, by concentrating on the system but remaining near silent on the actual topic. Only in this way, Wittgenstein was convinced, could ethics be safeguarded against being emptied: "...es wird aber doch nicht darin geschwefelt".

The *Tractatus Logico Philosophicus* treats of ethics by presenting a logical system of representation in a literary fashion. Maybe there is no combination more fascinating than this one, but perhaps there is also none that is more likely to be misunderstood. It was for good reasons that Wittgenstein expected his work to be accessible only to those who had had thoughts similar to his. The numbered theses of the *Tractatus* are much like markings along an unorthodox route, which force the reader to find his way from one to the other. Although more technical in appearance, the *Tractatus* is as much an album as the *Philosopische Untersuchungen* is.[5]

On reading the *Tractatus* for the first time, as I did when I was about

---

[4] 'It is quite strictly speaking the presentation of a system. And this presentation is *extremely* compressed since I have only retained in it that which really occurred to me – and how it occurred to me. (...) The work is strictly philosophical and, at the same time, literary, but there is no babbling in it.' Translation in *Letters to Ludwig von Ficker*, translated by B. Gillete, edited by A. Janik (Luckhardt, 1979, 92-94).

[5] At the conference 'Reading Wittgenstein', London 2007, Wolfgang Kienzler pointed out that the *Tractatus* has many features of a partially rendered dialogue, i.e., a form which is typical of Wittgenstein's later work.

eighteen years old, a main problem is that an overarching view is missing that gives all its aspects a proper place. Only after careful re-reading, extended with reading earlier sources and commentaries, I have come to see the book as inviting an ethically motivated, active reflection that brings one from everyday language, any language, to logic and ontology. Logic and ontology, in turn, are understood as a system showing how descriptive language has meaning, and with that how ethics is freed from descriptive, theoretical leanings.

This chapter aims to give a general overview of the *Tractatus*. It indicates how the text brings us from the world as the basis for logic and language to the world as that with which one has to reconcile in order to live the good life. The understanding reader, who is addressed in the preface to the book, sees from the beginning that the world can be viewed from both angles. Still, there is no easy and direct path to relating to the world in this way. The true nature of the world can only be uncovered via the philosophical method of analysis. Especially the analysis of descriptive language unveils the world as a totality of contingencies in a frame of necessities. It indicates what is essential to all descriptive symbolization, and how description relates to logic, language, and ethics. As long as philosophy is seen as descriptive theorizing the whole enterprise has the ring of paradox about it. This tension dissolves to a large extent if philosophy is taken as an 'ostensive' activity in which the system of representation shows what is essential and what is not.

1.1   Preface

According to the preface of the *Tractatus*, the point of the book is to show that what can be said at all can be said clearly, and what cannot be talked about must be passed over in silence. It aims to draw a limit to the expression of thought in language. Beyond the limit there is nothing but silence. Silence, however, has both negative and positive connotations. In silence there is no description, no theorizing, but in silence one may have a proper view on the transcendental aspects of the world, i.e. logic and ethics. Wittgenstein's correspondence with Von Ficker makes clear that the limits to the expression of thought are indeed made manifest to safeguard ethics from 'gassing'.

> The book's point is an ethical one. I once meant to include in the preface a sentence which is not in fact there now but which I will write out for you here, because it will perhaps be a key to the work for you. What I meant to write, then, was this: My work consists of two parts: the one presented here plus all that I have *not* written. And it is precisely this second part that is the important one. My book draws limits to the sphere of the ethical from the inside as it were, and I am convinced that this is the ONLY *rigorous* way of drawing those limits. In short, I believe that where *many* others today are just *gassing*, I have managed in my book to put everything firmly into place by being silent about it.[6]

To indicate the limits of the expression of thought clearly has the ring of paradox. But paradox mainly arises if the text would be intended as an explanation, and so would involve 'thoughts' and 'statements' that the tractarian philosophy indicates to be meaningless. The remarkable opening of the preface

---

[6] From McGuinness' Editor's Appendix, Engelmann (1967), 143.

discourages one to read the *Tractatus* in such a single minded way.

> This book will perhaps only be understood by someone who has already had the thoughts expressed in it himself – or at least similar thoughts.

Since nobody knows beforehand which thoughts the book will present, no reader will take for granted to qualify as being of the appropriate kind. Being challenged so early on, one may even feel put off or offended. The point of the preface is however to signal the reader not to expect a discursive treatise that will explain new scientific content. Instead, Wittgenstein indicates that he is about to show what according to his philosophy of language is essentially ineffable; namely, how the expression of thought is set a limit. Due to ineffability Wittgenstein takes resort to an ostensive philosophy – i.e., a series of elucidating acts rather than of lecture notes. Such elucidations require a setting in which the reader already has had the insights ('thoughts') to be pointed out, or at least insights that are akin. Once a reader is brought to recognize that the philosophy is indeed hinting at what can only make itself manifest, he may be left with the artistic pleasure of seeing it done so well.

1.2 World

The *Tractatus* has not one but two remarkable beginnings. Besides the first sentence of its preface, there is the opening of the sequence of theses.

1. The world is all that is the case.

Even a reader who has accepted the appropriate relationship between himself and the text to follow, may be struck by this powerful opening. It is as if a philosopher recalls the essentials of a genesis, but in his own way to put basic notions right. A reader of the kind addressed in the preface will sense that with this thesis the unfolding of the entire philosophy is started. In this philosophy the world is 'the world I find myself in' (5.631); it is the world that poses an ethical dilemma, the problem of life that needs to be dissolved (6.521). Like Wittgenstein, he will have considered a way out of this ethical dilemma, and will hold that intricate verbalization offers little hope in this respect. Instead a symbolic turn is needed that shows the symbolizing power of language to have its limit, and that this is what the book sets out to show.

The abstract presentation of the world with which the book begins, acts as a bridge toward the later theses on the nature of propositions and logic, and finally on ethics and mysticism. It lays the groundwork for the tractarian system in the light of which the interplay between logic, language and world should be seen. To demarcate ethics from within, any pluriformity in language and world must be understood somehow in terms of the unique system that captures the essentials of sense and truth. Indeed, in this book I will argue that a descriptive essentialism is core to the *Tractatus*, and that this essentialism is unveiled from its beginning: To be is to be essential to description.

1.3 Descriptive language

As I read the *Tractatus*, a considerable part of what is written concerns the questions: How is meaningful language possible, where meaning is prior to the facts? And how does meaningful language relate to logic? Answers to these questions should make clear that ethics is of a different nature than logic or the use of language. Ethics is non-theoretical and concerns acting just *sub specie aeterni*, in reconciliation with the world's contingencies. However, is there any hope of shedding light on ethics in this way? After all, the questions concerning the essentials of language and logic may appear pompous, grandiose, even impossible to answer. For sure there are many forms of language dealing, for example, with facts, beliefs, logic, arithmetic, probability, science, ethics, aesthetics, volition, the mystical..., and it is hard to believe that there is a common core to them all. In the *Tractatus* all these language forms are distinguished from each other and are understood in different ways. But to get a handle on the pluriformity, the text concentrates for a large part on descriptive language, i.e., the language that describes this contingent, ever changing world, my world (4.023).

To take descriptive language as most basic is quite natural. It is the part of language where the alluring idea of words having reference is vital, and where the interaction of language with reality is most promising when trying to get a hold on the notions of sense and truth. Also, descriptive language allows for a notion of proposition with which description and logic can be distinguished from each other without becoming incomparable. It is the part of language where the insights of predecessors such as Frege and Russell can be sharpened and refined.

1.3.1   The clothing of descriptive language

Reflecting on the nature of descriptive language we, humans, are well advised to start with the familiar: the perceptible signs that we utter, write, read, that we use to think. But the human capabilities to immediately grasp the workings of these signs are limited, as the apt imagery of 4.002 makes clear:

4.002 Humans are able to build languages capable of expressing any meaning, without having the slightest idea how each word has meaning or what this meaning is – just as people speak without knowing how the individual sounds are produced.

Everyday language is part of the human organism and is no less complicated than it.

It is not humanly possible to immediately derive from language what its logic is.

Language disguises thought. In fact so that from the outward form of the clothing the form of the clothed thought cannot be derived; since the outward form of clothing is designed for completely different purposes than to reveal the body's form.

The tacit agreements to understand everyday language are enormously complicated.

Early Wittgenstein held descriptive language to be in order as it is, but also that the form and the content of its meaning may be concealed. To see what is possible in this regard and so what the limits of meaning are, the essence of descriptive language should be made manifest.

5.471 The general propositional form is the essence of propositions.

5.4711 To indicate the essence of propositions means to indicate the essence of all description, and thus the essence of the world.

To show the general form of propositions, the perceptible signs of language can not be taken at face value. Wittgenstein advocates what could be called a Christo theory of language; our everyday signs wrap up the structure and content of meaning, if they have any, and in doing so conceal the general form of propositions from direct inspection. It is part of the tractarian reflection to make the general form manifest. Thus it is seen how descriptive language compresses its structure and content into a sign that is suitable for common use, and how analysis unveils its essentials.

1.3.2 The form and content of descriptive language

On a general level, the point of descriptive language is to see that we picture facts to ourselves (2.1). The semantics of descriptive language is pictorial, it is meaningful in the way pictures are. A picture consists of elements that are structured in a certain pictorial form. This form may vary per picture – it could be spatial, temporal, colored,... – and a picture is restricted to depict only those parts of reality that have the same form. But within the diversity there is a uniform essence: besides the restricted kinds of picture, there is a most general kind, the so-called logical pictures. Logical pictures, which are much like logical models, have the form of the world and can picture the world regardless. They are logical constructs of elementary pictures, and any picture – be it spatial, temporal, colored,... – is also a logical picture.

Two kinds of logical pictures can be distinguished: logically elementary and logically complex pictures. At the elementary level the crux of logical picturing consists of structures, i.e. configurations of elements, that are projected onto each other. The ontology of language gives the essentials of structure and so that of projection. According to the *Tractatus*, elementary pictures are independent of each other; an elementary picture may picture either correctly or incorrectly, but this independent of any other elementary picture. The independence is also at the heart of logically complex picturing. Logical complexity is truth-functional. That is, a logically complex sign is a list of elementary pictures together with an indication for each possibility of their correctness or incorrectness whether that possibility would make the complex picture itself true or would make it false. This general form of propositions – that of a truth-table sign, – is uncovered in analysis.

1.3.3 The analysis of descriptive language

Analysis is a philosophical method that starts with a perceptible sign in use. It concerns the general form of meaningful propositions, not so much the analysis of a specific proposition.

3.3421b And this is how things are in philosophy: again and again specific things will turn out to be unimportant, but the possibility of each specific thing gives us insight into the essence of the world.

In particular, analysis should be distinguished from an epistemological method to determine whether a sign is meaningful or not, and if so what its specific meaning is.

To show the general form of meaningful propositions, the first step of analysis consists in giving a sign a precise high-level structure.

3.251 A proposition expresses what it expresses in a determinate way that can be clearly indicated: a proposition is articulate.

Since the words of everyday language may be ambiguous – e.g., one and the same word sign may have different meanings, or different word signs may mean the same (3.323), – the step from a propositional sign to the first high-level analysis could be huge, and it is natural to assume that this first step does not have to be unique. But according to the *Tractatus* such apparent indeterminateness is a signal that certain aspects of the sign's significant use have been ignored (3.262, 3.326, 3.327). When both sign *and* significant use are taken in consideration, the process from the highest level analysis to the more detailed ones results in a unique, completely analyzed form.

3.25 There is one and only one complete analysis of a proposition.

In this way, each meaningful sentence can be shown to have a structure and content that fixes reality to 'yes' or to 'no' (3.31, 4.023a).

The principle that drives analysis is the same as the principle which started it: eliminate the suggestion of indeterminate meaning. If the meaning of a proposition appears to be indeterminate, this indicates that a seemingly simple expression in its sign goes proxy for a complex structure and content. The simple expression must be definiendum in a definition with the complex as definiens, and substituting the complex expression for the simple one will reduce the suggestion of indeterminate meaning (3.24).

3.261a Each defined sign has meaning *via* the signs that define it; and the definitions point the way.

It may appear that in spite of its unique result, the analysis itself could take many different forms, but the *Tractatus* states it must comply with its unique system of pictorial representation.

1.3.4 The fruits of analysis

Sense is determinate, so analysis must come to an end. An analysis is a sequence of signs σ_i and a corresponding sequence of definitions with expressions ε_{i+1} and ε_i, as in table 1. Here, $\varepsilon_1 = \varepsilon_0$ is used to transform $\sigma_0(\varepsilon_0)$ into

$$
\begin{array}{llll}
\sigma_0(\varepsilon_0) & & \varepsilon_1 = \varepsilon_0 & def. \\
\sigma_0(\varepsilon_1) & (\equiv \sigma_1) & & \\
\sigma_1(\varepsilon_2) & & \varepsilon_3 = \varepsilon_2 & def. \\
\sigma_1(\varepsilon_3) & (\equiv \sigma_2) & & \\
\vdots & & \vdots & \\
\sigma_m(\varepsilon_{2m}) & & \varepsilon_{2m+1} = \varepsilon_{2m} & def. \\
\sigma_m(\varepsilon_{2m+1}) & (\equiv \sigma_{m+1}) & &
\end{array}
$$

TABLE 1 Analysis of σ_0 using definitions $\varepsilon_{i+1} = \varepsilon_i$

$\sigma_0(\varepsilon_1)$, which is σ_1. In σ_1, say, the expression ε_2 may be discerned, so that more explicitly one has: $\sigma_1(\varepsilon_2)$. Now $\varepsilon_3 = \varepsilon_2$ yields σ_2, i.e., $\sigma_1(\varepsilon_3)$, and so on.

In all this, a definition $\varepsilon_{i+1} = \varepsilon_i$ should be read from right to left: ε_i can be replaced by ε_{i+1} (4.241). So each step in the sequence should be read as: σ_{i+1} is the result of applying the definition $\varepsilon_{2i+1} = \varepsilon_{2i}$ to a sub-expression ε_{2i} of σ_i. It is allowed that σ_i and ε_{2i} coincide.

A main principle for the analysis of language is: eliminate the suggestion of indeterminacy to show the proposition's unique and determinate sense (3.251, 4.023). The application of a definition, in particular, must show more clearly the determinate sense of a proposition (3.24). The analysis stops if the notation is perfect and maximal clarity is achieved. So each σ_i in the above example must suggest an indeterminateness of meaning that is reduced in σ_{i+1}; otherwise, applying the definition does not make sense. When made fully explicit – similar to e.g. the theory of definitions in Frege (1903a), III 1.a, – this principle will ensure that sequences of definitions resulting in circularity are precluded.

In analysis, there are logical and non-logical stages. The stages may be intertwined, but the first stage that should come to an end is the logical one, which lays bare the truth-functional structure of a proposition (5). At this point, the content of the proposition may not be analyzed completely, but no further truth-functional content can be found; otherwise the logical phase would of course continue.

The distinction between truth-functional and non-truth-functional structure returns in two kinds of definition. A definition may eliminate the use of a logical expression in context, much as in Russell's theory of definite descriptions. Table 2 has an example. Here the operation 'N' is of joint denial:$N(\varphi, \psi)$

$$\begin{array}{ll} \chi \vee (\varphi \wedge \psi) & \qquad N(N(\varphi), N(\psi)) \;=\; \varphi \wedge \psi \quad def. \\ \chi \vee N(N(\varphi), N(\psi)) & \end{array}$$

TABLE 2   Definition of logical structure

has the same meaning as $\neg\varphi \wedge \neg\psi$. See chapter 5 for details. A definition may also give a more meaningful structure to non-truth-functional content, as in table 3. The notation used here is explained in chapter 2.

$$\begin{array}{ll} \chi \vee (\varepsilon_1, \varepsilon_2)_c & \qquad (n_1, n_2)_{c'} \;=\; \varepsilon_2 \quad def. \\ \chi \vee (\varepsilon_1, (n_1, n_2)_{c'})_c & \end{array}$$

TABLE 3   Definition of elementary structure

Like the analysis of the logical parts of a proposition, the analysis of its non-logical parts must come to an end. The completely analyzed, truth-functionally simple parts of a proposition, are called 'elementary propositions'. They are the truth-arguments of the proposition's truth-functional structure.

5.01 Elementary propositions are the truth-arguments of propositions.

Elementary propositions consist of non-analyzable, non-logical, non-propositional parts, which are called 'names'.

4.221a It is obvious that the analysis of propositions must bring us to elementary propositions, which consists of names in immediate connection.

4.23 A name in a proposition only occurs in the context of an elementary proposition.

By definition, the genuine parts of elementary propositions must be non-propositional; this is what makes elementary propositions elementary. In this way the sequence of definitions in an analysis change the form of a meaningful sentence into a philosophically clarifying one.

As said, it is a basic assumption of the *Tractatus* that the resulting form must be a truth-function.

5.54 In the general propositional form a proposition occurs as part of a proposition only as basis of the truth-operation.

See also 5 and 6. The form of an analyzed proposition must be:

$$\sigma_n := \tau(p_0, \ldots, p_m, \ldots),$$

where σ_n is the truth-functional structure τ in which the elementary propositions p_j occur. The existence of elementary propositions is basic to the tractarian system.

4.411 It is immediately probable that the introduction of elementary propositions is basic to our understanding of all other kinds of proposition. Indeed, the understanding of general propositions depends *palpably* on that of elementary propositions.

As chapter 4 shows in detail, the elementary proposition is a picture, a model of reality, as we imagine it to be (4.01). In a picture, the configurations of names that constitute elementary propositions determine what reality must be for language to have sense.

1.3.5   Analysis is independent of truth

Although sense is determinate and analysis comes to an end, it is helpful to consider the opposite, an unending analysis, to clarify that sense and its analysis is independent of any truth or falsity. So, to derive a tractarian discord, lets us consider an infinite analytic process.

$$
\begin{array}{ll}
\sigma_0(\varepsilon_0) & \qquad \varepsilon_1 \;=\; \varepsilon_0 \quad def. \\
\sigma_1(\varepsilon_2) & \qquad \varepsilon_3 \;=\; \varepsilon_2 \quad def. \\
\quad\vdots & \qquad\quad \vdots \\
\sigma_m(\varepsilon_{2m}) & \qquad \varepsilon_{2m+1} \;=\; \varepsilon_{2m} \quad def. \\
\quad\vdots & \qquad\quad \vdots
\end{array}
$$

TABLE 4   Infinite analysis of σ_0

As said, a main principle for the analysis of language is: eliminate the suggestion of indeterminacy in a propositional sign to show the proposition's unique and determinate sense (3.251, 4.023). But besides this principle, there is a second principle driving analysis; namely that analysis is independent of the truth or falsity of any proposition. In other words: the sense of a meaningful sentence is prior to truth. See also Pears (1987), 117-18. The principle is stated clearly in 4.064.

4.064 Each proposition must *already* have a sense; it cannot obtain it from affirmation, because it is precisely its sense that is affirmed. And the same holds for denial, etc.

And it is assumed opaquely in 2.0211.

2.0211 If the world had no substance, then whether a proposition had sense would depend on the truth of another proposition.

The crux of 2.0211, or so I think, is the connection it highlights between the sense of a proposition and the essentially semantic structure of the world. We conclude from 2.0211 that the world must have substance, *because* sense is prior to truth. Substance and sense allow us to envision how the world may be, independent of propositional truth. Moreover, 2.0211 indicates that for analysis to show the inherent connection between sense and substance, it must come to an end. In a world without substance, a propositional analysis could continue indefinitely, entangling sense and truth in an undesirable fashion. To see this, let us continue the example in table 4.

The analysis must ensure that the more determinate ε_{i+1} helps elucidating the meaning of ε_i, and therefore it must be given independently of analysis that the ε_{i+1} contribute to sense. At this point Wittgenstein's version of Frege's context principle comes into play:[7]

3.3 Only propositions have sense; only in the context of a proposition does a name have reference.

Expressions are configuration of names that are part of propositions. See chapter 4 for details. Given 3.3, an expression may contribute to sense but only in the context of a proposition. The contribution to sense of a ε_i, in particular, is only guaranteed if there is a proposition $\psi_i(\varepsilon_i)$ in which it occurs. Now, if the world were to lack substance, the proposition $\psi_i(\varepsilon_i)$ could not describe it, its sense could not relate to it. Then, only the truth or falsity of $\psi_i(\varepsilon_i)$ would ensure ε_i to contribute to the sense of $\psi_i(\varepsilon_i)$, and the truth-value of $\psi_i(\varepsilon_i)$ should somehow be given independently of what the world may be like. Observe that precisely in this circumstance the *sense* of σ_{i+1} would depend on the *truth-value* of the $\psi_i(\varepsilon_i)$. This contradicts 4.064, much later in the text, stating that sense is prior to truth.[8] Moreover, due to the assumed lack of substance, the analysis continues indefinitely, and at no stage the determinate sense of the proposition would be manifest in its sign. This contradicts 3.251. Thus, for sense of propositions to be possible, the world must have substance.[9] Indeed, ontology captures the bare essentials of what the world must

---

[7]Cf. Frege (1884), 23. Dummett holds that the principle is absent in Frege's later work (Dummett, 1973, p. 7), but Hacker argues convincingly that Frege continued to adhere to it (Hacker, 1979).

[8]Cf. Stokhof (2002), pp. 117-120, for a similar argument. The argument is also discussed in Proops (2004), §4.

[9]One could argue that the two principles (i) that in each step of an analysis the resulting sign most make the determinate sense of a propositions more apparent, and (ii) that the

be like to enable sense (5.4711). The substance of the world, in particular, ensures that the sense of a proposition is determinate, as is manifest in its fully analyzed sign.

1.3.6 Analysis and ontology

Early Wittgenstein aims for a balance between the ontology of everyday language and the pragmatics of its analysis. Analysis is the process that uncovers the essentials of language and thought, which is normally hidden to enable human use. The balance is found in a semi-formal system of perfect notation that in certain respects departs radically from the more formal logics of Frege and Russell. For example, where Russell required the type of terms to be given prior to combination so that propositions could be defined in a bottom-up way, Wittgenstein insisted that the totality of elementary propositions is given holistically, all at once, so that the form of an object, which is like an abstract type, remains implicit. See chapter 2 for details. Variables, which are sequences of signs used as input for truth-operations, get a similar quasi-formal treatment with their specifics only resulting from analysis. Due to the semi-formal approach, Wittgenstein could hold on the one hand that ontology is given in the abstract as the totality of states of things (1), but hold on the other hand that the specifics of the totality is to be made manifest in the application of logic (5.55 and offspring). On this approach the exercise of trying to stipulate an arbitrary range of types prior to the analysis of language is circumvented.[10] Wittgenstein was against such *a priori* typing, since it may come with an unjustified limitation on the application of the system. Still, whatever outcome analysis may have, its result must comply with the abstract frame of ontology. This is how the *Tractatus* offers a descriptive essentialism. Perhaps there is more to reality than facts, states of things and objects; just as there is more to propositions than form and content or more to names than the ultimate simplicity of their semantic contribution. After all, propositions may be clothed in inessentials for human use (4.002), and names may appear structured when used in a sign. But the point is such appearances are semantically inessential, and do not constitute ontology. To be is to be essential to description.

1.3.7 *Interlude*: Russell's merit

Analysis is a philosophical method to make the determinateness of sense manifest, and with this the nature of logic and the essence of the world. According to Wittgenstein it was Russell who had shown convincingly that the surface structure of a proposition does not have to be its logical structure.

---

world has substance, do not suffice to enforce analysis to have finite depth; that is, they do not force analysis to be well-founded. In the context of the two principles, sense could still appear as the limit of a process of infinite depth. Perhaps this is as it should be, for even then the outcome of analysis could still be manifest if both the process and its outcome are regular enough. Cf. 4.2211.

[10]In the context of restricted domains, such as arithmetic or geometry, it may make sense to type basic expressions prior to their logical application, as Frege and Russell did. But when generalizing ontology and logic to provide for the semantic conditions of any language, as Wittgenstein does, such *a priori* typing is arbitrary and may block the application of the system.

4.0031 All philosophy is 'critique of language'. (Surely not in Mauthner's sense.) It is Russell's merit to have shown that the apparent logical form of a sentence does not have to be its real one.

The insight that the logical structure of a proposition reveals itself only after analyzing its superficial structure, is part of Russell's merit. The reference in 4.0031 is no doubt to Russell (1905) in which he presents his theory of descriptions as contextually defined expressions. See Black (1964), p. 161, or Anscombe (1959), chapter 2. But the praise should be taken broadly. The kind of analysis that the *Tractatus* advocates concerns all of any language. Wittgenstein generalizes analysis to show that each proposition is a truth-functional complex of elementary propositions. Russell's treatment of definite description, suitably adapted, is part of it at best.   [*End of interlude*]

1.4   Ontology

Within the tractarian system the analysis of a proposition results in a truth-functional complex of elementary propositions (5). The prominence of this logic is part of Wittgenstein's indebtedness to Frege. Frege held that the sense of a proposition, the objective core of its meaning, is the thought that its truth-conditions obtain. In the *Tractatus* the truth-conditions of a proposition are unveiled in stages. One stage is truth-functional, and truth-tables make especially clear under which conditions a proposition is either true or false. But in analysis the truth-functional stage does not continue indefinitely, and for the remaining logically elementary parts, truth consists in the structural identity with what they symbolize.

That the analysis of a meaningful sentence results in a unique structure and content is equivalent to holding that the analysis of a proposition comes to an end. The end-result of analysis shows what the world must be like for language to have meaning. However, the ontology of language contributes to the tractarian philosophy in various ways. It delimits the pictorial capabilities of description in giving the logical space of all possible structures. But in doing so it also shows what is contingent and what is necessary, and this is core to the transcendent, ineffable view on the world, where the nature of logic and ethics is to be found.

The logical space of ontology is unique and fixed.[11] Logical space does not wax or wane; it is as it is by necessity. It comes with a Siamese bond between the elementary structures called 'states of things' (*Sachverhalten*), and the objects that form states of things.[12] State of things are possibilia. They are the hallmark of contingency. They can be realized or can be not realized

---

[11] The adjective 'logical' in 'logical space' may be a bit misleading. The notion of logical space is akin to that of grammar in Wittgenstein's later work: it concerns a space of possible combinations, and indicates ways in which they can be used.

[12] I translate the German *Sachverhalt* as: state of things. There has been some dispute on how *Sachverhalt* should be translated. The two most common translations are: atomic fact (Ramsey & Ogden) and state of affairs (McGuinness & Pears). The beauty of the word *Sachverhalt* is its combination of the word *Sach* object, thing) with the suggestion of *Verhältnisse* (state), which makes it a natural candidate to express the philosophical concept: configuration of things (as substance). Neither 'state of affairs' nor 'atomic fact' retains this quality. 'State of affairs' is a natural phrase, but 'affairs' suggests a much coarser state than one of things as the ultimate constituents of the world. Perhaps for this

independently of each other. States of things may allow for substructure, but discerning substructures comes to an end. In the end it is clear that states of things are configurations of unstructured simples, called objects'. Objects are the substance of the world, which in a way determine logical space. Objects have form and content. Their form is the range of possible ways in which they may combine with other objects to form a state of things, and their content enables the same form to occur more than once. Yet, objects do not determine state of things in any independent way. Objects are bound to occur in the context of state of things, and have no existence outside such context. Fixed and unalterable, they constitute the necessary bedrock of contingency.

1.5 Logic

In this book I will argue that logical space encompasses both analyzed signs and their content. On this view, the difference between analyzed signs and their sense is functional. There is no necessary distinction between analyzed signs and their content that forces one to be symbolic, the other to be what symbols are about. Signs are state of things *used* as signs. The only restriction that applies is structural: the elementary parts of a symbol should be isomorphic to their content.

Logical complex signs come with non-material extensions of state of things that show their truth-conditions in a uniform way. The logical notation used does not refer (4.0312), so it must be purely formal itself. See chapter 6 for discussion. Reflecting on the possible forms of logically complex signs, two borderline cases can be discerned. Firstly, there are the elementary signs, which are truth-functions of themselves. From a truth-functional point of view they are identity functions, which show the triviality that an elementary proposition is true if its is true, but false otherwise. Since this is empty, their content rather comes from projecting the elementary sign onto the state of things it describes. Only in the context of projection an elementary sign constrains the world by having fixed what its truth or falsity consists of. A second borderline case are logically complex signs that are true or false unconditionally; i.e. regardless of whether their content is realized or not. They are the logical propositions, the tautologies and contradictions whose elementary content is dissolved, as it were, and are about nothing. Being about nothing, the nature of logical propositions itself cannot be described. Logic is transcendental.

1.6 Ethics

Wittgenstein's urge to safeguard ethics and the mystical from pompous prate and his desire to find a way to acting just, brings the analysis of language to the fore, and with it logical space with its delicate balance of contingency and necessity. Like logic, ethics pervades the world's contingencies in its frame

---

reason, Wittgenstein disliked the translation, and with some hesitation suggested the Latin *status rerum* as an alternative (Letters to C.K. Ogden, 21 (13)). This, however, is hardly an improvement: even if accurate, the phrase is very uncommon. The translation 'atomic fact' is more accurate than 'state of affairs' but has two disadvantages: (i) it is a newly invented, technical term, and (ii) it suggests the fact to hold, whereas I take a *Sachverhalt* to be a logical possibility. I prefer 'state of things': an ordinary phrase with about the same associations as *Sachverhalt* that is suitable to indicate unrealized possibilities.

of necessity. Both are transcendental: they can become manifest but are impossible to describe (cf. 6.4 and its offspring). This makes plain that ethical value, which must be absolute, cannot be among the world's contingencies.

6.41 The sense of the world must lie outside it. In the world everything is as it is, and everything happens as it happens; *in* it there is no value, – and if something were to give it value, it would be without value.

If there is a value that has value, it must lie outside all what happens and is the case. For all what happens and is the case is accidental.

That what makes it non-accidental cannot lie *within* the world, since then it would be accidental again.

It must lie outside the world.

Ethics cannot be spoken of, but shows itself. Its nature is revealed in a holistic view on the world *sub specie aeterni*, but from a different perspective than in case of logic. What is shown is perhaps best hinted at in negative terms: no thinking, no judgment, no imagination, no longing. Still, there is not just indifference. Suicide is the ultimate sin. For in the limit of description this world is still my world, and my life should be reconciled with all that is. It makes me act so as to dissolve all its problems, which in the limit encompasses the problems of all others.

Seemingly a defect, the ultimate virtue is silence. In silence – not distracted by verbal constructs concerning God, Categorical Imperatives, Kabbalah, The Prophet, Compassion, Love, – one lives the good live in the eternity of timeless now.

Wittgenstein's ethical view on ontology naturally expands his earlier logical insights; both for logic and for ethics it is crucial that ontology gives a balanced account of necessity and contingent change. The resulting system has a distinct Kantian flavor, in that it shows the conditions for meaningful language to be possible, with logic and ethics, each in its own way, appearing in the limit of sense.[13]

With Stokhof (2002) I hold that from a systematic point of view ethics is on a par with logic and semantics. In the last stages of writing Wittgenstein seems to have realized that the demarcation of ethics is well suited to sharpen his insights on language, logic, and ontology. This at least is the pattern that

---

[13]In this chapter I just highlight some core aspects of tractarian ethics. See also McGuinness (2001), chapter 14, or Stokhof (2002), chapter 4, for more detail. Stokhof (2002) has a careful reading of how Wittgenstein's view on ethics leads to those on language, logic and ontology, and how his early philosophy relates to the philosophies of Schopenhauer and Kant. Other Kantian readings of Wittgenstein are Stenius (1960), Hacker (1984), Pears (1987) and Proops (2004). Wittgenstein is a Kantian philosopher to the extent that he formulates the semantic conditions of language and uses them to distinguish between what can be said and what we must remain silent about. This is similar to Kant's approach, who gives the conditions of pure knowledge to delimit it from what can not be known. Apart from similarities, there are of course vast differences. For one, Kant's logic comes with a rich structure of intuitive and categorial concepts to understand the possibility of the synthetic a priori and different forms of judgment. See Van Lambalgen's recent work on this topic. By contrast, Wittgenstein based his philosophy on the abstract logical structure of ontology, which separates the necessary from the contingent and brings all propositions and concepts on a par.

emerges from McGuinness' reconstruction of how the *Tractatus* is composed (McGuinness, 2001, 259-269). Clearly, the demarcation would have been impossible without the single-minded unity of the system. (This not to say that the demarcation determines the system uniquely. E.g., the historical context of Frege's and Russell's work has surely helped to give it its final shape.)

1.7 Paradox?

On the symbolic turn there is a threat of paradox in delimiting ethics via the semantics of everyday language. For how can insight in the workings of meaningful language be obtained using a language that might be unsuitable for this task? Whether paradox does arise depends on whether language is reflexive or not, i.e., is able to describe its own sense conditions. Only if language is *not* reflexive paradox will result.[14]

Suppose for the sake of argument that language is reflexive. There are two ways in which this may be so: firstly the sense conditions of language consists of a self-referential web, or secondly they are reduced to basic parts whose sense is self-evident.

As to the first option, in line with Russell's vicious circle principle (Whitehead and Russell, 1910, introduction chapter II), the view on language as a self-referential web of sense conditions is far removed from early Wittgenstein's thought. In the *Tractatus* he explicitly rejects the possibility of simple forms of self-reference: "No proposition can make a statement about itself, because a propositional sign cannot be contained in itself." (3.332, also 2.173) On the basis of this stark rejection one may safely assume that more complex forms of self-reference, if considered at all, are rejected as well. Chapter 10 shows in detail that for the tractarian system self-reference and describing its own sense-conditions is impossible.

On the second option sense conditions are reduced to basic parts whose sense is self-evident; they require no further elucidation, since their sense conditions are immediate. Here we should recall that Wittgenstein had little respect for an appeal to self-evidence: "Russell would say: 'Yes! that's self-evident.' *Ha!*" (NB 3.9.1914)[15] Indeed, what is gained with an appeal to self-evidence? In case of names Wittgenstein does consider such elucidations of their reference, but points out that they presume knowledge of the sense of the propositions in which the names occur (3.263). For language as a whole self-evidence is blocked. The parts of language that are assumed to provide the sense conditions of the remaining parts must themselves have sense, and how could their sense conditions become apparent? We will have learned how the sense of language depends on the sense of some of its parts, but for the basic parts the questioning stops, while the point of the initial question remains open.

---

[14] In 'The Big Typescript' of the middle period, Wittgenstein did consider language to be its own foundation: 'Investigating grammar is fundamental in the same way in which we may call language fundamental, as if it is its own foundation.' (p. 412 in the original, or p. 279 in Wittgenstein (1933). Cf. also Schulte (2005), 80.

[15] An appeal to self-evidence plays an important rôle in Russell (1913), which grew out of Wittgenstein's critique on Russell's theory of knowledge in Russell (1912). Wittgenstein fierce reactions to the new manuscript made Russell decide to leave it unfinished.

1.8   Ostensive philosophy

In case language is not reflexive – is not able to describe its own sense conditions, – its sense conditions must be based on the non-linguistic and ineffable. Then, if access to sense conditions is granted at all, they can at best show themselves but cannot be spoken of. It is this position which is paradoxical. It is not paradoxical to assume that language cannot *provide* for its own sense conditions. But if the semantics of descriptive language is grounded in necessities, while all meaningful language consists in describing the contingent, it is impossible to *describe* the semantics of descriptive language. Still, this is what the *Tractatus* seems to attempt when approaching ethics at the boundary of the system of descriptive language.

Wittgenstein's solution of the paradox consists in abandoning theorizing concerning the transcendental, i.e., the context in which the paradox arises, and to highlight that philosophy is an activity rather than a theory (4.112). This position is unacceptable for the scientific mind that strives to understand and control the basic assumptions of language, logic and life. But a returning theme in Wittgenstein's philosophy is that he seeks alternatives for the scientific attitude without getting soft in the head.[16]

There are many traditions, both philosophical and mystical, in which a search for the sense of life is felt to result in paradox that, after having experienced an insight into the nature of the world, is resolved by taking action, that is: to lead the good life with all traces of paradox dissolved (6.521). What sets early Wittgenstein apart from mystics and philosophers such as Tolstoy and Kierkegaard, whose company he sought when writing and trying to publish his work (Janik, 1979), is the highly technical nature of the tractarian paradox: the key to good life is found in the boundary of meaningful language, essentially a logical system that meaningful language itself cannot describe. This paradox forces thought to stretch to its limit before giving up.

In the *Tractatus* philosophical activity consists of using language in a nondescriptive, ostensive way to gain insight into the essentials of descriptive language. Clearly, such non-descriptive ostensions will only have the desired effect for a reader 'who has already had the thoughts expressed in it himself – or at least similar thoughts' (*Preface*).

The ostensions are meta-linguistic in nature, and use quotation to sharpen a reader's understanding of the system as a whole. The basic approach is in 4.012.

4.012 It is obvious that we perceive a proposition of form 'aRb' as a picture. Here the sign is obviously a simile of what is signified.

In section 10.6.2, I give a detailed choreography of ostension, showing that its success is based on the extreme uniformity of the tractarian system. This allows philosophers to quote just a few sentences to show an understanding addressee the essence of all description.

---

[16]Wolfgang Kienzler drew my attention to the correspondence between Schlick, Wittgenstein and Carnap, from the Innsbruck Edition of Wittgenstein's letters, which clearly shows that round about 1930 Wittgenstein abhorred the single-minded scientific approach to philosophy.

3.3421 A particular mode of signifying may be unimportant, but it is always important that it is a *possible* mode of signifying. This is anyway how things stand in philosophy: time and again the individual case turns out to be unimportant, but the possibility of each individual case clarifies something about the essence of the world.

This is the crux of ostensive philosophy. It allows engendering the temporary make-believe that language can be a meta-language describing its own semantics. But the finger pointing toward the moon is not the moon. In the end, the make-believe has to be recognized as such: the workings and boundaries of descriptive language must be shown, cannot be said (6.54). Once seen, the nonsense of make-believe is restricted to the meta-linguistic route taken, not to where it has taken us, for the world *is* changed completely (6.43). Regardless of how it is arrived at, the symbolic turn is genuine. It stops one from pursuing ethical theories and illusions, but rather pursue *living* the good life.

1.9   *Interlude*: ethics in real life

For young Wittgenstein ethics was not an abstract, academic exercise; it had an immediate impact on his everyday life. This is particularly clear from the ciphered parts of his diaries (MS101-3). They come with the remarkable suggestion that philosophy may help to reconcile oneself with such devastating events as happened to his brother Paul. Also, they have examples of ethical principles that Wittgenstein strived to adhere to. Still there is a direct link with the more restrained and polished theses in the *Tractatus*. Except perhaps for some background information given shortly, this brief anthology should speak for itself.

> Time and again I must think of poor Paul, who has LOST HIS VOCATION SO SUDDENLY. How awful. Which philosophy would be needed to come to terms with this! If at all there is another option than suicide!![a] [17]

> Read Tolstoi with great gain.[b] Time after time I repeat Tolstol's words to myself: 'Man is POWERLESS in the flesh, but FREE through the Spirit.' Let the Spirit be in me![c]

> Do not loose yourself!!! Keep yourself together! And do not work for pastime but devoutly in order to live! Do nobody wrong![d] Understand people! Each time you want to hate them, instead try to understand them. Live in inner peace. But how to attain inner peace? ONLY if you live piously! ONLY so it is possible to endure life.[e]

> I should find indifference with regard to the difficulties of EXTERNAL life.[f] I feel to be dependent on the world, that is why I should fear it even if there is no acute danger. I see myself, the I in which I may certainly rest as on a remote island seen from afar that has drifted away from me.[g]

> Man should not depend on the accidental; neither on the favorable, nor on the unfavorable.[h] How should I live to exists in this moment? To live in Goodness and in Beauty until life of itself ceases to be.[i] Dead really gives life

---

[17]The references for *a-l* below are as follows: a: MS101 28.10.14; b: MS101 3.9.14; c: MS101 12.9.14; d: MS102 12.11.14; e: MS103 6.5.16; f: MS101 11.10.14; g: MS102 9.11.14; h: MS102 6.10.14; i: MS101 7.10.14; j: MS103 9.5.16; k: MS101 13.10.14; l: MS103 6.4.16.

its meaning.[j]

I am Spirit and so I am free.[k] Life is one.[l]

The 'Paul' mentioned in the first paragraph is Ludwig Wittgenstein's older brother, who was a pianist. In 1914, his right arm had to be amputated, after he was shot in the elbow by the Russians during an assault on Poland. In the years to follow, he found new techniques to play with his left hand, and arranged or commissioned works to continue performing. Most famous among these works is Maurice Ravel's Piano Concerto for the Left Hand in D major.

The Tolstoy that Wittgenstein read 'with great gain' is *The Gospel in Brief*. In August 1914, as volunteer in the Austrian army, Wittgenstein obtained a copy of this book. It had a profound influence on him, then and in the years to follow. July 24th, 1915 he wrote to Von Ficker:

> Are you acquainted with Tolstoy's *The Gospel in Brief*? At its time, this book virtually kept me alive [...] If you are not acquainted with it, then you cannot imagine what an effect it can have upon a person.

[End of interlude]

2

Ontology

The ontology of the *Tractatus* – its world, reality, logical space, its states of things and objects, – is so abstract that it leaves much room for interpretation. In the literature there are about three ways in which ontology has been taken: as concerning logical categories, sense data, or physical atoms.[18] In this chapter I start pursuing the first, symbolic interpretation. Ontology concerns what must be in order for description and its symbolization to be possible. To be is to be essential to description.

As shall become clear in the course of this book, in my view ontology concerns not just what sign and symbols are about, it includes signs, sense, and symbols. Ontology is a most abstract, semi-formal alternative to the type theories of Frege and Russell, which offers Wittgenstein's solution to Russell's paradox. And when Wittgenstein's work started to extend from the foundations of logic to the essence of the world (Notebooks, 1.8.16), ontology also provided for the world's substance, which comes with the contingent and the necessary that delimits language and, finally, ethics.

This chapter still presents ontology largely in a sign-independent way. After having linked ontology to the approaches to types of Frege and Russell – i.e., part of the context to which the *Tractatus* reacted, – I continue with detailing the niceties of logical space. Here the main task is to come to grips with its twin-notion of states of things and objects.

In subsequent chapters, the apparent sign independence of ontology is quickly left behind. In chapter 3 on projection and in chapter 4 on elementary propositions, I shall argue that logical space is the realm of elementary propositions; it includes both their signs and their sense. Indeed, the current study of ontology is best seen as a first step in capturing the essentials of descriptive language: its meaning and its limit.

2.1 Ontology and types

As we have seen in the previous chapter, the analysis of descriptive language should make clear that propositions are truth-functional structures of elementary propositions. In this way, the end-result of analysis shows what the world must be like for language to have sense. This insight is captured in the ontology at the beginning of the book, which, apart from its ethical and logical

---

[18]I do not given an overview of the arguments pro and con, but refer to Stokhof (2002).

dimensions, gives the essentials of sense.

Tractarian ontology is based on the notions of object (*Gegenständ, Ding*) and state of things (*Sachverhalt*). State of things and object are like siamese twins: they can be discerned from one another but hardly separated. Wittgenstein changed his mind a few times concerning their nature (Pears 1987, ch. 6, 137-8), but finally he did not bother about their specifics. In my opinion this is not a weakness, but a deliberate choice. When composing the final version of the *Tractatus*, Wittgenstein realized that his view on logic and the semantics of descriptive language should not constrain its application in any way.[19] The only types that are essential are those of objects and states of things, and their linguistic variants: names and (elementary) propositions. Wittgenstein went so far as to see further specifics given prior to analysis as an arbitrary and unjustified constraint. Instead, he presents a generic framework of logical combination. In this framework only the most abstract features of object-combination are given; namely that objects have a form that delimits their range of combination, and a content that allows multiple occurrences of the same form. Although names are presented as having a form ('type'), the specifics of these forms are inessential and are left to analysis to make manifest.[20]

I begin with recalling the main aspects of Frege's and Russell's views on logical types, which are core to the context that elicited Wittgenstein's early philosophy. Next, I shall argue that Wittgenstein was against the typing of names and objects prior to analysis. This position comes with the question: how much detail should a more formal approach attempt to give? After having weighed some pros and cons, I decide to stay close to the text and keep the treatment of the tractarian ontology rather abstract. In chapter 10, I show how the abstract ontology still allowed Wittgenstein to solve Russell's paradox.

2.1.1 Frege on function and object

Frege analyzed propositions as consisting of functions and objects, combined in a certain way.[21] The distinction is the basis of an informal theory of types, which arose from a study of mathematical functions. E.g., Frege analyzed the term (1a) as (1b).

(1) a. $2^3 + 2$
   b. $(2)^3 + (2)$
   c. Object: 2
   d. Function: $(\ )^3 + (\ )$.

In (1b) the object 2 merges with the function $(\ )^3 + (\ )$ to result in the object 10. But if the function were to merge with the object 3, the object 30 would result. Frege observed that numbers should be taken as complete, in the sense

---

[19] According to McGuinness' (McGuinness, 2001, 259-269) the theses on the application of logic are added last.

[20] Thus I agree with Ishiguro (1981, p. 47) where she writes that 'For Wittgenstein, a theory of types was a necessary truth about symbolism and language: something which could be grasped as evident, if we correctly understand the nature of the symbolism. It was not a philosophical position which we can argue for or against in a non-circular manner. Nor on the other hand is it something we are free to decide by stipulation.'

[21] This section is based on Frege (1891). But cf. also Laan (1997), chapter 1.

that they lack empty places, while functions should be taken as incomplete, they need objects to be completed.

Based on his analysis of mathematical functions Frege came to posit all objects as complete (also: saturated), and all functions as incomplete (also: unsaturated), and this independent of whether they are mathematical or not. The incompleteness of functions is always due to empty places that indicate where they are to be saturated with objects. Objects are always saturated in this regard as they lack such empty places.

According to Frege, there are different kinds of object and different kinds of function. Indeed, objects must be taken broadly, in an abstract manner. They include numbers, points, material objects, places, times, and the truth-values True and False. Also, for each function $\Phi(\xi_1, \ldots \xi_n)$ its course-of-values (*Wertverlauf*) is an object:

$$\acute{\varepsilon}_1 \ldots \acute{\varepsilon}_n \Phi(\varepsilon_1, \ldots, \varepsilon_n),$$

Course-of-values somehow capture the 'class' associated with the term Φ, the association of the possible arguments of a function with the corresponding value.

With regard to propositions Frege held that concepts and relations should be seen as functions. A concept is a function with one empty place that, when filled with an object of the appropriate kind, yields a truth-value. For example, (2a) is analyzed as (2b).

(2) a. Hadewijch wrote 'Visions'.
    b. (Hadewijch) wrote 'Visions'
    c. Object: Hadewijch
    d. Concept: ( ) wrote 'Visions'

In the analysis the name under (2c) denotes an object, which is complete as it lacks empty places. The phrase (2c) denotes a concept, which as the parentheses indicate is incomplete. Filled with the object Hadewijch the concept yields the object True, but filled with Elisabeth Anscombe it yields the object False. Relations are like concepts but they have more than one hole, which we distinguish using different parentheses:

(3) a. Hadewijch wrote 'Visions'.
    b. (Hadewijch) wrote ['Visions']
    c. Objects: Hadewijch, 'Visions'
    d. Relation: ( ) wrote []

Frege was well aware of the fact that functions are of different types; they vary not only in the number of their arguments, but also in their kind and level. In *Funktion und Begriff* he puts it thus:

> Now just as functions are fundamentally different from objects, so also functions whose arguments are and must be functions are fundamentally different from functions whose arguments are objects and cannot be anything else. I call the latter first-level, the former second-level. [...] In regard to second-level functions with one argument, we must make a distinction, according as the rôle of the argument can be played by a function of one or of two arguments. Frege (1891), pp. 27, 29

The examples given until now are all of first-level holding of objects. There are also second-level functions, which hold of (the course-of-values of) first-level functions. For example, the existential quantifier is analysed as a second-level concept requiring the course-of-values of a first-level concept to yield a truth-value:

(4) a. There is a mystic.

    b. ⊢⌐ᵥ⌐ mystic(𝔳)

    c. First-level concept: mystic( )

    d. Second-level concept: ⊢⌐ᵥ⌐ Φ(𝔳)

If (4a) and hence (4b) is true, the second-level concept ⊢⌐ᵥ⌐ Φ(𝔳) – or in modern notation: $\neg\forall\neg x.\Phi(x)$, – and the object $\acute{\varepsilon}.\text{mystic}(\varepsilon)$ obtained from the first-level concept 'mystic', combine into the True. This is the case if and only if there is an object h such that 'mystic(h)' is the True.

Frege's imagery of functions as incomplete and objects as complete is helpful, but at the same time a bit misleading. A logical view on functional expressions seems to have inspired the image. As a consequence one may be tempted to think that the structure of how function and argument combine is still available at the level of reference – as in case of expressions, – but it is not. When objects complete an unsaturated function, both function and argument merge into an object, a complete, structureless whole (Frege (1891), 21). Indeed, if the structure of combination were preserved in reference, different functions could not be identical to each other. For instance:

$$3^2 + 3 = 3(3 + 1)$$

is true, because function and argument at both side of the identity merge into the object 12. In this object the structure of how it is attained is lost. The same holds in case of propositions. E.g., if the concept $C(\ )$ combines with the object o into the proposition $C(o)$, the result is either the True or the False, and at this level any trace of how the truth-value is arrived at has disappeared. 'One could say that to judge consists in distinguishing parts within a truth-value.' But 'the distinction proceeds via a return to the thought.' (Frege (1892b), 50) At the level of reference, there is no such structure.

2.1.2 Russell and *a priori* typing

The distinction between function and object is part of Frege's approach to reduce arithmetic to logic. Independently, Russell started working on a similar project round about 1900. In 1902 Russell began his study of the impressive system in Frege (1893), and noted that it was susceptible to about the same paradox as the one that he had found earlier in the theory of classes.[22]

Having noted the inconsistency in Frege's system, Russell continued analyzing the paradox to find a consistent reduction of mathematics to logic. His idea was to use typed terms for this purpose. Much as Frege had done before him, he categorized terms as belonging to different simple types prior to their

---

[22]See chapter 10. Recent insight has shown that when appropriately constrained Frege's system still gives an awesome reduction of arithmetic to a few basic principles, though not the reduction to logic that he strived for. See Burgess (2005) for an introduction and overview.

being combined into complexes.[23] So there are individuals, propositional functions, properties, relations, propositions, etc. But in Russell's opinion simple types do not suffice. For on an informal approach there is nothing to preclude, e.g. a relation to hold of itself. And if propositions of the form $\neg r(r)$ are allowed, paradox would result much as before. Russell thought a more involved notion of typing was required. The so-called ramified types that Russell proposed, resulted from his idea that all paradoxes involve a certain kind of vicious circle, which can be avoided by means of a ' vicious circle principle'.

> The vicious circles in question arise from supposing that a collection of objects may contain members which can only be defined by means of the collection as a whole. [...] The principle which enables us to avoid illegitimate totalities may be stated as follows: "Whatever involves *all* of a collection must not be one of the collection." (Whitehead and Russell, 1910, ch. 2, §1)

In other words, objects should not only be categorized in terms of simple types, the types should also be ramified into different orders to keep track of the complexity of their definition. For instance, the definition of a class is one level above the definition of its elements, which clearly precludes classes to have themselves as a member. See Laan (1997), chapter 2 and 3, for a clarifying formalization.

2.1.3 Against *a priori* typing

On both Frege's and Russell's approach, the type of an entity is an explicit characteristic of the entity itself: (i) the type is fixed prior to any combination, and (ii) the type is independent of the type of other entities. Today, in the line of Frege, a similar approach can be found, in the implicit typing of constants, functions, and terms of predicate logic. Such a notion of entity-type allows a system to be defined in a bottom-up fashion, in which the type of an entity determines which complexes are possible.[24]

Quite late in the *Tractatus* – namely, in 5.55 and its offspring, – Wittgenstein pronounces himself against *a priori* typing.[25] Each logical question must be answered without more ado, but to specify all possible elementary propositions and the forms of objects that constitute them, is *not* part of the enterprise.

5.553a Russell said there were simple relations between different numbers of things (individuals). But between what numbers? And how will this be decided? – By experience?

5.554 To give any specific form would be completely arbitrary.

---

[23] For want of a better phrase this book uses 'a priori typing' in a logical sense; namely: typing prior to combination or analysis. Thus 'a priori' is not used in its epistemological sense of prior to being experienced. Similarly, 'a posteriori typing' means: types abstracted from given combinations.

[24] Instead of 'bottom-up' the term 'recursive' could be used. I choose not to, because of its connotation with computability. In my opinion, neither Frege, nor Russell, nor Wittgenstein were concerned with decidability or algorithmics. Cf. section 11.4.

[25] But already in his *Notes dictated to G.E. Moore* of April 1914, 109–110, Wittgenstein criticizes the theory of types in that it tries to say what can only be shown.

Similarly, although the *Tractatus* starts with presenting ontology as the to-tality of states of things, it deliberately does not indicate the specific states of things that make up the totality.

5.555 It is clear that we have a concept of elementary propositions regardless of their particular logical form.

When symbols can be formed systematically, it is the system which is logically important and not the individual symbols.

And how would it be possible that I had to deal in logic with forms that I can invent? What I have to deal with is that which makes it possible for me to invent them.

Instead of inventing forms (types) prior to analysis, we should concern our-selves with the system that gives the general form of propositions, and with this the form of each end-state of analysis.[26] It is the pragmatics of applied logic that elucidates which specific elementary propositions and names there are, and so which specific states of things and forms of objects. This is how logic should be in harmony with its application (5.557).[27]

It is no wonder that the *Tractatus* offers a quite different analysis of propo-sitional structure than that of Frege or Russell. Instead of making the types (forms) of expressions more formal and explicit, as Russell did to circum-vent paradox, Wittgenstein made them more abstract and less specific. In the analysis, all *a priori* typing in elementary propositions is absent, even Frege's abstract version that distinguished between function and object. The ontolog-ical theses about objects and states of things deliberately refrain from giving a more specific formal approach to logical combination, simply because it is inessential to understanding how descriptive language has meaning and what logical propositions are.

Analysis makes clear that there must be elementary propositions, the small-est possible units of meaning, which consist of names, the smallest possible units that contribute to meaning in the context of a proposition. Their on-tological analogues are states of things, the smallest possible units that can be described, and objects, the smallest possible units that can be referred to in the context of description. On this reading, names and objects on the one hand, and elementary propositions and states of things on the other should be seen as giving a highly abstract, tractarian framework of 'types'. The question is: how far should we go in speculating about the specifics of these 'types'. My position is: since according to the *Tractatus* it is impossible to give the specifics prior to analysis and since they are not crucial to see how language has sense, speculation should be kept at a minimum. See 2.3 for an overview of pros and cons.

---

[26]The conditional formulation of 5.5571 relaxes this position somewhat: If it is impossi-ble to give the elementary propositions *a priori*, trying to do so should result in obvious nonsense. Recall that we noted earlier that types for restricted domains, such as arithmetic or a specific kind of geometry, may be useful. Wittgenstein's position concerning types is related to the universality of his system.

[27]Ishiguro (1981, pp. 55-56) writes as if Wittgenstein did assume elementary propositions to result from typed propositional functions and expressions. I think the abstract setting of the *Tractatus* allows for the circumstance that the form of a name– its space of possible combinations, – cannot be captured by a type. Cf. section 2.4.

With a view to Russell's paradox, it is quite remarkable that Wittgenstein opted for making the forms (types) of objects more abstract instead of more formal, as Russell did. But in chapter 10 I will argue that he must have seen that this abstract setting still allows one to solve the paradox, but in a structural way that avoids vicious circles much as Russell had required.

Compared to Frege, the tractarian system is based on a radical shift of ideas: all tractarian propositions are structured and saturated, while all tractarian objects are unstructured and unsaturated. See also Anscombe (1959), 98. The concept of a tractarian form (type) must shift accordingly. As we shall see in detail shortly in logical space: (i) *no* form is known prior to combination, and (ii) *each* form depends on the form of other expressions.

2.2 Logical space

A detailed account of logical space requires finding a proper view on the interdependent notions of objects and states of things that leaves their abstractness in tact. As a first step in this direction we just define logical space as the unique totality of all contingencies, the states of things.

Definition 1 *Logical space – notation: Λ, – is the totality of states of things, which are configurations of objects. We use α, with or without an index, to vary over states of things.* □

Observe that logical space does not have to be 'Cartesian'. The space Λ gives all possible combinations of objects, but not in the sense that it allows the combination of any object with all other ones. The phrase 'logical' in 'logical space' indicates that the space contains all possible, logically independent configurations. In this regard it resembles the phrase 'grammar' in Wittgenstein's later work. It does *not* indicate that the possibilities in the space are logically complex. In chapter 7, I will argue that only signs have overt logical structure and that the logic of non-pictorial reality just consists in states of things being independent from one another: the existence of one state of things is entirely independent of the existence of another state of things.

Is logical space a class or a set? Strictly speaking it is neither: it is a totality that should not itself be seen as an entity or thing. As such, it is closer to a modern class than to a set. Anyway, the *Tractatus* advocates a classless philosophy (e.g., 6.031), and I will try to stay close to this assumption. In a similar vein one may wonder whether logical space allows for multiple copies of a state of things? I will not assume such multiplicities here. The ontology is a descriptive essentialism, and from this point of view multiple copies are inessential. Cf. interlude 2.6.

Clearly, definition 1 requires the notions of state of things and object to be developed further. Whatever else they may be, states of things are configurations of objects, and objects are fully specified by the logical notions of form and content (2.025):

- the *content* of objects enables the same form to be instantiated more than once;[28]

---

[28]The tractarian notion of content is ambiguous. In 3.13 'content' ('*Inhalt*') has a different meaning; namely, the states of things a proposition is about. This meaning is not the one

- the *form* of objects are their possibilities to combine with other objects in states of things.

The notion of form is analogous to that of type and the notion of content to that of token, the instantiation of a type. In this regard the tractarian ontology is a creative response to Frege's notions of function and object and to Russell's theory of types. But there are vast differences. For one, Wittgenstein's presentation of ontology is even more abstract than those of Frege and Russell.

2.3 Structure

Logical space consists of states of things. States of things, in turn, are configurations of objects. Thus in a way objects determine logical space.

2.0124 With all objects, all *possible* states of things are given as well.

It makes one wonder *how* logical space results from the combination of objects, and *what* the possible structure of states of things is. In line with the view that objects cannot be typed prior to analysis, Wittgenstein says next to nothing about the structure of states of things, or how it results from combining objects with one another. There is just a metaphor.

2.03 In a state of things objects fit into one another like the links of a chain.

But how serious should the metaphor be taken? How far should one go in attempting to specify the form of objects and the ways in which they fit into one another? For example, the chain metaphor suggests linearity. Yet one feels that state of things should be left so general as to encompass all kinds of configuration. To see what is at stake, it may be clarifying to consider an example. How to view figure 1 in line with the chain metaphor?

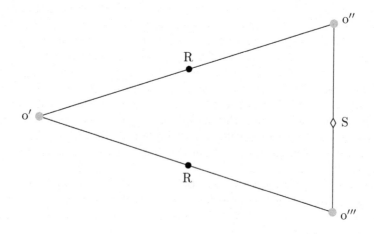

FIGURE 1 State of things?

---

intended for objects as the substance of the world (2.024). Cf. also page 62. If required, I distinguish the two by using 'sign-content' or 'object-content' for content as the instantiation of form, and 'sense-content' for the content an expression is about.

If the structure is seen as a single 'saturated' state of things, a linear view seems to require nesting, say, as in:

$$(r(o',o''),r(o',o'''),s(o'',o'''))_c.$$

On this view the form of its objects is given via the parts:

$$r(o',o''),r(o',o'''),s(o'',o''')$$

of the larger configuration $(-)_c$. From the serialized structure, which may well be invariant under the order in which its parts are presented, the specifics of the configuration and its objects can be gleaned.

An alternative would be to prefer a LEGO® view on objects that exhibits their form more directly. Then, the structure of the state of things is seen as obtained from the combinatorial possibilities of objects, possibilities they wear on their sleeves. On this view, each combination of objects is bisected in a positive and a negative pole that gets saturated in object combination. See figure 2.

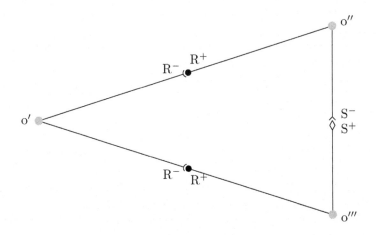

FIGURE 2 Lego view on state of things

In figure 2, o' has two R^- connectors that must combine with R^+ connectors to contribute to a state of things; similarly for the other objects.

Yet another alternative would hold that Wittgenstein had a tendency of taking structure as logically defined, and that logical definition is based on configurations that are so simple they may as well be captured in a semi-formal, linear notation to highlight what is essential to their contribution to sense. Then, Wittgenstein would prefer viewing the complex structure in figure 1, not as a single saturated whole, but as a logical complex of much simpler structures. On this approach, the serialized:

$$(r(o',o''),r(o',o'''),s(o'',o'''))_c$$

is replaced with a conjunction of elementary propositions:

$$(o',o'')_r \wedge (o',o''')_r \wedge (o'',o''')_s.$$

What were unsaturated parts of a whole, now become saturated states of things themselves. To support this view, one could point to the fact that the chain metaphor may be a residue of an earlier observation of Wittgenstein; namely, that we are right in being confident that any sense can be expressed in a two-dimensional notation, even if we do not yet see how or why. Cf. MS101, 26.9.14. The second dimension is used, e.g., in logical tables, graphs or tableaux. The view also complies with Wittgenstein's early experiments to find a notation for structural complexes that attempts to push complex structure as much as possible to logical combination. Cf. Wittgenstein (1914-16), 5.9.14, where the definition of structure is compressed into a quasi-elementary proposition:

$$\varepsilon[aSb] = \varepsilon(a) \wedge \varepsilon(b) \wedge aSb \qquad Def.$$

For instance, $n'R[n''Sn''']$ is short for: $n'Rn'' \wedge n'Rn''' \wedge n''Sn'''$ describing figure 1. In this way, a proposition about a 'complex' is internally related to the propositions about its constituents (MS102, 26.4.15).[29]

With three alternatives at hand, how to choose among them? Although the first and the third alternative differ in important respects, I think they are sufficiently alike to treat them on a par. In this book the third alternative is followed, but only after having weighed some pros and cons of the LEGO®-view.[30] Advantages of the LEGO® approach are:

+ It comes with a clear view on the kind of object configuration a states of things consists of.

+ It allows states of things a wide range of complex configurations *without* having to assume any logical structure.

+ It presents the form of an object as a local notion independent from the configurations in logical space.

The advantages mainly speak for themselves. We shall see that the third advantage has a bearing on the creativity of language, discussed in section 7.7.4, which is easier to understand on a local than on a holistic view on objects.

As a model for tractarian ontology, the LEGO® approach has the following disadvantages:

---

[29]Potter (2009), chapter 11, argues convincingly that in the early stages of his logical work Wittgenstein distinguished between complex and fact. A complex may come with many facts. Potter's example is of a salad as a complex of objects – such as parts of leaves, eggs, cheese, nuts, grapes, tomatoes,... – related to one another in different ways. Then, there may be many facts that are grounded in the single complex: it may be a fact that the salad contains two eggs; that it contains more tomatoes than cheese, etc.

Does the *Tractatus* still assume the same distinction? Potter seems hesitant, but I think it allows for logical complexes only. The distinction between complex and fact appears to be inspired by Russell's approach to have meaning as part of the world. This strongly suggests that meaning may have spatial or temporal aspects to it. In the *Tractatus*, however, the complex is essentially logical in nature, and the one-many relation between complex and positive fact can only be understood logically; e.g, as conjunctions of elementary propositions.

[30]Advantages and disadvantages are relative to what one is aiming for. It is surely interesting to study how object configurations can be had if all objects are assumed unsaturated. See Visser (2011) for a sophisticated model. My point is rather that such a formalization runs counter to a main tenet of the *Tractatus*: do not speculate about the specificities of objects prior to analysis.

- It attempts an abstract typing of the kind that is countered in 5.55 and its offspring.
- It gives object configurations a regularity for which there is hardly any foothold in the *Tractatus*.
- It posits objects and elementary propositions to have a general form.
- It presents logical space as the closure of operations to combine objects, but the *Tractatus* lacks such operations.

Let me elaborate. Firstly, a LEGO® approach gives objects a specificity they lack in the text. The connectors that objects are presumed to have, amount to a kind of *a-priori* typing that Wittgenstein was opposed to. According to him, it is not up to logic to restrict the possible kinds of states of things. In spite of its generality, there is no reason to assume that all object configurations fit the LEGO® mould.

Secondly, the LEGO® approach induces a regularity of combination that tractarian objects do not support. The form of an object gives its possibilities to combine with other objects, but the abstract way in which objects are presented leaves it open for this form to be quite irregular. That is, it may well be *im*possible to view the states of things in logical space as obtained from 'typed' objects in a bottom-up way. In case of such 'typing', the form of an object would be constant across all its occurrences in states of things. Then, if an object *o* occurs in a state of things, it combines in this state with a fixed number of objects that are adjacent to it and have forms that are permissible according to the form of *o*. There may be good reasons to favor such regularity of object-combination; e.g., to come to grips with the creativity of language (cf. Hacker (1979) p. 230 ff). Perhaps for this reason Wittgenstein relaxes his radical position by holding that 'in analysis the sign of a complex does not dissolve itself arbitrarily' (3.3442). Nevertheless, the abstract presentation of objects in the *Tractatus* leaves it open for objects to combine in less restricted ways than 'typed' objects do.

Thirdly, it is a main theme of the *Tractatus* to highlight the general form of propositions. The variable of its sixth main thesis captures it. It is telling there is no such variable to generate elementary propositions. If elementary propositions were generated like on the LEGO® view, Wittgenstein would surely have presented a variable for elementary propositions. But he did not. According to 5.472, the variable in 6 is unique in this regard.[31]

Fourthly, if objects had a general form, there should be operations that generate the space of all object combinations in a way analogous to propositions, but the *Tractatus* has no trace of such operations. For all we know, logical space may consist of a subtotality of all possible combinations that the purported types of objects incur, or may not be generated in this way at all.

To prevent speculating about the specificities of objects, for which there is no ground in the *Tractatus*, I take resort to a semi-formal, linear notation

---

[31] It seems that at the level of elementary propositions, where the core of picturing is to be found, Wittgenstein started to sense the diversity that is so typical of his later work even prior to having noted the color-exclusion problem. Indeed, truth-functionality is the most formal aspect of the tractarian system.

capturing the main characteristics of states of things. The notation presents states of things as configurations of object occurrences, – but leaves the specific configurations open as much as is possible.

2.4   Objects and holism

In the work of Frege and Russell there is the idea that only propositions have meaning, and there is at least the suggestion that propositions are obtained from entities whose type is given independently. Such *a priori* typing could be feasible in restricted areas, such as arithmetic or geometry, but according to Wittgenstein it is arbitrary as soon as all of language is taken into consideration. In the *Tractatus* 'object', with its form and content, is a holistic notion. The form of an object is not given prior to combination; it shows itself relative to the states of things in logical space (2.0141). That objects lack an independent 'type' and an independent existence is rooted in a radical version of Frege's context principle. Frege held that one should ask for the meaning of a word in the context of a proposition, not on its own (Frege, 1884, 23). In the *Tractatus* this becomes: names only have reference in the context of a proposition (3.3). But also: the referents of names must occur in states of things (2.012 and its offspring).[32]

Interestingly, Russell argues against the position that typing is only given with analysis. This is clear, e.g., from the beginning of his third lecture on The Philosophy of Logical Atomism, where he talks of the self-subsistence of particulars. (Russell, 1956, 204 ff.)

> ...each particular has its being independently of any other and does not depend upon anything else for the logical possibility of its existence.

In this respect Wittgenstein's holism is in stark contrast with the bottom-up approach to logic that Russell advocates in his version of logical atomism:

> ...propositions concerning the particular are not necessary to be known in order that you may know what the particular itself is. It is rather the other way round. In order to understand a proposition in which the name of a particular occurs, you must already be acquainted with that particular. (*Ibidem*)

In the same lecture Russell states: 'A very great deal of what I am saying in this course of lectures consists of ideas which I derived from my friend Wittgenstein. But I have had no opportunity of knowing how far his ideas have changed since August 1914, nor whether he is alive or dead, so I cannot make any one but myself responsible for them' (*ibidem*, 205). Wittgenstein explicitly denies that objects have existence independent of states of things (2.0122): objects must be considered in the space of possible combinations within which they occur (2.013). The notions of object and logical space are interdependent on each other, one cannot be had without the other. But in logical space, as the totality of what is possible, state of things, not object, is the basic notion.

---

[32]Dummett held that the context principle plays no rôle in Frege's later work, but Hacker (1979) argues convincingly that it does. Early Wittgenstein's version is quite radical: it concerns expressions in signs *and* what the expression signify.

2.5 State of things

A state of things is a configuration of objects. We have observed already the possibility of a state of things is interdependent on the form of its ᵕbjects, a holistic notion that is abstracted from logical space. From a moᵕ formal perspective this means that states of things must be seen as configurations of object-*contents*. States of things are instantiations of possible object combinations, but lack a full indication of the form of its objects.

In section 2.1.3, I have argued that it is not given prior to analysis which specific combinations in logical space are possible. Only the most abstract features of states of things are at hand – that they are objects in immediate combination. Further details need not be anticipated, since they are inessential to see how descriptive language is possible. Still, to indicate what logical space consist of we do need a notation that presents their main characteristics despite the possible variation in form. A discussion of the notation follows shortly.

Definition 2 *A state of things (Sachverhalt) is an immediate combination of object-contents:*

$$(o_0, \ldots, o_n)_c$$

The lower-case o's indicate object-contents, and the indexed parenthesis '$(-)_c$' indicate how they are configured. The sequence $o_0, \ldots, o_n$ is kept finite with $n \geq 2$. Its object-contents may have multiple occurrences. □

In the *Tractatus* the notion of object-configuration is as abstract as the notions of object and object-form. This means the notation should remain abstract, too. Be this as it may, it is important to think of the seemingly distinct features of the notation as inseparable. I now consider some apparent shortcomings of the notation.

2.5.1 Aspects of notation

In spite of its simplicity, the status of the notation in definition 2 should be clear. States of things are basic to about all that follows, and the notation may be misleading in some respects. In particular the following comes to mind:

1. How formal is the notation?
2. How does the notation capture that configuration is due to internal relation?
3. Does a certain type of configuration require a specific number of objects?
4. Does an object occur more than once in a state of things? And if so, must the same hold for the notation?
5. Can the same group of objects combine in different ways, or is their combination unique?

The topics are discussed in the given order.

Semi-formality States of things may allow for a diverse array of structure types. This appears to be in conflict with the uniform notation in definition 2, but it is not: the notation is semi-formal.

For a modern logician a semi-formal notation may be deeply unsatisfactory. But the point is: it suffices to indicate the essentials of states of things

– that they are configurations of object occurrences, – but it should be left open which configuration types are possible. One should not attempt to formalize what cannot be formalized. Cf. 5.55 and its offspring. In particular, the notation does *not* capture the general form of elementary propositions, and it does *not* indicate how the configurations $(-)_c$ result from operations based on the forms of its objects. Precisely this early Wittgenstein held not to be given prior to analysis, and so inessential to seeing the true nature of descriptive language.

Notation and configuration In the notation for states of things '$(-)_c$' indicates its configuration. This notation comes with a few warnings. Firstly, the notation may suggest 'c' indicates an external relation 'c'; instead it indicates an internal relation. To preclude an infinite regress, the configuration is immediate of the objects involved. Also, in the ontology the configuration is not an entity itself! It must therefore be stressed that $(-)_c$ does *not* name an object, and it is *not* a material function or concept applied to the arguments $o_0, \ldots, o_n$. The type of combination is rooted in the *form* of objects. However, since object-form cannot be determined on the basis of a specific state of things – it requires all possibilities in logical space instead, – the forms of the objects cannot be described.

Perhaps a comparison of the notation with modern logic clarifies its nature. In modern logic terms *are* implicitly typed. E.g., $R(c_0, \ldots, c_n)$ is a proposition, since it combines an n-ary relation with n individual constants. In the system the formula could be written: $(R, c_0, \ldots, c_n)_{lo}$ indicating that the configuration '$R(c_0, \ldots, c_n)$' is a linear order of the objects $R, c_0, \ldots, c_n$.[33]

The current notation does not keep track of configurations of objects beyond the level of a state of things. In principle this could be done, but it leads to a formality that is disproportional to the abstract, semi-formal notion of object as it is given in the text. There is also a more fundamental objection to such an approach. If object-occurrences can be seen as related to each other *in* a purported state of things, one should wonder whether the configuration is better seen as a state of things in itself, with the remaining structure resulting from logical combination. This would comply with the principle that the complexity of states of things must be reduced as much as possible in favor of logical definition.

Does a configuration have an arity? One may wonder whether a configuration $(-)_c$ always requires the same number of objects? The *Tractatus* is silent on the issue, so it is best to stay as abstract as possible. Here we think of the notation as indicating an instance of a linear order. Consequently, a configuration may come with different numbers of objects. For example, $(o_1)_{lo}$, $(o_1, o_2)_{lo}$, $(o_1, o_2, o_3)_{lo}$, $(o_1, o_2, o_3, o_4)_{lo}$ may all be instances of linear orders.

Rôles Apart from the status and the arity of the notation '$(-)_c$' there is another subtlety that needs attention. Should it be allowed, as in definition 2, that an object-content o_i occurs more than once in a sequence $o_1, \ldots, o_n$? The notation can be used as in $(o, o)_c$ and $(o', o')_c$ to indicate that o and o' c-relate

---

[33] Although the discussion on the topic is a still open, I assume relations and properties, etc. to be objects. Cf. Wittgenstein (1914-16), 16.6.15.

to themselves.

The aim is to find a proper notation to indicate configurations of objects. But what would it mean for an object to have multiple occurrences in a state of things? This question, however, puts us on the wrong track. That an o_i occurs more than once in the notation does not mean that the object itself does, it rather indicates that the object has several rôles in the configuration. Objects occur uniquely in a combination, or so I think, but in it they may have different rôles. The rôles are crucial when it comes to one state of things modeling another, and so should be captured in the notation. Also, due to the perfect balance between language and its ontology these aspects of representation should be aspects of both.

The position of an object-content in the sequence $o_1, \ldots, o_n$ is a notation for the rôle of an object in the configuration $(-)_c$. Both in formal and in natural language such rôles are readily available; e.g. in $(o, o', o)_c$ the two rôles of o could model that o connects to o' and to itself. Indeed, if we were to allow ontology to be of coarser grain than language – that is, if only single occurrences of object-contents occur in formalized configurations, no multiple ones, – it would imply that propositions which should be different, are identified via their senses. For example, it would be impossible to distinguish between the propositions $(n, n, n')_c$ and $(n, n', n)_c$ if n refers to o and n' to o' and ontology were only to have, say, the combination $(o, o')_c$.

Functionality Can the same group of objects, say, with contents $o_1, \ldots, o_n$, form multiple states of things or, if any, do they form just one? Black argues against multiple combinations (Black, 1964, 82-83). He reads 2.1514, 3.318, and the offspring of 3.4 as saying that the reference of the names in an elementary proposition should fully determine its sense. Among the theses he uses in his argument, the crucial ones, it seems to me, are 3.318 and 3.42a.

3.318 I regard the proposition – like Frege and Russell, – as a function of the expressions it contains.

Thesis 3.42a states that a proposition determines a unique location in logical space. The question is: how should the functionality of a proposition be taken? For now we concentrate on elementary propositions. See section 7.7 on compositionality for a discussion of the general case.

Elementary propositions are configurations of names in immediate combination (4.221). In the absence of *a priori* typing, it may not be clear without further ado what kind of object a name refers to. In particular, none of the names need to go proxy for a function taking the other names as arguments. E.g., a configuration $(o_0, \ldots, o_k)_c$ does not have to be of form $o_0(o_1, \ldots, o_k)$. Using Frege's imagery, all the objects named may be unsaturated, just to be completed in mutual combination. This seems to imply that the functional character of an elementary proposition can only concern the uniqueness of combination of the given names, as Black argues.

Still, are not the theses consistent with a more Fregean reading, too, namely that an elementary proposition is determined by the reference of its names and *the way in which the names are combined*? On the current interpretation, names refer to instances of forms. Suppose for the sake of argument that analysis has shown, say, the name n_1 to refer to a relational form taking as argument the remaining names n_2, n_3, which are both of the same form. Then, to the extent the possibilities are available in logical space, there is no reason to assume that $(o_1, o_2, o_3)_c$ is the sole possible combination of the objects referred to; $(o_1, o_3, o_2)_c$ may occur as well; cf. the now current notation: $o_1(o_2, o_3)$ and $o_1(o_3, o_2)$. On this reading, it is not just the names that figure in the propositional function, but also the way in which the names can be combined given the range of possibilities due to their forms. In the notation, I use $(-)_c$ to indicate the type of configuration. The configuration $(o_1, o_2, o_3)_c$ is an instance of type c, but so is $(o_1, o_3, o_2)_c$. It is the sign of the proposition that indicates which of the possible combinations is intended (see 3.21, 4.031, and interlude 2.11). Contra Black I assume that the same group of objects may form different combinations. That the notation indicates the functionality in an abstract way is strictly in line with the abstract way in which states of things are presented in the *Tractatus*.

2.6   *Interlude:* names and objects

Quite early on, Wittgenstein considered to what extent a distinction should be made between configurations of names and of objects (things). For instance, in the Notes on Logic (1913) he wrote:

NL.102(5) It is to be remembered that names are not things, but classes: 'A' is the same letter as 'A'. This has the most important consequences for every symbolic language.

In the Prototractatus this becomes:

PT 3.201221 'A' is the same letter as 'A'. For our language this is of the utmost importance.

And finally thesis 3.203 of the *Tractatus* just ends with the parenthetical remark:

3.203a3 ('A' is the same sign as 'A'.)

It is interesting to trace the changes. Compared to the Notes on Logic, the Prototractatus no longer makes an explicit distinction between objects and names, and 'symbolic language' is generalized to (daily) 'language'. In the *Tractatus* only the observation on multiple occurrences of the same name remains; a point of attention in a descriptive essentialism, which should mainly be concerned with a name's essentials.

I surmise that especially the first change is based on the insight that the suggested difference between objects with single occurrences and names with multiple occurrences, should be refined. The picture theory of meaning, first formulated in 1915, requires isomorphic projection between elementary propositions and states of things, and so does not allow for a difference in granularity between ontology and language. In the current notation the friction is resolved by noticing that despite its unique occurrence in a fact, an object still has different rôles in its configuration. It are the rôles that enable the isomorphism, which therefore should be present in the notation, not only in the elementary signs but also in the states of things.   [*End of interlude*]

2.7   Object form and objects

Objects have content and form. The form of an object is a holistic notion that cannot be described, but should be abstracted from logical space. The tractarian system introduces so-called formal concepts and relations for this purpose, which capture abstractions using propositions with a common form.

4.122a,d In a certain sense we could speak about formal properties of objects and states of things, or about structural properties of facts, and in the same sense about formal relations and structural relations. [...]

So, the existence of such internal properties and relations shows itself in the propositions that describe the relevant states of things or that are about the relevant objects.

The form of an object, in particular, should be given as a formal concept. Strictly speaking, in Wittgenstein's descriptive essentialism this can only be done via the corresponding sign-element; viz. the form of a name. But for now I indicate the abstraction directly in ontological terms, and use it to introduce the totality of objects.

Definition 3 *The* form $F(o)$ *of an object* o *is defined by:*

$$F(o) :\equiv \overline{(\alpha(o))}.$$

Here $\overline{(\alpha(o))}$ *is the totality of all states of things* α *in* Λ *in which the object-content* o *occurs.*   □

The form of an object is defined using a so-called bracket-expression, which indicates a variable. As said, this is not quite correct, for variables are sequences of propositional signs (5.501) and so sign-based in nature. Consistent with its claim that classes are superfluous (6.031), the *Tractatus* has no notation for arbitrary ontological totalities. Also, in the full system none is needed, because only those totalities are considered that are defined by a variable. Later I shall be more precise, but for now the current notation will do. In the notation the brackets '()' indicate that the order of the elements in the sequence is immaterial, and the bar '$\overline{}$' over the sequence indicates that all elements complying with the given form – here: that $\alpha(-)$ is variable but o constant, – are taken into consideration.[34]

Definition 4 *An object* $\hat{o}$ *is a pair* $o, F(o)$, *with* o *the object's content and* F *the object's form as specified in definition 3.*   □

Like in case of state of things, the notation highlights the main aspects of objects that as such are inseparable:

• An object $o, F(o)$ has a unique content o.

---

[34]The notion of form-sequence is reminiscent of sets or modern classes, but there are differences. Like classes a form-sequence is not an abstract object, as sets are. Also, it seems that abstractions over abstractions are not allowed; there are no form-sequences of form-sequences. Further, according to the *Tractatus* classes are based on an accidental kind of totality (6.031), whereas form-sequences are based on an essential kind of totality (form is an inherent property). Finally, note that form-sequences do seem to permit the repetition of elements. However, such repetitions may as well be disregarded – as I will do, – since they are inessential to showing a common form.

- An object $o, F(o)$ has a unique form $F(o)$: given that both its content o and the logical space Λ are unique, so is $F(o)$.

This presentation of objects – in particular of their form, – gives a detailed interpretation of the main theses concerning objects:

2.012 In logic nothing is accidental: If an object *can* occur in a state of things, the object should already predetermine its possibility.

2.0123a If I know an object, I also know all its possible occurrences in states of things.

2.013 Each object is, as it were, in a space of possible states of things. This space I can imagine empty, but I can not imagine the object without the space.

Here, space is a metaphor for the form of an object that captures all its possible occurrences. On the current reading this form is indeed given with the object, be it as an abstraction over logical space.

2.8 *Interlude*: Do states of one object exist?

The detailed notion of object sheds light on the question whether there are states consisting of a single object (thing)? The following text-fragments seem contradictory. According to 2.0121d, there is *no* object that can be imagined without the possibility of combining with others, but 2.013 states that each object is, as it were, in a space of states of things that could be imagined empty. To see that there is no contradiction, observe that if the space surrounding an object o is empty, there is *no* state of things in which it occurs; not even the fact consisting of just itself. In the current formalization $F(o)$ would be empty, and hence o can only be a superfluous element that does not occur in logical space. This non-realization might be taken as a borderline case of the possible combinations that 2.0121d requires. Anyway, 1.1 implies that an object constituting an entire fact should still be distinguished from this fact: objects are no part of the world. The formalization does just this: the object $\hat{o}$ is different from the state of things $(o)_c$, and only the latter would be part of the ontology. Be this as it may, I think 2.0121d should be taken at face value. Contingency concerns the realization or the non-realization of genuine complexity, so states of things contain at least two object-contents (whence $n \geq 2$ in definition 2). This should suffice to block any suggestion that an elementary proposition may consist of just one name that 'names' the object in the corresponding state of things, much like a Fregean proposition naming a truth-value object. However, the contents could still be of the same object, as in $(o, o)_c$. There is no need to eliminate cases like this. [*End of interlude*]

2.9 Identity of object-form

The form of an object is unique. But this should of course not be read in the strong way as saying that each form occurs as of just one object. It remains to be specified, therefore, what identity of object-form means. We are on our own here. Although the *Tractatus* does state that different objects can have the same logical form, the identity is not elucidated further:

2.0233 Apart from their external properties, two objects of the same logical form are just distinguishable from each other because they are different.

2.02331a Either an object has properties that no other object has, than a description can be used directly to distinguish it from the others and refer to it; or there are several things that have all of their properties in common, in which case it is quite impossible to single out one of them.

The question is: what does it mean for two object-forms to be identical to each other? Given that form is the possibility of combination, two forms should be considered identical if there is a sense in which they enable the same combinations. Assume, for example, that logical space Λ consists of just two states of things: $(o, o)_c$ and $(o', o')_c$. Then there are two objects: $o, (o.o)_c$ and $o', (o', o')_c$. These objects are different, as their contents are, but their forms are the same: $\hat{o}$ and $\hat{o}'$ allow the same combination with themselves. Notice that in the example $F(o) \neq F(o')$, so identity of form cannot be captured as identity of the totalities $F(o)$ and $F(o')$. What the example suggests instead is that identity of form requires an abstraction from content. This can be done as follows.

Definition 5 *The forms $F(o)$ and $F(o')$ are identical, notation:*

$$F(o) \overset{\sim}{=} F(o'),$$

iff[35] the sequences $F(o)$ and $F(o')$ can be one-one related so that a bijection ι between object-contents can be abstracted with:

i) $\iota(o) = o'$, *and*

ii) $(o_1, \ldots, o_n)_c$ *in* $F(o) \Longleftrightarrow (\iota(o_1), \ldots, \iota(o_n))_c$ *in* $F(o')$ □

Some examples may give a feel of what identity of form amounts to.

Example 1 Consider two objects: $\hat{o}$ and $\hat{o}'$ with $F(o) \equiv (o, o)_c$ and $F(o') \equiv (o', o')_{c'}$. Then $F(o) \not\overset{\sim}{=} F(o')$, for due to the difference in configuration the bijection $o \mapsto o'$ does not satisfy definition 5 ii).

Example 2 Suppose Λ consists of just two states of things: $(o, o', o)_c$ and $(o', o, o')_c$. Then $F(o) \overset{\sim}{=} F(o')$ via the map $o \mapsto o'$.

Before proving that identity of form has the properties one would expect, let us first check to what extent the notion is tractarian in nature.

Identity of form is tractarian At first sight, identity of form appears to be non-tractarian. Its definition seems of a meta-level that cannot be expressed within the system. However, observe that although the system has no bijections as maps from sets of object-contents onto each other, bijections can still be taken as a matter of speech. Definition 5 indicates that the bijection shows itself once the sequences of the given forms are arranged in an appropriate one-one relation. In chapter 3 we shall see that such bijections are basic to sense.

What remains to be clarified is that the sequences of states of things are available within the system. Purely ontologically this is not so: the *Tractatus* requires a sequence to be defined by a common form of propositions. But as

---

[35] The phrase 'iff' is short for: if and only if.

soon as elementary propositions are at hand the relevant sequences are defined respectively by the forms $\alpha(\underline{o})$ and $\alpha(\underline{o}')$, with $\underline{o}$ and $\underline{o}'$ the names of o and o'.

Form as congruence Both the form of an object and the identity of object-form are available within the system (as soon as elementary propositions are). But proofs about the formal properties and relations could be out of reach. Such proofs contain statements in which the formal concepts occur, and in general there is no procedure to reduce such statements to admissible ones: propositions, or sequences of propositions. Be this as it may, the following proposition can still be seen as part of the system's elucidation, because it is based on presenting the given variables jointly as pairs or triples with the same form.

Proposition 6 *Identity of form $\tilde{=}$ is an equivalence relation satisfying formal congruence.*

Proof. It is clear that $\tilde{=}$ is an equivalence relation: it is reflexive, symmetrical and transitive.

- *Reflexivity*: Every object has the same form as itself.
- *Symmetry*: If $\hat{o}$ has the same form as $\hat{o}'$, $\hat{o}'$ has the same form as $\hat{o}$.
- *Transitivity*: If $\hat{o}$ has the same form as $\hat{o}'$ and $\hat{o}'$ has the same form as $\hat{o}''$, $\hat{o}$ has the same form as $\hat{o}''$.

Moreover, identity of form satisfies formal congruence.

- *Formal congruence*: If o has the same form as o', then: o has a certain rôle in a configuration of type $(-)_c$ if and only if o' has the same rôle in a configuration of type $(-)_c$.

Formal congruence is immediate from definition 5 ii): o has rôle i in a configuration of type $(-)_c$ if and only if o' has. $\qquad\qquad\square$

In the following I shall not always be as precise, and may leave it to the reader to check whether a notion or proof is tractarian or not.

2.10 *Reflection*: Objects determine logical space

Objects and states of things in logical space are twin-notions; one cannot be had without the other. The current reconstruction does justice to this. Although logical space encompasses all possible configurations of object-content, objects themselves cannot be given independent of it: the form of an object requires *all* states of things in which it occurs. Yet there is a sense in which objects determine logical space.

2.0124 With all objects, all *possible* states of things are given as well.

This can be understood as follows: once the objects are given, they determine a space of configurations that is essentially the same as the space from which their forms are abstracted. The next proposition shows in which sense this is so.

Definition 7 *Let $\hat{o}_1, \ldots, \hat{o}_n$ be objects and $(-)_c$ a configuration type:*

$$(\hat{o}_1, \ldots, \hat{o}_n)_c \Leftrightarrow_{df} \text{For all } i \text{ with } 1 \leq i \leq n: (o_1, \ldots, o_n)_c \text{ in } F(o_i).$$

We use $\hat{\Lambda}$ to indicate the space of configurations $(\hat{o}_1, \ldots, \hat{o}_n)_c$. $\qquad\square$

It is worth to notice that the construction is non-tractarian. Within logical space Λ an object is a totality of states of things in which a particular content o occurs (and so defined by the tractarian variable $\overline{(\alpha(\underline{o}))}$). But a sequences of such totalities – as used for the states of things $(\widehat{o}_1, \ldots, \widehat{o}_n)_c$ in $\widehat{\Lambda}$, – is not a formal notion over Λ.

Proposition 8 *The map $\widehat{-}$ from the totality of object-contents to the totality Ω of objects is an isomorphism between Λ and $\widehat{\Lambda}$.*

Proof. We have to show that $\widehat{-}$ is a bijection that preserves configuration. To show that $\widehat{-}$ is a bijection, first assume for arbitrary o and o' that $\widehat{o} \neq \widehat{o'}$. Then either $o \neq o'$ or $F(o) \neq F(o')$. If $o \neq o'$, we are done. But if $F(o) \neq F(o')$, $o \neq o'$ too, for an object cannot have two forms. Thus, $\widehat{-}$ is injective. As to its surjectivity, take object ω. The object ω is the pair $o'', F(o'')$ for some object-content o''. Therefore $\widehat{o''}$ is ω, as required. It remains to show that configurations $(-)_c$ are preserved:

$$(o_1, \ldots, o_n)_c \text{ in } \Lambda \Leftrightarrow (\widehat{o}_1, \ldots, \widehat{o}_n)_c \text{ in } \widehat{\Lambda}.$$

For $(-)_c$ arbitrary, assume $(o_1, \ldots, o_n)_c$ in Λ. Then $(o_1, \ldots, o_n)_c$ in $F(o_i)$ for all i with: $1 \leq i \leq n$. So $(\widehat{o}_1, \ldots, \widehat{o}_n)_c$ in $\widehat{\Lambda}$. The converse implication holds by definition. $\square$

What does proposition 8 learn us? Apart from giving a detailed interpretation of 2.0124, it also sheds some light on Wittgenstein's view on *a priori* typing. Apparently, contra *a priori* typing, the proposition shows that each object-content o_i in a states of things $(o_1, \ldots, o_n)_c$ can be regarded as being typed by its form $F(o)$. On this view the logical space Λ is not altered essentially, because with typing an isomorphic copy results. It should be noted however that this kind of typing is rather superfluous. In defining $F(o_i)$ it is impossible to indicate for a state of things $(o_1, \ldots, o_n)_c$ to which o_j the object $\widehat{o}_i$ is immediately connected and to which it is not. We may as well assume that $\widehat{o}_i$ is connected to all. Thus, the types do not learn us much about the combinatorial possibilities of an object. In the construction the type $F(o_i)$ of $\widehat{o}_i$ is holistic and not really local to it. They remain abstractions over logical space in which the form of an object is given implicitly. To make the same point from the other side of the equation: given that explicit typing results in a isomorphic copy, the type of an object in the logical space Λ may as well remain implicit, until analysis explicitly shows which forms are at stake.

2.11 *Interlude:* notation

When using the notation $(o_1, \ldots, o_n)_c$ for states of things, I have stressed that $(-)_c$ is *not* an object or a material function applied to the arguments $o_1, \ldots, o_n$. The connection among the objects $o_1, \ldots, o_n$ is immediate, based on their inherent properties, and $(-)_c$ is just a notation for the resulting configuration-type. In particular, $(-)_c$ does not refer. This interlude shows that there is a striking similarity between the notation used here and that in the diaries and in the *Tractatus*.

On 23rd September 1914 (MS101, 25r), just a few days before the picture theory of meaning was formulated, Wittgenstein made a drawing in which two relational facts are assigned to each other via the objects related:

$$
\begin{array}{ccc}
a & R & b \\
| & & | \\
c & S & d
\end{array}
$$

Even if the idea of immediate connection of objects is not articulated yet, this drawing only makes sense if R and S indicate the configurations of the objects as it results from such a connection. The same assumption clarifies the enigmatic 3.1432.

3.1432 Not: "the complex sign 'aRb' says that a stands in the relation R to b", but: that "a" stands in a certain relation to "b", says *that aRb*.

Here 'a certain relation' is best read as pertaining to the immediate connection between the objects a and b, while R just indicates the structure of the resulting complex without itself referring to an object. So R is used as $(-)_c$ is in the notation introduced in definition 2. The functions in 4.24 can be understood in a similar way:

4.24 Names are simple symbols, I indicate them by single letters ("x", "y", "z").

   I write the elementary proposition as a function of names that has the form "fx", "$\varphi(x,y)$", etc.

   Or I indicate them by the letters 'p', 'q', 'r'.

If the f and φ were material functions, they would be configurations of names themselves (see section 4.5 below). Then, the φ and the f in 4.24b would hide these names, whereas by contrast there is the strong suggestion that all names occurring in the elementary proposition are listed. A complete listing of names is fully compatible with taking f and φ as indicating the type or their configuration, without any reference to parts of this structure.

   Finally, in the Notes on Logic the notation used here is mentioned explicitly. I quote two examples:

NL Sum. Consider a symbol "xRy". To symbols of this form correspond couples of things whose names are respectively "x" and "y". The things xy stand to one another in all sorts of relation, amongst others some stand in the relation R, and some not (...).

NL 2nd MS Whatever "x" and "y" may mean, "xRy" says something indefinable about their meaning.

Note in passing that the first example explicitly states that names and the objects they refer to may be combined in different way. This is in line with how functionality of configurations was understood on page 33. As to notation, the example suggests that something like Tarskian assignments for object-contents might do. But this would be too weak for we need explicit structure, which is indicated by the configuration-type R. Their configuration must show itself, cannot be described. Thus, the indefinable mentioned in the second example gives a first glimpse of what developed later into the idea that objects are immediately connected. *[End of interlude]*

3

Projection

To safeguard ethics from dilution, the system of the *Tractatus* demarcates language from within, by pointing at the symbolic essentials of descriptive language. The ontology with which the text begins is a first step in this direction. The next step is based on the Hertzian insight that our ability to picture facts to ourselves is essential to description and its symbols (2.1). So, a view is required on how the pictorial part of logical space is projected onto what is depicted.

Based on the ontology developed in chapter 2, the present chapter starts giving details of elementary pictures. Since the notions of state of things, object, and identity of object-form are already in place, two things are still required: more detail on the elementary signs projected, and more insight into the notion of projection itself. Both aspects are covered here. They prepare the ground for chapter 4, which is on elementary propositions and their proper parts (expressions, material functions). For now I start with an overview of what picturing is all about.

3.1 Picturing in overview

The main point in coming to grips with the sense of descriptive language, is to see that we picture facts to ourselves (2.1). Picturing involves the projection of a picture onto what is depicted, a notion that is elucidated in 3.11–3.13, and is discussed shortly.

Projection is based on the so-called pictorial form of a picture. There are many kinds of picture – such as spatial installations, movies, photographs, maps, different kinds of notation – and there are as many kinds of pictorial form and projection. Thesis 4.0141 gives the example of music, where score, performance, sound and record-groove are projected onto each other, since somehow they share the same form.

It is a distinctive feature of the tractarian philosophy to hold that the plethora of pictures have a common core. Each picture, of whatever kind, is a picture because it is a specific instance of the most abstract kind of picture: logical pictures, which consist of a truth-functional complex of elementary pictures that are projected into logical space. Logical picturing is the essence of all picturing. Indeed, the core of each proposition is a logical picture.

Just as there is a most general kind of picture – the logical picture, – so there is a most general kind of projection. I call it: logical projection. It is the kind of projection that is based on the form of a logical picture.

In logical projection there is a strict relationship between projection and reference; they are siamese twins, like the related notions of states of things and objects are. Logical projection concerns propositions as logical pictures, and reference concerns the rôle of names in logical projection. The contextual principle states that names only have reference in the context of an elementary proposition (3.3). Phrased in terms of projection the principle becomes: only in the context of logical projection names have reference. But the converse holds, too: without reference there is no projection. One is not prior to the other. The reference of names depends on the propositions in which the names occur, and referents can only be named in the context of a state of things as it is depicted in projection.

Theses 3.11–3.13 elucidate how picturing involves the projection of a picture onto what is depicted, and the offspring of 3.2 clarifies reference. Combining the theses in a systematic way, we arrive at the following overview of how an elementary proposition is projected onto what it depicts:

- The elementary proposition is a projected sign (3.12).

- The elementary proposition is projected onto a state of things (follows from 3.201, 3.21).

- The simple elements of a sign's structure are called 'names' (3.202), those of states of things 'objects' (2.02).

- The names of the sign represent the objects in the depicted state of things (3.2). Names have objects as their reference (*Bedeutung*); they stand proxy for objects (3.203, 3.22).

- The sign exhibits the form of its projection, not its content (3.13).

- The pictorial form of the propositional sign, its structure, is the same as that of the state of things it is projected onto (3.21). The sign indicates that the objects its names refer to, can be configured in exactly the same way as the configuration of its names (3.13).

So an elementary proposition is an elementary propositional sign in its projective relation to the world. It gives the possibility of what is projected, since its sign provides a possible configuration of the referents of its names, but the proposition does not include what is projected. In chapter 4, I discuss the complete notion of elementary proposition, with special attention to its sense and truth. But before doing so, we first have to consider the status of elementary propositional signs as facts, and also what logical projection amounts to.

3.2 Propositional signs as facts

When formalizing elementary propositions as logical pictures, an interesting difference presents itself between the tractarian system and the model-theoretic approach to logic that Tarksi has defined. The model-theoretic approach assumes an absolute distinction between syntax and semantics, between language and its meaning.[36] By contrast, the *Tractatus* is ambivalent as to what extent propositional signs – roughly: interpreted syntax, – should be distinguished from the state of things they describe – roughly: semantics. On the one hand, both signs and sense are presented as facts: the world is the totality of facts (1) of which pictures and propositional signs are part (2.141, 3.14, 3.1431). An apt example can be found in the *Notes on Logic* from 1913: 'Propositions are themselves facts: that this inkpot is on this table may express that I sit in this chair.' NL(97). On the other hand, if pictures and propositions were mere facts, one would expect their analyzed forms to be described in the ontological terminology: elementary sentences would be configurations of objects just as states of things are. By contrast, Wittgenstein speaks of 'elements of pictures' and calls the simplest such elements not 'objects' but 'names'. This suggests a rather strict distinction between language and the world described. If language is to be part of logical space, the distinction between propositional signs and their senses is not entirely clear. What does it amount to to have sign and sense as of one category? I will argue that the difference between sign and sense is not absolute but functional: the sign of an elementary proposition is a state of things *used* as a sign, and names are the objects in a state of things so used. It is logical projection in use that determines whether a state of things is a sign or is depicted.

To take both pictures and the depicted as part of logical space is a way of indicating that for Wittgenstein descriptive language is a worldly matter throughout. But 'worldly' should be taken in an Aristotelean fashion: it is the essential form and content of language and world that enables one to describe the other, the other to provide meaning of the one (Pears, 1979, 202). Only facts can express sense (3.142), and the tractarian urge for unification makes a unique logical space that encompasses all possible sense and signs to be preferred over a system with two distinct realms: one for signs, one for sense. This is not to say that all semantics can be part of logical space. A projection of one form onto another, which turns a configuration of elements into an elementary proposition, is not itself a state of things or an object occurring in logical space. A projection is a formal (internal) relation between configurations: it shows itself in abstraction over specific instantiations of form, but it is not part of reality.

The reading that elementary signs and their senses are of one logical category has to be argued for in detail. To this end I shall respectively consider two characteristics: their configuration and their elements. Since configuration is due to the forms of elements, elements require most of our attention.

3.2.1 The configuration of elementary signs

It is not hard to see that if there is any difference in the category of sign and sense, it cannot be due to configuration. Just recall that the *Tractatus* assumes its system to be descriptively complete: the world can be described completely using the basic ingredients of analysed language.

---

[36]I refer to model theory in general, and not to specific kinds of model. E.g. for arithmetic or set theory there are models where syntax can be thought of as being part of their domain.

4.26 A specification of all true elementary propositions describes the world completely. The world is described completely by specifying all elementary propositions together with an indication which of them are true and which false.

Since elementary description involves identity of the form of an analyzed sign and of its sense, it follows that the kinds of configuration that signs and sense allow must be the same. For each type of configuration in a sign there is the same type of configuration in its sense, and conversely. Therefore, any categorical difference between them should be due to their elements. I will argue however that there is no essential difference.

3.2.2 The elements of elementary signs

The *Tractatus* has approximately twice as many theses on objects as on names. The theses on names are about how they represent objects and how their reference (*Bedeutung*) differs from the sense (*Sinn*) of elementary propositions. By contrast, the theses on objects are about what enables them to form the totality of contingent configurations. Although both objects and names are positioned as the most basic elements of configuration, only objects are explicitly presented as having content and form. Yet, it can be argued that within the system names, too, have both content and form.

To begin with, the simplicity of names, stated in 3.26, must be as absolute as that of objects.

3.26 A name cannot be analyzed any further by means of a definition: it is an ursign.

It may seem that 3.26 allows a name to be a ursign just to the extent that it goes proxy for a simple object, while from any other perspective it may be complex. Since the propositional sign is itself a fact (3.14) it is hard to comprehend what this complexity could consists of, but anyway it cannot be essential to description. It cannot consist of object configuration, because with such complexity the name ceases to be a name. Although Wittgenstein's interpretation of Occam's maxim does allow for unused sign-elements (3.328, 5.47321), such frills and fringes are not essential to the end-state of analysis, and it is this context we are interested in. This means that the only 'complexity' a name could have is accidental, inessential to meaning, and the same holds of the object it refers to. Again: to be is to be essential to description.

We picture facts to ourselves (2.1, italics added). Since elementary signs are facts, names can only be objects *used* as names. Thus, names like any object have content and form. This interpretation is supported further by the fact that expressions *are* explicitly stated to have form and content (3.31). Expressions as part of elementary propositions are configurations of names (cf. section 4.5). Clearly, the content and form of an expression can only result from the content and forms of the names it consists of; strictly analogous to the form and content of a state of things being due to the content and form of its objects. The content of a name is its occurrence in a sign, and its form is implicit in the totality of all elementary signs in which it occurs. In this regard there is no difference between object and name. Names without form could not go proxy for objects, for which form is essential, and names without content could not instantiate form.

Some may object to a reading in which the distinction between signs and the depicted is non-absolute. It is however in line with the strong a-psychological tendency of Wittgenstein's early work, in which even tables, chairs and books may express the sense of a proposition, its 'thought' (3.1431, 4). Although Wittgenstein did once suggest that thought may have its own constituents (Wittgenstein, 1995, 125), this did not interest him much, as the assumed mental nature of such constituents would be inessential to description. Since an absolute division between language, world and their constituents would be necessary in a way that according to the *Tractatus* only logic is, I see no reasons for positing a quite arbitrary schism between the symbols of 'language' and the 'worldly' facts. But those who think otherwise can easily adapt the system presented here to their insights.

In sum, signs and what they are about are of one category: both belong to logical space. The difference between the two is functional. An elementary propositional sign is a state of things *used* as a sign by projecting it onto another state of things. Names are the objects of such signs.

3.3 *Interlude*: names and the coherence of a sign

It is interesting to observe that the combination of names in an elementary proposition is presented using the same image as is used for the combination of objects in states of things. According to 2.03, objects in a state of things fit into one another like the links of a chain (*Kette*), and 4.22(1) states that an elementary proposition is a connexion, a linking-up (*Verkettung*) of names in immediate combination. This suggests that elementary propositional signs are similar to state of things, and names to objects. In my opinion they are essentially the same.

On this reading one wonders why Wittgenstein asks how the coherence of propositional signs comes about (4.221b)? Is not the connection of names in elementary propositions as immediate as the connections of objects in states of things? Precisely for this reason, I think 4.221b is not about the coherence of elementary signs, but about the coherence of logically complex signs. In the *Tractatus*, the coherence of such signs is indeed left unaccounted for. E.g., logical space does not seem to allow for the 'interpunction' of complex signs. See chapter 6 for further discussion.   [*End of interlude*]

3.4 Logical projection

In reflecting on how meaningful language is possible, the notion of picturing (modeling) comes to the fore, which is to be understood, among other things, in terms of logical space. Logical space provides the framework of picturing: its structure and content give the essentials of what is required for one configuration to be a logical picture of another one. Yet, logical space is presented in the abstract, not in full detail, because only the analysis of language makes clear which specific states of things there are.

In the previous section I have argued that logical space includes elementary signs and what they are about. The next step is to clarify how one can be projected onto the other. Here, we shall be mainly concerned with what I have called 'logical projection', i.e., the projection of logical, fully analyzed pictures. Logical projection is a subtle concept that is central to the tractarian system. To indicate pros and cons, I let the concept evolve in a few steps corresponding to some natural interpretations. I will distinguish between possible projection and projection in use; closely related concepts that are not quite the same.

3.4.1 Possible projection

Isn't logical projection a local matter of somehow assigning the names in a particular elementary proposition to the objects in a state of things so as to do justice to both configurations? Even on this basic assumption not just any local assignment will do. Firstly, the configuration of the names must correspond to the configuration of objects in the state of things (3.21), and this in a strict sense: the elementary sign must have the form of what it depicts (3.13). Secondly, the names in the proposition go proxy for the objects in the states of things (3.22), and for each object in the state of things there must be a name in the proposition (3.21). Finally, the tractarian treatment of identity requires that different names refer to different objects (5.53). All in all, this results in the following notion of projection.

Definition 9 *Write α for $(o_0, \ldots, o_m)_c$ and α' for $(o'_0, \ldots, o'_n)_{c'}$. The state of things α possibly projects onto the state of things α' iff a one-to-one relation $o_i \mapsto_\pi o'_i$ between object-contents can be abstracted from α and α' so that:*

$$(\pi(o_0), \ldots, \pi(o_m))_c \text{ is the same configuration as } (o'_0, \ldots, o'_m)_{c'}.$$

That π shows α to project onto α' is also written as: $\alpha \xrightarrow{\pi} \alpha'$. Here α is the sign and α' what the sign is about. □

It should be stressed that the bijective map π is *internal*; it is not an object itself but a formal abstraction from the structural identity of the states of things involved. If α possible projects onto α', the form of the states of things are the same but not the object-contents occurring in them. The abstracted π indicates which object-contents are associated with each other one-to-one to highlight the structural identity.

The 'space' of all possible projections has some well-known properties. For any α, α' and α'' one has:

- *Symmetry*: If α projects onto α', then α' projects onto α. For if $\alpha \xrightarrow{\pi} \alpha'$, then the inverse π^{-1} of the abstracted π shows $\alpha' \xrightarrow{\pi^{-1}} \alpha$.

- *Transitivity*: If α projects onto α' and α' projects onto α'', then α projects onto α''. For if $\alpha \xrightarrow{\pi} \alpha'$ and $\alpha' \xrightarrow{\pi'} \alpha''$, then the composition $\pi' \circ \pi$ of the abstracted π and π' shows $\alpha \xrightarrow{\pi' \circ \pi} \alpha''$.

From a logical point of view it seems natural to add reflexivity: α projects onto itself: $\alpha \xrightarrow{\pi} \alpha$. But Wittgenstein would surely object to this manner of speech. Self-projection would involve projecting each object-content onto itself, which says as little as stating of an object that it is identical to itself (5.5303). Further, the content of projection is explicitly stated to be no part of the projection, which in case of self-projection is impossible (3.13). Cf. section 3.4.3 for more discussion on self-picturing.

If for each α we consider the totalities consisting of α itself and all the states of things onto which α projects, one gets a partition of logical space into subtotalities of states of things with the same form (or: 'type'); the elements of each of these sub-totalities can therefore depict each other.[37] I will argue it is precisely this aspect of possible projections which should be constrained to get the more restricted notion of projection in use. Sense is determinate, and identity of form is unambiguous. So a π showing the structural identity of α and α' should be fixed in a logical way to be unique.

3.4.2   Projection in use

There is an interesting difference between the logical space of all possible states of things and the space of all possible projections. Since states of things are independent of each other they may all be realized, but for philosophical reasons not all possible projections can be in use together. If projection is used in (elementary) description, the states of things used as a sign is itself a fact (3.14), hence realized. And, as we shall see, in such a context projections should not be, say, symmetrical or transitive. But before showing this in detail, let us first define what projection in use is.

Definition 10 *A world is given as a pair* Λ, R, *where R gives the states of things that are realized and which not. We use (variants of) w for worlds.*

In definition 10, the R used to indicate whether a state of things is realized or not is purely notational: existence is not a property assigned to states of things.

Definition 11 (Projection in use) *A projection* π *is in use in a given world w, iff* π *is a fixed possible projection* π *of which each sign* α *is a realized state of things in w. We employ the notation:*[38]

$$\alpha^\circ \xrightarrow{\pi} \alpha',$$

if α *is used to project onto* α'. *The* $^\circ$ *indicates that* α *is is realized.* □

A possible projection need not be in use or thought (3), but if the projection $\alpha^\circ \xrightarrow{\pi} \alpha'$ is fixed to be used we leave the space of pure logical possibility. Firstly, the projection itself is transcendental: it shows a structural identity that cannot be described itself. Secondly, in the context of use α° is a propositional sign, so a fact, a realized possibility.

Projection in use helps capturing what is essential to the meaningful use of a sign. Meaning does not result from analysis. Analysis is a philosophical method that brings to the fore what is inherent in a sign used to convey meaning (3.326). The *Tractatus* posits logical projection at the heart of use. In a way, the use is non-empirical – as sense consists in employing a transcendental identity of forms, – and should thus be located in the metaphysical subject.

---

[37]A partition of logical space divides it in non-overlapping totalities that cover the space entirely.

[38]The notation is a bit misleading: it could also indicate a merely possible projection involving a realized α. In what follows no confusion is likely.

In a similar vein, names only acquire reference in relation to my metaphysical will that fixes one possible projection rather than another one.[39]

On the current approach projection is still a logical variant of the idea that projection is mainly a matter of relating names in a given sign with objects, while doing justice to the configurations of sign and depicted. Despite the naturalness of the idea, it happens to be too liberal. This is especially clear when moving from the local view that considers just two states of things to a more holistic view in which several projections are involved.

3.4.3   Self-picturing

Can a picture be a picture of itself? Clearly it is logically consistent to assume that there are pictures picturing themselves: $\alpha \xrightarrow{\pi} \alpha$. But self-pictures have some non-tractarian features that exclude them from the system.

Even as a purely possible projection – i.e., a projection in which α does not have to be realized, – a self-picture would be a projection that includes its own content, contradicting 3.13. And in case of projection in use – where α is realized, – self-picturing contradicts the basic principle that picturing is contingent. The realized elementary sign should be true in one world and false in another. By contrast self-pictures have the characteristic that they are realized iff they are true! To see this note that if, pace 2.141 and 3.14, the sign α is allowed not to be realized, then the content of its sense, namely α itself, is not realized either, and so α is false. But if α is realized, as 2.141 and 3.14 explicitly state, then the content of its sense is, and so α is true. In other words, if the sign of a self-picture in use is a fact, the self-picture is non-contingent: in use it is always true. *Quod non.*

The realization of a self-picture is conditional for its truth. This means that as soon as a self-picture is thought, no further comparison with the world is needed to see its truth (3.05). This would be an example of an *a priori* truth, and according to 2.225 such *a priori* truths do not exist. Self-pictures are, however, not *a priori* in the stronger sense of 3.04 where a thought is said to be *a priori* if its possibility is conditional for its truth. A purely possible self-picture could be false as long as it is not realized.

Thesis 2.173 prohibits self-picturing in stating that a picture represents its topic from the outside. In the notebooks, Wittgenstein is even more outspoken on the impossibility: The description of the world by propositions is only possible because what is described is not its own sign! (MS 101, 19.10.14.) To preclude self-picturing, definition 11 should be adapted to require that the domain and the range of a projection π must have no object-content in common.

3.4.4   Univocality

According to definition 11 projection is a local matter between two states of things. Thus it allows a sign to be projected differently onto different states

---

[39]Thus I agree with Ishiguro (1969, 20-50) and McGuinness (2001, 87-88) that both in the *Tractatus* and in the *Philosophische Untersuchungen* meaning in use is most basic. Of course, in the later work use is no longer restricted to structural identity in a unique logical space ('grammar'). It allows near endless variation in the kinds of use; use in which sign and what is represented may be loosely coupled, if at all.

of things. For instance, the sign α could be projected onto α' and onto α'':

$$\alpha' \xleftarrow{\pi'} \alpha^\circ \xrightarrow{\pi''} \alpha''.$$

Although it follows that α' and α'' are instances of the same configuration type – for α' can be projected onto α'' via $\pi'' \circ \pi'^{-1}$ – the elementary proposition α would be ambiguous. This is unwanted. As the states of things α' and α'' can be realized independently of each other, dual projection may result in α being true and false simultaneously, which is impossible. To preclude ambiguous projection, each state of things α in logical space is either not in use as a picture, or it is a picture of at most one other state of things. In other words, the internal relation $\xrightarrow{\pi}$ of projection in use must be a partial function: in use, each α projects onto at most one other α'.[40]

3.4.5 Naïve projection

The next question is whether projective chains are allowed? If pictures are facts, may pictures picture pictures picturing facts without constraint? This is what I call: naïve projection. The inkpot-example from the *Notes on Logic* invites one to reflect on how naïve projection may be.

NL(97) Propositions are themselves facts: that this inkpot is on this table may express that I sit in this chair.

If the inkpot on this table expresses that I sit in this chair, could then at the same time that I sit in this chair express that the inkpot is on this table? Or in a similar vein: if the inkpot on this table expresses that I sit in this chair, could then at the same time that I sit in this chair express that the envelop is lying on the doormat?

Even if projection in use is restricted to be irreflexive and partial, it still allows for finite, infinite and circular projective chains. For example, assume that the states of things $(o_i, o_i)_c$ are all available $(0 \leq i \leq n+1)$, then projection π can be as follows:

$$(o_0, o_0)_c \xrightarrow{\pi} (o_1, o_1)_c \xrightarrow{\pi} \ldots \xrightarrow{\pi} (o_{n+1}, o_{n+1})_c$$

Since logical space is infinite, infinite chains would also be possible. In a similar vein, circular projection could result:

$$(o_0, o_0)_c \xleftrightarrow{\pi'} (o_1, o_1)_c.$$

To what extent do these examples fit into the tractarian philosophy of meaning? It is clear that if projection in use is transitive, projective chains can not be allowed. For instance, in the above finite chain, say, $(o_0, o_0)_c$ would not just picture $(o_1, o_1)_c$, but due to transitivity it would picture all of: $(o_1, o_1)_c, \ldots, (o_{n+1}, o_{n+1})_c$, and *mutatis mutandis* the same holds in case of infinite and circular projective chains. I have already argued such multiple projection results in an ambiguity that is unacceptable from a tractarian point of view. This means that if the system were to allow projective chains or circles, the projection in use must be atransitive:

If $\alpha \xrightarrow{\pi} \alpha'$ and $\alpha' \xrightarrow{\pi} \alpha''$, then *not* $\alpha \xrightarrow{\pi} \alpha''$.

---

[40]The function is called 'partial', because not all states of things α need to be used as a picture of another state of things.

Observe that atransitivity implies irreflexivity: If for any α, $\alpha \xrightarrow{\pi} \alpha$, then $\alpha \xrightarrow{\pi} \alpha$ and $\alpha \xrightarrow{\pi} \alpha$, and from atransitivity: not $\alpha \xrightarrow{\pi} \alpha$, which is contradictory. So, not $\alpha \xrightarrow{\pi} \alpha$, for all α. With projection in use constrained to be an atransitive partial function, the chains of projections in use that still remain possible will be quite harmless. Yet, as I will argue, there is good reason to ban even those.

3.4.6 Describing signs?

That chains of projections in use are better disallowed, has to do with the curious nature of describing fully analyzed signs (pictures). There are two ways in which, say, an elementary sign can be described: from the outside in terms of properties that are inessential to its sense, and from the inside in terms of essential properties. I will argue that in the end the system supports neither.

External description of sign An example of describing the externals of a sign would be to describe 'Jo runs.' as a linear sequence of the marks:

$$\text{'J', 'o', ' ', 'r', 'u', 'n', 's', '.' .}$$

The externals of an unanalyzed sign, taken as a fact, have a structure of their own, but the description of this structure would tell us nothing about the sense of the sign. E.g., the above description of the sign 'Jo runs' is silent about its sense.

The idea of describing the externals of a sign is harmless, but does the tractarian system leave room for such descriptions? Description is only possible, if the unanalyzed sign as a fact has structure of its own. But I think it is non-tractarian to allow the externals of a sign to have real structure. It is basic to the *Tractatus* that propositions are *not* intermediate entities between a sentence and what it describes (like in case of Fregean senses). Analysis makes it manifest that the proposition is given *in* the sentence. After analysis all that is essential to what the sentence describes is laid bare. As a consequence, its structure in unanalyzed form, to the extent that it departs from its analyzed form, must be *in*essential to what it describes. Now, if it were allowed to describe this inessential structure 'from the outside', its structure would become essential for that description. Thus, what is or what is not essential would be relative to the perspective from which a fact is described. Although an interesting idea, this line of thinking is incompatible with the tractarian philosophy.[41] The semantic essentialism of this philosophy is absolute, not relative. For instance, objects as the ultimate constituents of structure, the substance of the world, are clearly absolute, not relative to any description. If this is correct, it means that the inessential aspects of signs cannot be described. Signs in use are the only means available to effect description, but its inessential clothing is no topic for description.

Internal description of sign A sign's clothing cannot be described, but the same goes for its essential structure. E.g., consider a logical analysis of the sign 'Jo runs.', which is, e.g., $(o_1, o_2, o_3)_c$. (The deviation from the stan-

---

[41] Perspective does have a rôle in the work of Husserl; see Husserl (1925/26) or see Lambalgen, M. van, and J. van der Does (2000), section 5.3, for a logical elaboration.

dard analysis R(j) is on purpose. Within the tractarian system one does not know prior to analysis how subject and predicate are represented.) The analyzed proposition states that the possibility $(o'_1, o'_2, o'_3)_c$ – i.e., an instance of the logical form of 'Jo runs', – is realized. The state of things $(o_1, o_2, o_3)_c$ projects onto $(o'_1, o'_2, o'_3)_c$, and if we were just concerned with possible projection, identity of form would be the only constraint. However, for projection *in use* the situation is different. Then $(o_1, o_2, o_3)_c$ is itself a fact, hence realized, and what would describing this fact add to its use as a sign? I hold there is no effective way to describe the essentials of pictures *in use*. The point is that description is grounded in structural identity. So for a picture in use either the description fails completely, in case the attempted description has a different configuration type, or the description is non-contingent, since the picture to be described is a fact, hence realized. Since in $\alpha° \xrightarrow{\pi} \alpha'$, α is a picture (sign) in use, all pictures in such a descriptive chain will be true except perhaps for one. For instance, in the chain:

$$(o_0, o_0)^°_c \xrightarrow{\pi} (o_1, o_1)^°_c \xrightarrow{\pi} \ldots (o_n, o_n)^°_c \xrightarrow{\pi} (o_{n+1}, o_{n+1})_c,$$

$(o_{n+1}, o_{n+1})_c$ will lack a truth-value, since it is not a picture; $(o_n, o_n)_c$ will be either true or false, depending on whether $(o_{n+1}, o_{n+1})_c$ is realized or not; and $(o_0, o_0)_c, \ldots, (o_{n-1}, o_{n+1})_c$ will all be true, since $(o_n, o_n)_c$ is a picture in use and so realized. Similarly, in the circle:

$$(o_0, o_0)^°_c \xleftrightarrow{\pi} (o_1, o_1)^°_c,$$

both $(o_0, o_0)_c$ and $(o_1, o_1)_c$ must be true, since both are realized pictures. This makes clear that chains and circles of projection *in use* come with redundant copies of identical structures that do not add anything to our descriptive capabilities. Everything that is worth to be described in a picture (sign) is apparent from the picture itself. Any attempt to describe the essentials of the picture as a fact can only succeed with a picture that is essentially the same as the picture to be described. This is clearly an ineffective redundancy that is inessential to semantics and is better shunned. Also, as in case of self-picturing, pictures describing the essentials of pictures in use do not have the contingency required, simply because pictures in use are realized.

It is best to disallow naïve projection completely: all projection is immediate, there are no redundant chains of projections *in use*. This insight leads to the further constraint that if projection in use involves multiple states of things:

$$\alpha_0, \ldots, \alpha_n, \ldots,$$

they are partitioned into two totalities, one of pictures and one of the depicted, such that each α_i is either a picture or a depicted, but not both. In non-tractarian terms: projection *in use* is a partial function whose domain and range are disjoint. Projection in use is best seen as a non-chained, immediate relation between a picture and what it depicts. In consequence, this kind of projection is irreflexive, asymmetric and atransitive.

The conclusion that signs cannot be described seems to contradict theses 3.317, 3.33, 5.472, and 5.501 in which Wittgenstein talks about the description of signs. Although I hold that in the final analysis elementary propositions are state of things used as signs, I think the theses use '*beschreiben*' (to describe) in a different way than in case of describing states of things. A sign could be 'described', e.g., by means of quotation to highlight some of its features in a way independent of its sense. Thesis 4.012 has a typical example. Indeed, according to Wittgenstein the way in which a sign or rule is 'described' is inessential (3.317).

Starting from a local notion of possible projection we have arrived at a more holistic variant. Just one step remains to see that projection in use must be truly holistic.

3.4.7   Descriptive completeness

In section 3.2.1 we recalled the world can be fully described using elementary language; that is, Wittgenstein assumes the tractarian system to be descriptively complete.

4.26 A specification of all true elementary propositions describes the world completely. The world is described completely by specifying all elementary propositions together with an indication which of them are true and which false.

Descriptive completeness serves several purposes. Hintikka (1986), p. 99, highlights the most important ones: it ensures that each name has reference, and that each elementary sign describes a possible state of affairs. That is, descriptive completeness ensures that basic presuppositions of determinate description are fulfilled. Cf. also section 9.2. In this regard, descriptive completeness has a rôle analogous to Frege's principle that each expression has a sense and a reference.

As I have argued in the previous section, the list of 4.26 should not include a description of pictures as facts: it is hard to believe that each description would induce a possibly infinite chain of descriptions describing other essentially identical descriptions. Instead, description is immediate of non-depicting facts. Therefore, each projection in use comes with a totality of sign facts that are not described themselves.

From a more formal point of view descriptive completeness has the consequence that each state of things in logical space must either be a picture or a state of things that is depicted. We have already seen that each picture depicts at most one state of things, but due to descriptive completeness each state of things that is not a picture must be depicted. Using the previous terminology, descriptive completeness comes with a partition of logical space: the pictures and the depicted are disjoint totalities that cover logical space completely.

3.4.8   *Interlude*: Projection, partition and solipsism

Projection in use is holistic in the sense that it induces a bi-partition of logical space. It is worth to observe that this reading reconciles the apparent tension noted in section 3.2; namely that on the one hand the *Tractatus* implies propositional signs and their senses to be of the same category – states of things in logical space, – while on the other hand the distinct terminology used suggests sign and sense to be vastly different. The reconciliation consists in holding that signs and sense are indeed both states of things, but that in each context of use there is also a strict but non-absolute distinction between them, i.e., the distinction determined by projection in use.

Although the approach remains implicit in the *Tractatus*, it must have been quite natural for early Wittgenstein to let projection determine a strict division between language and reality. Particularly telling in this respect is a figure in the notebooks, which Wittgenstein drew on November 15th, 1914. It presents the projection of a model (*Modell*) onto reality (*Wirklichkeit*) using two lines:

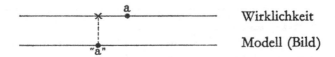

Wirklichkeit

Modell (Bild)

If in this picture both model and reality consists of facts, it can only be projection that effects their partitioning.

There may be several options to partition logical space, a projection does not have to be unique. Moreover, there is no constraint on how logical space is partitioned. But doesn't this mean that we are too general? After all, signs are basic elements of thoughts, and on the current reading projection is so liberal that one and the same state of things could be part of a thought in one projection and of a non-thought in another.

I have already indicated that Wittgenstein's notion of thought is a-psychological: inkpots and tables could show thoughts just as effectively as 'mental' signs can. Thoughts rather concern the essence of logic and symbolism, and Wittgenstein makes no claim about the elements of our cogitative capabilities. His indifference to the mental nature of thought is particularly clear in the letter to Russell dd. 19.8.19.:

> 'I don't know *what* the constituents of a thought are but I know *that* it must have such constituents which correspond to the words of language. Again the kind of relation of the constituents of thought and of the pictured fact is irrelevant. It would be a matter of psychology to find it out.'
>
> Wittgenstein (1995, letter 68)

A strict division between thoughts and what they are about would indeed be necessary in a way only logic is. Instead, language is a 'worldly' matter throughout.

In the *Tractatus* the world is shown in the metaphysical subject as my world, but this self retreats, so to speak, from the world without leaving a trace.

5.64 Here one sees that solipsism, strictly pushed to its limit, coincides with pure realism. The self of solipsism contracts into an expansionless point, and the reality that is co-ordinated with it remains.

As I am my world (5.63) and the limits of my language are the limits of my world (5.6), language and world – i.e., elementary signs and facts described, – must be essentially the same. This chapter introduces a radical version of this insight. Projection is not related to the mental, but to the limit of experience in which realism and solipsism are unified. The essence of meaning is based in the metaphysical rather than in the empirical subject.

Projection is transcendental. It is a trace of the metaphysical self in meaning and world. There is no fixed projection in logical use. In the context of one projection a state of things may partake in a picture, a thought, while in the context of another projection it may partake in the depicted world outside thought. Indeed, if realism and solipsism are unified as indicated in 5.64, there is no need to have an absolute distinction between thought and the depicted. But, as the descriptive completeness of 4.26 phrases, in any context there must be a perfect match between the two.[42]   *[End of interlude]*

3.4.9 Holistic projection

Up till now the discussion has moved from a local notion of possible projection, inspired by a purported human assignment of objects as the referents of names, to projection with a coarser granularity at the level of state of things, which keeps sense determinate by considering language as a whole. To arrive at a detailed view on holistic projection, we have to return to an approach that involves names and objects.

According to the *Tractatus* the sense of a sign is no part of its projection (3.13). This suggests it would be better to have a concept of projection that just uses the sign and the reference of its names, and to show that having both suffices for the possibility of the sign's sense. It turns out that to arrive at such projection, names and objects should be involved in full: both their content and their form must play a rôle.

It is not difficult to see that local projection based on just object-content is too shallow.

Definition 12 (Projection, tentative) *Let α be the state of things $(o_0, \ldots, o_n)_c$. A projection π is a injection from $o_0, \ldots, o_n$ onto other object-contents $o'_0, \ldots, o'_n$ disjoint from $o_0, \ldots, o_n$.* □

Projection as specified in definition 12 is unacceptable, simply because for given α and π the complex $(\pi(o_0), \ldots, \pi(o_n))_c$ does not have to be in logical space. This is due to the fact that it is phrased in terms of content, disregarding form. To eliminate this drawback, we should look for a richer, holistic version of projection.

From a holistic perspective, the projections of definition 9 and 12 are too meagre. If names are objects used as names – as I think they are, – projection should not only involve the content of name and object but also their form. Tractarian objects are neither like the infinitesimal, independent balls free-

---

[42]Like Ishiguro (1969) and McGuinness (2001), my interpretation stresses the logical nature of the tractarian system, and it is based on a radical form of contextuality. But in line with Stokhof (2002) I think tractarian ontology requires more than a metaphorical status. Language is itself a 'worldly' matter – its propositions are facts. One is hard pressed to give language a favored status while denying one to its ontology. In my opinion the difference is functional: language uses facts to describe states of things that lie outside it. On this reading, it is as difficult to think of language as 'answerable' to the structure of its ontology (Hacker's term). The *Tractatus* offers a descriptive essentialism: both language and its ontology are on a par, in essence both are part of the one universal logical space ('grammar').

floating in empty space, as the vexing image of Democritus has it, nor like the points in set-theoretic domains. Rather, each object is essentially dependent on all the objects with which it may combine to form a state of things. Thus, as Pears (1987) has rightly observed, the reference of names should be 'intensional' in that it is consistent with all relevant possibilities of combination.

On a holistic approach, the holism of objects and states of things will surface again. For if projection must comply with object-form, this cannot just hold for the object-contents involved immediately in a description, but it should hold recursively, so to speak, for all objects in the object's form. Before giving definitions, let us consider some examples to get a feel for what the holistic aspect of projection amounts to.

First suppose logical space Λ consists of:

$$(n_0, n_1)_{c^0}, \ (n_1, n_2)_{c^1}, \ (n_2, n_3)_{c^2},$$
$$(o_0, o_1)_{c^0}, \ (o_1, o_2)_{c^1}, \ (o_2, o_3)_{c^2}.$$

Now abstract from these states of things the projection π_0:

$$\pi_0 : \quad n_0 \mapsto o_0,$$
$$n_1 \mapsto o_1.$$

This projection is in line with definition 9 and seems sufficient to project $(n_0, n_1)_{c^0}$ onto $(o_0, o_1)_{c^0}$. But note that o_1 combines with o_2 to form c^1. So, π_0 does not do justice to the form of o_1 and should be extended to π_1:

$$\pi_1 : \quad n_0 \mapsto o_0,$$
$$n_1 \mapsto o_1,$$
$$n_2 \mapsto o_2.$$

The projection π_1 is in accord with the form of o_0 and o_1. Moreover, it projects $(n_0, n_1)_{c^0}$ onto $(o_0, o_1)_{c^0}$ and $(n_1, n_2)_{c^1}$ onto $(o_1, o_2)_{c^1}$. But again, π_1 disregards the form of o_2 which combines with o_3 to form c^2. Thus we see that π_2 is required:

$$\pi_2 : \quad n_0 \mapsto o_0,$$
$$n_1 \mapsto o_1,$$
$$n_2 \mapsto o_2,$$
$$n_3 \mapsto o_3.$$

For logical space Λ, π_2 is a projection doing justice to the combinatorial possibilities of the objects involved (its converse is another one). And to do so for the elementary proposition $(n_0, n_1)_{c^0}$, it must include such objects as o_2 and o_3, which at first sight do not appear to be involved at all. In particular, o_3 is not immediately related to an object referred to in $(n_0, n_1)_{c^0}$, but due to the holism of form it has to be taken into account all the same. By contrast π_3 with:

$$\pi_3 : \quad n_0 \mapsto o_0,$$
$$n_1 \mapsto o_1,$$
$$n_2 \mapsto o_3,$$
$$n_3 \mapsto o_2.$$

should not be considered a projection. For although π_3 seems to project $(n_0, n_1)_{c^0}$ onto $(o_0, o_1)_{c^0}$, it attempts to map $(n_1, n_2)_{c^1}$ onto $(o_1, o_3)_{c^1}$ which is not a possibility in Λ.

As a further example, let logical space Λ' be infinite and consist of:

$$(n_0, n_1)_{c^0}, \ (n_1, n_2)_{c^1}, \ (n_2, n_3)_{c^2}, \ldots$$
$$(o_0, o_1)_{c^0}, \ (o_1, o_2)_{c^1}, \ (o_2, o_3)_{c^2}, \ldots$$

The map π_0' given by:

$$\pi_0' : \quad n_i \mapsto o_i$$

is again a projection, but the map π_1' with:

$$\pi_1' : \quad n_i \mapsto o_{i+1}$$

is not. Although it maps possibilities onto possibilities, it fails to capture that o_0 combines with o_1 to form c^0.

Finally, let logical space Λ'' be as simple as:

$$(n_0, n_1)_{c^0}, (o_0, o_1)_{c^1}, \ (o_1)_{c^2}.$$

Then, although there is a bijection from $(n_0, n_1)_{c^0}$ to $(o_0, o_1)_{c^0}$, no projection is possible, because the form of o_1 can not be captured.

The above examples should suffice to introduce a holistic kind of projection that is in harmony with the holism of object-form. The idea is to base logical projection on names having the same form as their referents. Since we have already argued that projection induces a partition of logical space, the form of the name and that of the object will be abstractions, respectively, over the pictorial and the depicted part of logical space. And due to descriptive completeness the forms of names on the one hand and of depicted objects on the other will cover these parts of logical space entirely. Therefore, the definition of holistic projection can remain simple. As before, projection that is merely possible is defined first, and then projection in use.

Definition 13 *A possible projection π is an abstraction over logical space Λ so that:*

1. *π induces a bi-partition consisting of N, the occurrences of names in pictures, and O, the object-contents occurring in depicted states of things;*
2. *If an n in N occurs in picture α, all other object-contents in α are in N. If an object-content o in O occurs in $\pi[\alpha]$, all other object-contents in $\pi[\alpha]$ are in O;*
3. *π maps the object-contents in N one-to-one onto the object-contents in O;*
4. *for each n in N: $F(n) \stackrel{\sim}{=} F(\pi(n))$, with identity of form as in definition 5 but via π.*

If no confusion is likely, we use 'projection' to indicate holistic projection. $\square$

As it happens, there are several way in which holistic projection could be presented. Here, I have opted for the version that shows most clearly that π is holistic in the way that preserves form. The next section shows that holistic projection and isomorphism are equivalent to each other.

In holistic projection both content and form of names and their referents are represented. Each name $\widehat{n}$ is mapped onto an object $\pi(\widehat{n})$, or in full:

$$n, F(n) \quad \mapsto \quad \pi(n), F(\pi(n)).$$

Because names stand proxy for objects, name and object are both instances of the same form. As before Wittgenstein's view on identity is captured, too: in the context for projection different names refer to different objects. Also, holistic projection guarantees that what is projected is part of logical space.

Proposition 14 *The content of projection $\pi[\alpha]$ is a state of things in logical space Λ.*

Proof. Let α be $(n_1, \ldots, n_k)_c$. Then α is in $F(n_i)$ $(0 \leq i \leq k)$. Since $F(o_i) \overset{\sim}{=} F(\pi(o_i))$, $\pi[\alpha]$ is in $F(\pi(o_i))$ and so in Λ. $\qquad\square$

In definition 13, projection is possible projection. But projection *in use* can be specified as holistic projection within a world so that signs are facts.

Definition 15 *Call a pair $\mathcal{F} \equiv \Lambda, \pi$ a frame. A holistic projection can be used in a world w, Λ, π, R, iff it is a holistic projection π for Λ with all its pictures realized (in R).* $\qquad\square$

If no confusion is likely, I use 'projection in use' or even 'projection' to indicate a fixed holistic projection in use. Figure 3 gives a visualization of a projection in use. In the figure, world w is indicated as the light-gray part

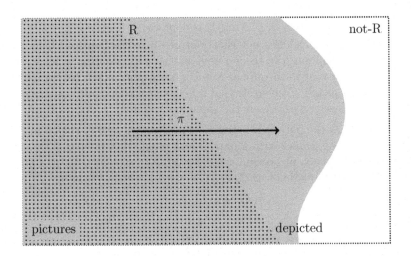

FIGURE 3 Holistic projection in use

of logical space it realizes (R) and the part it does not realize (not R). Also, a projection bisects logical space Λ in two areas: pictures (dotted pattern) and the depicted (the rest). Since the projection is in use, all its pictures are realized. But among the depicted states of things, some are realized and some are not.

This finishes the discussion of projection. Before using projection to detail elementary propositions and their sub-expressions, we first chart some consequences of the notion of projection just given.

3.5 *Interlude*: Projection and isomorphism

Holistic projection as presented in definition 13 is a rich notion involving the form of names and their referents. As we saw in chapter 2 the form of an object is an abstraction over logical space, and identity of form is shown using partial isomorphisms between these abstractions. It is therefore no surprise that holistic projection ensures the state of things an elementary proposition is about, to be possible. In projection all partial isomorphisms are chosen uniformly, and so determine a bijection between elementary signs and the states of things they are about.

Proposition 16 *In a frame* Λ, π, *holistic projection is equivalent to* π *being an isomorphism between the pictures and the depicted:*

$$(o_1, \ldots, o_n)_c \text{ is a picture} \iff (\pi(o_1), \ldots, \pi(o_n))_c \text{ is the depicted.}$$

Proof. Since the pictures and the depicted have no object-content in common, for each object $\hat{o}$, $F(o)$ consists only of pictures or only of the depicted. So, if $F(o) \doteq F(\pi(o))$ holds via π, the following are equivalent for object-content o_i in a picture:

$$(o_1, \ldots, o_n)_c \text{ is a picture}$$
$$(o_1, \ldots, o_n)_c \text{ in } F(o_i)$$
$$(\pi(o_1), \ldots, \pi(o_n))_c \text{ in } F(\pi(o_i))$$
$$(\pi(o_1), \ldots, \pi(o_n))_c \text{ is depicted}$$

The first and third equivalence use that pictures and the depicted have no object-content in common, the second equivalence uses identity of form via π. Therefore, identity of form implies π to be an isomorphism between the pictures and the depicted. Conversely that π being such an isomorphism shows identity of form, is seen with about the same argument. □

Next observe that the partition between elementary signs and the states of things they are about, comes with a partition of names and referents. Due to their holistic nature, each projection π can be lifted to a one-to-one mapping $\hat{\pi}$ between names and referents so that:

$$\hat{\pi}(\hat{n}) \mapsto \widehat{\pi(n)},$$

or in full:

$$\hat{\pi}(n{:}F(n)) \mapsto \pi(n){:}F(\pi(n)).$$

Proposition 17 *In the context of projection, each object is either a name or a referent, not both. Names* $\hat{n}$ *are one-to-one related to objects* $\hat{o}$ *via the map:* $\hat{\pi}\colon n, F(n) \mapsto \pi(n), F(\pi(n))$.

Proof. To show that names and referents are different from each other, we use both their content and form. As to content, for each o definition 13 gives that o must be either in N or in O, but never both. Also, for each $(o_1, \ldots, o_n)_c$ in Λ, all $o_1, \ldots, o_n$ are in N or all $o_1, \ldots, o_n$ are in O. It follows that each form $F(o)$ is either a subtotality of N – if o is in N, – or of O – if o is in O: names and referents are different from one another.

Since π is a bijection between contents and since each content comes with a unique form, the mapping is bijective: $\hat{\pi}$ maps each name $\hat{n}$ to a unique referent. To show $\hat{\pi}$ to be injective, assume the referent $\hat{\pi}(\hat{n})$ is identical with

$\widehat{\pi}(\widehat{n}')$. Then $\pi(n){:}F(\pi(n))$ is the same as $\pi(n'){:}F(\pi(n'))$. So $\pi(n)$ is identical with $\pi(n')$, hence n with n', and $\widehat{n}$ with $\widehat{n}'$, as required. As to surjectivity, let $\widehat{o}$ be a referent. Since π is a bijection, there is a name-content n with $\pi(n) = o$. It follows from $F(n) \stackrel{\sim}{=} F(o)$ via π that $\widehat{\pi}(\widehat{n})$ is $\widehat{o}$. □

The elegant relation between holistic projection and isomorphism is based on the assumption that description is immediate and complete. As soon as descriptive completeness is dropped, parts of the form of a name may be left out of the bijective relation, and so some of its object-contents may fail to have reference. If projection is holistic, such indeterminacy of sense is absent, as it should be. This is seen clearly from the examples that started the discussion on this kind of projection in section 3.4.9.

Up till now we have assumed logical space to be available. The consistency of the notion is obvious. But as a corollary to proposition 16 it can also be checked in a more formal sense.

Proposition 18 *Logical space is a consistent notion.* □

In view of proposition 16 it suffices to join two isomorphic totalities of states of things. But of course more complex structures are also possible.

The tractarian system is absolute: it is based on a unique logical space Λ satisfying definition 1. Still, as 'there are no prominent numbers in logic' (4.128b), Λ should not be forced to have a particular size. The space Λ is part of the framework to understand the possibility of meaning, but at best its specifics are known via the pragmatics of analysis. The current approach does justice to this philosophy.   [*End of interlude*]

4

Elementary propositions

Insight into the nature of descriptive language is insight into its sense. This holds for elementary propositions, which embody the core of description, just as well as of logically complex propositions.

In this chapter I will use the notions developed up until now – logical space, state of things, object-content and form, projection, – to get a clear view on elementary sense and truth. I also discuss subparts of elementary proposition, such as expressions and material functions.

The general notion of sense requires more information on the logical structure of propositions. This is given in chapters 5-7. The perfect notation of sense is discussed in chapter 9.

4.1   Elementary sense

In the tractarian system propositions are truth-functional compounds of elementary propositions, which suggests a highly extensional notion of meaning. This suggestion is to a large extent mistaken. Even in case of an elementary proposition, its truth-functional structure is used to project its sense – an intension concerning a possible state of things, – into logical space.

It is a basic principle of the *Tractatus* that the sense of a proposition is prior to its truth or falsity. Also, sense is determinate. It leaves reality a 'yes' or 'no', but no other options. A proposition consists of articulate structures of names combined into an articulate truth-functional structure, which is projected onto the articulate structure of the world. If its sense – the content of its projection, – is realized, the match is perfect and the proposition is true, but if its sense is not realized, the mismatch is perfect and the proposition is false. In this way having a determinate sense is prior to being either true or false.

The elucidation of sense is an annex to 4, which presents thoughts as meaningful propositions.

4.022 A proposition *shows* its sense.

> A proposition *shows* how things relate *if* its true. And it *states that* they do so relate.

In 4.022 we find the tractarian analogue of a Fregean idea: the sense of a proposition is its content (*Inhalt*), which as such is presented in a neutral way; it may be true, it may be false. Only the *use* of a proposition comes with the statement that the sense is realized, and it is the world that determines whether this claim is correct or not.

4.2 The elementary proposition

A projection in use determines which states of things in logical space function as pictures and which as the depicted. A more detailed view on sense is found in 3.13 about projection.

3.13 A proposition includes all that the projection includes; but not what is projected.

Therefore, the possibility of what is projected but not what is projected itself.

A proposition, therefore, does not yet contain its sense, but does contain the possibility of expressing it.

('The content of a proposition' means the content of a proposition that has sense.)

A proposition contains the form of its sense, but not its content.

At this point we should note an ambiguity in the tractarian uses of 'content'. Objects and expressions have form and content (2.025, 3.31). This content constitutes the substance of the world, and allows the form of an object to occur more than once. In 3.13 'content' is used in another way. 'Content' is intended here as the states of things a proposition is about.

In projection, whether a state of things functions as an elementary proposition or not, is given holistically all at once. In this regard, with each elementary proposition the entire logical space is given (3.42); if only because the form of the names occurring in an elementary proposition and the form of the objects they stand proxy for may expand well beyond the single elementary proposition. See section 3.4.9.

In this chapter we are just concerned with the essentials of elementary propositions; i.e., with the elementary proposition as a logical picture. A logical picture exhibits in its sign the form of its sense, which is also the form of the state of things it depicts. The configuration of names in its sign indicates that the referents of the names may configure in the same way. This state of things is what the proposition is about, its content, which strictly speaking is no part of the proposition itself (3.13). The sense of an elementary proposition presents its content as realized, and in using the proposition it is stated that this content is realized. Whether this is true or not is for the world to decide. Figure 4 visualizes the idea.

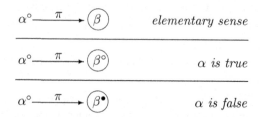

FIGURE 4 Elementary proposition

Although we focus on a single elementary proposition α, the projection π concerns the entire logical space. Projection turns the state of things α into a logical picture of β. Projection π is in use, so α is realized as the circle $\circ$ indicates. The circle around β indicates that the sense of α shows β as realized. If β is in fact realized (β°), the statement of α is true. But if in fact β is not realized ($\beta^\bullet$), the statement of α is false. The different aspects of elementary propositions can be summarized as follows.

Definition 19 *Let π be a projection in use in a world w: Λ, R. An elementary proposition is an elementary sign α together with the projection of α via π. The sense of α has the form of its sign $(n_0, \ldots, n_k)_c$. The content of its sense is $(\pi(n_0), \ldots, \pi(n_k))_c$. The proposition shows its content as realized.* □

The idea is that the content of the sense of an elementary proposition, the state of things it is about, is not part of the proposition (3.13), but that the referents of its names are. This is precisely how projection is given in definitions 13, 15. Projection should ensure that the sense of an elementary proposition is about a possibility in logical space. This we already know to be the case from proposition 14: the content of its sense $(\pi(n_0), \ldots, \pi(n_k))_c$ is a state of things in Λ.

4.3 Elementary truth and falsity

Elementary propositions come with an intensional interpretation of truth-table signs. According to thesis 5, elementary propositions are truth-functions of themselves, so an explicit sign of α has form:[43]

$$\begin{array}{c|c} & \alpha \\ \hline T & T \\ F & F \end{array}$$

In a truth-table sign we may restrict our attention to the truth-grounds of a proposition, which are the rows marked 'T' at the left-hand side. This is due to 4.022: 'A proposition *shows* how things relate *if* its true'. The sign for α, in particular, indicates that α is true (left-hand side 'T') iff the content of its sense is realized (right-hand side 'T'). Thus in projection the sign shows this content *as* realized, independent of whether it is actually realized or not. For the negation of α the situation is reversed:

$$\begin{array}{c|c} & \alpha \\ \hline F & T \\ T & F \end{array}$$

Then, α is true (left-hand side 'T') iff the content of its sense is not realized (right-hand side 'F'). Thus in projection the sign shows this content *as* not realized.

Apart from showing their sense as a possibility in logical space, elementary propositions are actually true or false when used in a world.

Definition 20 *Let w be the world Λ, R in which the projection π is used, and let α be an elementary proposition.*

---

[43]In section 5.2, I discuss in detail that truth-tables are signs rather than graphs of Boolean functions.

- α *is* true *in w iff* $\pi[\alpha]$ *is in R.*
- α *is* falsity *in w iff* $\pi[\alpha]$ *is not in R.*

We write 'w $\Vdash \alpha$' for α being true in w, and 'w $\nVdash \alpha$' for α being false in w. $\square$

Definition 20 makes plain, once more, that the sense $\pi[\alpha]$ of proposition α is prior to its truth or falsity. Truth or falsity presupposes sense, whose content can be realized or not. In general there is an important difference between being realized and being true. States of things in reality can be realized, but if they are not a sign they lack sense, and cannot be true or false: truth and falsity concern propositions only. Observe, too, that the *use* of α is a precondition of its truth or falsity. On this reading, only statements that are actually made are true or false. Possible propositions, which are not in use, may have a truth-value, but they need to be used in a statement to actually have it.

It is a basic feature of elementary propositions that they are independent of each other; the simultaneous statement of elementary propositions can not result in tautology or contradiction.

Proposition 21 *Elementary propositions have the following properties:*

   i) *they are bivalent: either true or false* (4.023).

   ii) *they are logically independent* (5.134).

   iii) *they cannot contradict each other* (4.211), *or be tautologous.*

Proof. As to i), in a world w $\pi[\alpha]$ is either realized or not, so α is either true or false. As to ii), let α_1, α_2, $\pi[\alpha_1]$, and $\pi[\alpha_2]$ be in logical space Λ. Clearly there is:

a world w with $\pi[\alpha_1]$ and $\pi[\alpha_2]$ in R;

a world w' with $\pi[\alpha_1]$ in R' and $\pi[\alpha_2]$ not in R';

a world w'' with $\pi[\alpha_1]$ not in R'' and $\pi[\alpha_2]$ in R'';

a world w''' with neither $\pi[\alpha_1]$ nor $\pi[\alpha_2]$ in R'''

These alternatives show iii) to be a consequence of ii). In the above α_1, α_2 are logically independent. Therefore they do not imply or contradict each other, and they are not tautologous. $\square$

This finishes our discussion of elementary propositions. After an interlude on the mutual influence of Russell and early Wittgenstein on each other, we shall focus on propositional parts: expressions and material functions.

4.4   *Interlude*: Russell on judgments and propositions

With the treatment of elementary propositions in place, it should be interesting to compare Wittgenstein's notion of statement with Russell's theories of judgment. In the *Tractatus*, there are but a few traces of admiration for Russell's work. Still, it is fair to say that Wittgenstein's early work profited much from the work of Russell it opposed to.

To compare the work of Russell and Wittgenstein from the period 1910 until 1918 is a tricky business. When in the autumn of 1911 Wittgenstein decided to study with Russell, the relation of pupil and teacher quickly changed into that of equals, and it is not always clear who has influenced whom. Also,

Russell often adapted new editions of his numerous publications to the latest insights. This makes it difficult to keep track of how his insights developed – e.g., as is done in Potter's excellent history of ideas (Potter, 2009), – or to assess the similarities in the work of Russell of Wittgenstein, which may be due to either. Despite these misgivings, one is on fairly save grounds when discussing the ways in which their work differ. This is what I will concentrate on here, profiting much from Pears (1979) and Potter (2009).

For the sake of comparison, let us engage in a counterfactual. If Russell had written a text like the *Tractatus*, it could have started thus:[44]

#1. The universe consists of objects having various qualities and standing in various relations.

#1.1 Some of the objects which occur in the universe are complex.

#1.11 When an object is complex, it consists of interrelated parts.
   Broadly speaking, a *complex* is anything which occurs in the universe and is not simple.

#1.12 We will give the name of "a *complex*" to any such object as "a-in-the-relation-R-to-b" or "a has the quality q" or "a and b and c stand in the relation S".

#1.13 The complex object "a-in-the-relation-R-to-b" may be capable of being *perceived*; when perceived, it is perceived as one object.
   Attention may show that it is complex; we then *judge* that a and b stand in the relation R. The judgment is a relation of four terms a and b and R and the percipient.

#1.131 The judgment is *true* if there is a complex *corresponding* to the discursive thought which is the judgment.
   That is, when we judge "a-in-the-relation-R-to-b," our judgment is said to be *true* when there is a complex "a-in-the-relation-R-to-b," and is said to be *falsity* when this is not the case.

#1.132 The relation which constitutes judgment is not a relation of two terms, namely the judging mind and the proposition, but it is a relation of several terms, namely the mind and what are called the constituents of the proposition.

#1.2 When a judgment occurs, there is a certain complex entity, composed of the mind and the various objects of the judgment.

#1.21 It follows that 'proposition' as *the* object of judgment is a false abstraction.
   A judgment has several objects, not one.

#1.22 We call a judgment *elementary* when it merely asserts such things as "a-in-the-relation-R-to-b" or "a has the quality q" or "a and b and c

---

[44]The compilation is based on the second edition of the *Principia Mathematica* (to be precise: Whitehead and Russell (1910), introduction, chapter II, section III, 43-44). The 1910 edition has the same text. The numbering, like that of the *Tractatus*, is inspired by the *Principia*.

stand in the relation S".

#1.23 Then an *elementary* judgment is true when there is a corresponding complex, and false if there is no corresponding complex.

There are striking similarities between this counterfactual, tractatus-like text based on Whitehead and Russell's *Principia* and Wittgenstein's real *Tractatus*; e.g., the use of the term 'elementary proposition', which Wittgenstein seems to have borrowed from Russell. There are also striking differences.

One of the crucial points in comparing Russell and Wittgenstein is their treatment of complexes. Russell calls a complex "a-in-the-relation-R-to-b" a complex object or thing. This complex is based on a certain simple typing, with in this case a and b individuals that stand in the relation R to each other. In opposition Wittgenstein held that the world consists of facts, i.e., structured complexes, *not* of objects or things (1.1). The objects that form these complexes are presented in the abstract, in that all have content and form. Any differentiation in these forms, any specific 'typing', is presented as inessential to the system, and at best to be made manifest as the result of analysis.

To call the complexes 'objects' made up of other, simply typed objects (Russell) or to call them 'states of things' made up of objects *simpliciter* (Wittgenstein) is not just a point of conceptual purity. According to Russell (1913), 144, in which some ideas of the introduction to the *Principia* are developed further, a judgment like (R.a) should be analyzed as (R.b).

(R) a. S judges that Socrates precedes Plato.

b. J(S,F,'Socrates','precedes','Plato').

That is, the judgment is a five-place relation between a subject S, a logical form F, and three simple objects 'Socrates', 'precedes', 'Plato'. The logical form F is a *summum genus* resulting from utmost generalization. Earlier in the text Russell gives the example xRy, where the logical difference between the relation R and its terms is indicated using different kinds of letter (Russell, 1913, 98).

That Russell argued for the involved analysis seems due to his resistance to adopt intensions such as Fregean senses; cf. also Russell (1905). For without sense or intension, what could it mean to form a judgment about what does not exist? But once the constituents of a proposition, whether realized as a complex or not, are assumed to subsist, like in Russell (1903), the problem does not seem to arise: the subsisting components just combine when realized as such. On this route, however, a much more troublesome objection comes to the fore than the philosophical status of sense: how to ensure that the judgment concerns the proper combination of objects? After all, their combination does not have to be unique. Given the logical form xRy and the objects 'Socrates', 'precedes', 'Plato', 'Plato precedes Socrates' can be formed just as well as 'Socrates precedes Plato'. The drawback was most problematic for an earlier sketch of Russell's theory, in which the form xRy was still lacking, and thus allowed for nonsensical combinations, like 'precedes Plato Socrates' (Pears, 1979, 191-2). In the 1913 installment, such unwanted combinations are eliminated. It held that the objects of a judgment should respect the different logical types in its logical form. In particular, 'Plato' and 'Socrates' are of the object type that may associate with either x or y in xRy, and 'precedes' is of the relational type that may be associated with R. This does rule out nonsensical combinations, but still requires a cumbersome, inconclusive argument to deal with the permutation of terms. Cf. Russell (1913), part II, chapter V.

The theory of judgment is a key element of Russell's 1913 manuscript. As is well known, Wittgenstein criticized it so fiercely that Russell felt he 'could not hope ever again to do fundamental work in philosophy' (Russell, 1987, 282). Wittgenstein phrased his main objection in a letter of June 1913 (abbr. preserved):

> I can now express my objection to your theory of judgment exactly: I believe it is obvious that, from the prop "A judges that (say) a in the Rel R to b" if correctly analyzed, the prop "aRb .$\vee$. $\neg aRb$" must follow directly *without the use of any other premiss*. This condition is not fulfilled by your theory. (Wittgenstein, 1995, 29)

Given Russell's theory as just explained, Wittgenstein's point must have been that though judging nonsense is excluded, the resulting relation of judgment may still be permutative. Both 'Plato precedes Socrates' and 'Socrates precedes Plato' are allowed. E.g., on Russell's theory from the judgment either "aRb .$\vee$. $\neg aRb$" or "bRa .$\vee$. $\neg bRa$" may follow, and without further premisses there is no principled way to choose which.

As Pears (1979) argues convincingly, another objection of Wittgenstein to Russell's theory of judgment was his use of Platonic logical forms, such as xRy. There are some observations against the introduction of such forms:

1. They do not give a full explanation of judgments;
2. When separating the logical form and the constituents of a complex so radically, it is hard to see how they are to unify again.
3. To unify the constituents of a judgment with the elements of the logical form, the constituents need to be 'typed', and when properly understood such typed constituents make the logical form superfluous;
4. Logical forms leave in the dark how complexes formed with logical constants such as 'not', 'and', 'some', 'all' should be understood. As Russell (1913), 99, puts it: 'I do not know what the logical objects involved really are'.

Each objection must have been known to Wittgenstein, and in the *Tractatus* he offers principled solutions to all of them.

Eliminating the second and third objection, I surmise, has led to the tractarian ontology. As we have seen in chapter 2, the ontology is based on the notions of form and content, which abstract from specific types and their instantiations. Form is the possibility to combine, and logical space encompasses all possible combinations. Also, the combinations of parts cannot be separated from their logical forms. At best logical form is shown in abstraction using a tractarian variable. Thus the need for a Platonic realm of logical forms is eliminated in favor of an Aristotelean ontology, but taken in an abstract logico-semantic manner rather than in a material way.

Apart from making form essential to all instantiations in logical space,

Wittgenstein made some further steps to perfect the theory that Russell had started to develop. Firstly, other than Russell he made a principled distinction between fact (complex) and object (simplest part of a complex), and related both using a variation on Frege's context principle: objects must occur in the context of states of things. But Wittgenstein did retain Russell's idea that the basic constituents of a complex subsist, and associated his notion of object with the traditional concept of substance. Secondly, Wittgenstein varied on Frege's notion of sense to turn the logical space of possible combinations into an intensional realm where the form of signs is projected onto what they depict. This step removed the need to make the elements of a complex the constituents of a judgment instead of the complex itself. Since logical forms are internal to the complexes, the difficulties surrounding form in Russell's theory of judgment are eliminated. On Wittgenstein's view the complex suffices, as it is a logical possibility that may be realized in some worlds, and not realized in others. Non-existing states of thing can be judged as well as existing ones. Finally, with ontology understood in a logico-semantic way, Wittgenstein saw, too, how to cope with Russell's search for logical objects: there are none.[45] Logical constants do not refer, but indicate rules to combine propositions into propositions with logically complex truth-conditions. From a logical point of view, this is the prime locus where Wittgenstein's originality is to be found. As will be discussed extensively in chapters 5-7 and 9, his philosophy of logical constants is different from both Frege and Russell, as is their intensional treatment in logical space.   [*End of interlude*]

4.5   Expressions

An expression is a part of a proposition that contributes to its sense in a characteristic way. Elementary propositions are prime examples of expressions. They are the smallest expression having independent sense. Proper expressions do not have sense themselves, but as long as they have structure, they contribute systematically to the sense of a proposition. Cf. the offspring of 3.31 and 4.126.

In a way, proper expressions generalize names. Other than names, expressions are complex, but like names:

- expressions have form and content (3.31);
- they characterize the sense of propositions (3.31);
- they may occur in different propositions (3.311-3);
- they comply with a context principle (3.314).

The *Tractatus* lacks detail on how expressions occur as part of a proposition. It mainly states that they can be seen as a constant factor in a class of propositions, and that the variable of all propositions in which they occur characterizes their form. See the discussion on prototypes and variables in sec-

---

[45]See Hintikka (1986), chapter 4, section 7, for a similar observation. Hintikka's interpretation of the *Tractatus* is based on the assumption that Wittgenstein was by and large in agreement with Russell's 1913 manuscript. By contrast I think the manuscript inspired Wittgenstein only because he was opposed to the theory of judgment it put forward. On this reading it is quite natural to interpret the tractarian ontology logico-semantically.

tion 11.3.1. Here they are interpreted as invariants in the analysis of complex phrases.

3.3442 In analysis the sign of a complex does not dissolve itself arbitrarily, so that, say, its dissolution in each propositional structure would be different.

On the current reading, expressions can be taken to correspond to proper names, which are best thought of as abbreviated complexes of logical names, or to the proper objects of everyday life, which are best seen as complexes of tractarian objects. I should stress, though, that the complexity is of a symbolic nature, and is hence neither material nor psychological. People and material objects can be described at different levels of logico-semantic granularity, until the semantic rock-bottom of named objects has been reached. Expressions codify the semantic complexity.

Definition 22 starts presenting my understanding of expressions in a more detailed way: an expression is a configuration of names as part of a proposition. Like propositions, expressions have content and form, but their content and form is parasitic on that of propositions. We begin with specifying the sign-content of expressions.[46]

Definition 22 *Let α be an elementary propositional sign $(o_0, \ldots, o_n)_c$. The content of an expression of α is a sub-configuration $(o_{i+1}, \ldots, o_j)_{c'}$ ($1 \leq i \leq j \leq n$). We let ε vary over expression-contents and write $\alpha(\varepsilon)$ in case ε is the content of an expression in α.* □

The idea is that expressions are 'after the fact': they abbreviate or highlight a subpart of the entire configuration, which strictly speaking is not made up of them. Expressions are a pragmatic device to compress the complex information of a logical picture.

The notation c' for the configuration type of an expression should be interpreted in strictly the same fashion as the configuration type c of a state of things; see the discussion on definition 2. It should also be noticed that the type c of the main configuration does not change, for highlighting a sub-configuration does not alter the configuration as a whole. This interpretation justifies writing $\alpha(\varepsilon)$ with:

$$\varepsilon \ = \ (o_i, \ldots, o_j)_{c'} \quad def.$$

in less abbreviated form as:

$$\alpha((o_i, \ldots, o_j)_{c'}), \text{ or as}$$
$$(o_1, \ldots, o_i, \varepsilon, o_{j+1}, \ldots, o_n)_c, \text{ or as}$$
$$(o_1, \ldots, o_i, (o_{i+1}, \ldots, o_j)_{c'}, o_{j+1}, \ldots, o_n)_c.$$

These are all different presentations of one and the same state of things:

$$(o_1, \ldots, o_n)_c.$$

It remains to specify the form of expressions. Like the form of an object, the form of an expression is a formal concept. According to 3.312 this form must be presented using the variable of propositions in which the expression

---

[46]Recall the distinction between sign- and sense-content on page 25.

occurs. In the variable the expression is constant, while the remaining parts of the configuration may vary:

Definition 23 *The form of an expression ε – notation: $F(\varepsilon)$, – is given by the variable $\overline{(\alpha'(\varepsilon))}$ of all states of things α' that contain ε.* □

Definition 23 uses the bracket notation for variables; cf. the explanation following definition 3. Now that both content and form of an expression is clear it remains to introduce expressions themselves.

Definition 24 *An expression – notation: $\widehat{\varepsilon}$, – is given by its content and form: $\widehat{\varepsilon} = \varepsilon, F(\varepsilon)$.* □

As in case of objects, it should be stressed that the current notation is a bit misleading: expressions are not pairs with content on the one hand, and form on the other. Content and form of an expression are two aspects of one and the same part of a proposition. Form is inherent to an expression in the Aristotelean way; it gives its 'type', its range of possible combinations. Content is the logical notion of form instantiation; it ensures that a form can occur more than once.

At this point we are in a position to introduce an expression's contribution to sense.

Definition 25 *Let π be a projection in use and let $\widehat{\varepsilon}$ be an expression. The contribution of $\widehat{\varepsilon}$ to sense-content – notation: $\pi[\widehat{\varepsilon}]$, – is the unique sub-configuration of objects in all projections $\pi[\alpha]$ for α in $F(\varepsilon)$ that is determined by π and the content ε. The contribution of $\widehat{\varepsilon}$ to sense-form is the configuration of ε itself as it occurs in the α in $F(\varepsilon)$.* □

It is obvious that $\widehat{\varepsilon}$ characterizes the sense of all propositions in which it occurs; it contributes to their form. But one should verify that expressions contribute to the content of sense *within logical space*. Also, the uniqueness of an expression's contribution to content should be verified.

Proposition 26 *Let π be a projection in use for logical space Λ, and let $\widehat{\varepsilon}$ be an expression. The contribution $\pi[\widehat{\varepsilon}]$ of ε to sense-content exists and is unique.*

Proof. Let ε be the content $(o_h, \ldots, o_j)_{c'}$. For any α in $F(\varepsilon)$, it holds that α is in $F(o_i)$ $(h \leq i \leq j)$. Since $F(o_i) = F(\pi(o_i))$, $\pi[\alpha]$ is in $F(\pi(o_i))$, and hence in Λ. Also, π is a bijection and ε is part of each α in $F(\varepsilon)$. Therefore, the sub-configuration $\pi[\varepsilon]$ occurs in all $\pi[\alpha(\varepsilon)]$ in Λ, and is unique. □

Proposition 26 shows that relative to projection the sense-content of (elementary) propositions is rigid, and the same holds for names and expressions. What is projected, is 'rigidly' the same across possible worlds.

It is also worth noticing that according to the proof of proposition 26 just one name in an expression suffices to establish its contribution to sense-content. The reason is that on the abstract approach of the *Tractatus* there is no way to detail the dependencies among objects within a state of things. For all we know, each object in a state of things should be assumed to depend on all others.

Definition 24 captures the essentials of an expression, but leaves some of its aspects open. For one, may two proper sub-expressions of an expression overlap? The *Tractatus* is silent on the issue. I think that across analyses overlapping expressions are allowed. But such expressions cannot be abbreviated simultaneously, and should hence be used in different analyses of a proposition. For example the propositional sign:

$$(n_0, n_1, n_2)_c$$

may well allow the expressions:

$$\varepsilon' = (n_0, n_1)_{c'} \quad def.$$
$$\varepsilon'' = (n_1, n_2)_{c''} \quad def.$$

But to abbreviate $(n_0, n_1, n_2)_c$ as $(\varepsilon', \varepsilon'')_c$ would be mistaken. Yet, both $(\varepsilon', n_2)_c$ and $(n_0, \varepsilon'')_c$ lead to the same result $(n_0, n_1, n_2)_c$. The sense of propositions is determinate: in the end-result of analysis the sense of proposition is one and the same.

In a similar vein one may ask whether the sub-expressions of an expression should have a unique structure? Cf. the unique readability of formula's and terms in modern logic. As the analysis of everyday language involves replacing expressions by their definitions (see chapter 1), the absence of a unique structure would imply that several analyses of the same piece of language may be possible. This appears to contradict 3.25.

3.25 There is one and only one complete analysis of a proposition.

The contradiction is apparent at best. One should distinguish between the process of analysis and its end-result. Thesis 3.25 is, I think, about the completed analysis, the end-result, not about the process. And the end-result is unique, since it is fully determined by the elementary propositions and their truth-functional combination. Sub-expressions only play a rôle at the intermediate stages of analysis prior to reaching the end-state. Here, expressions are seen as a pragmatic device to compress the complex information of a logical picture. Elementary propositions set boundaries to the contribution of expressions to sense. Only elementary propositions figure in the end-result of analysis.

4.6 Material functions

Expressions are instrumental in formalizing material functions. At a few places Wittgenstein speaks of such functions – cf. 5.44, NB37(1), – but what can a function be, given an ontology consisting of interdependent objects that form elementary facts? In particular, the classless philosophy of the *Tractatus* lacks the point-sets and other abstract entities that are used in a set-based treatment of functions. But to view functions along these lines is misleading: material functions are propositional functions adapted to the tractarian system. See Hylton (2005), chapter 7, for an overview of propositional functions in *Principia Mathematica*.

Although there is no definition or thesis clarifying what material functions are, they must provide for a notion of function that is adapted to the tractarian ontology. My understanding of material functions is based on combining the insights that (i) propositions are functions of the expressions they contain (3.318); that (ii) these functions correspond to material properties or concepts as opposed to formal ones (4.126d); and that (iii) such material properties are formed of configurations of names (2.0231). Reasoning along these lines, one arrives at material functions as a specific kind of propositional function; they are functions from expressions to elementary signs.[47]

Definition 27 *Let* $(\overline{\alpha(E)})$ *be the variable of expressions in which* $\alpha(-)$ *is constant. The* material function $\alpha(-)$ *is the function from the expression-contents* ε' *occurring at the place that* E *indicates to the signs* $\alpha(\varepsilon')$. □

In a way, the material function $\alpha(E)$ is dual to the form of an expression. In the form of an expression ε, ε is kept constant, whereas for the material function the context $\alpha(-)$ is the constant factor. One could say that the range of the material function $\alpha(-)$ is the variable $(\overline{\alpha(\varepsilon')})$ of all such propositional signs.[48]

A function $\alpha(-)$ comes with a strict requirement on the 'type' of its arguments: only those expressions ε' are admitted that result in a propositional sign $\alpha(\varepsilon')$. In this regard ε is the prototype of $\alpha(-)$'s argument (3.333), but it is unclear how strict the idea of prototype should be. May the argument be replaced only by expressions that are isomorphic to ε? Or should we allow all values in which the context $\alpha(\ )$ is constant? For instance, let ε be the name e. Then on the first interpretation $\alpha(e)$ could determine, say, the forms:

$$\alpha(e), \alpha(e_1), \ldots, \alpha(e_n), \ldots$$

But on the second interpretation it could determine, say:

$$\alpha(e), \alpha(e, e), \ldots, \alpha(\underbrace{e, \ldots, e}_{n \text{ times}}), \ldots$$

Whichever interpretation is favored, the main point is there will always be a rigid context that, given the possibilities in logical space, puts strict requirements on what the admissible values of the function are.

Definition 28 *Let* π *be a projection in use and let* $\alpha(-)$ *be a material function. The* contribution to sense-content *of the material function* $\alpha(-)$ *is the unique sub-configuration of objects in the sense-content* $\pi[\alpha(\varepsilon)]$ *of the* $\alpha(\varepsilon)$ *in the range of* $\alpha(-)$ □

As in case of expressions one has:

Proposition 29 *The contribution to sense-content* $\pi[\alpha(-)]$ *of* $\alpha(-)$ *exists and is unique.* □

Based on this interpretation, we see an important difference between material functions and truth-functions (5.44, NB 37(1)). The configuration of names that make up a material function is rigid, and so its content is fixed relative

---

[47] Cf. Anscombe (1959), page 101 ff. for a similar analysis.

[48] Expressions are implicitly assumed to be coherent parts of an expression, but there is no such assumption for material functions. If one were to allow discontinuous expressions, material functions would be expressions of a particular kind.

Material functions obtained form a strict sub-expressions ε_m of elementary propositional signs lead to a kind of parameterized variant. The range of such a material function $\varepsilon_m(-)$ would be the variable $(\overline{\alpha'(\varepsilon_m(E))})$ in which both α' and E may vary. The admissible values of $\varepsilon_m(-)$ would depend on the context α'. In what follows there is no need for this kind of generality.

to a projection in use. It is a complex grounded in the reference of names, which is not interdefinable in terms of another complex of names. By contrast, truth-functions are interdefinable in terms of each other, which is an indication that they lack reference (5.254, 5.44d). In chapter 6, I will argue in detail that truth-functions are no part of the non-pictorial world, and other than material functions will disappear when interpreting language. Here we finish with an interlude on the extent in which sense relates to human capabilities.

4.7 *Interlude*: setting up sense?

The interpretation of elementary propositions involves a cluster of notions concerning names, objects, propositional signs, states of things, projection in use. At this point it is natural to ask: 'If even elementary propositions are as complex as described here, how does the use of language ever come of the ground?'[49] In chapter 3, I have discussed the local forms of projection quite extensively, since they seem to capture a kind of human feasibility, but in the end it was clear that the local notions should be rejected. For similar reasons we shall see here that the tractarian system is little to do with such human capabilities as the creation of sense. As has also been noted in Hacker (1984), there is a tension between the holistic transcendental view on language and its human use that the *Tractatus* does not resolve properly.

In attempting to answer the above question, let us start with reference as involved in logical projection. A tempting image concerning reference is that we, humans, first correlate names with the objects they refer to, and next use the correlation in elementary description. Sometimes the image is even held to be asymmetrical, in that the object is taken to be leading instead of such linguistic elements as names and propositions. Pears makes the point thus:

> Wittgenstein did always qualify the contact between name and thing by making it conditional on the correct use of the name in sentences. [...] The condition imposed by Wittgenstein *qualifies* name-thing contact, but does not *replace* it. The thing, with its independent nature, is the dominant partner in the association, and if the name does not remain faithful to the possibilities inherent in the thing, the association is annulled. So representation (*Vertretung*) requires an initial correlation followed by faithfulness to the possibilities intrinsic to the thing with which the initial correlation was made. (Pears, 1987, 75)[50]

Although I fully agree with Pears' holistic treatment of reference, which requires faithfulness to all possible occurrences of an object, I do not think objects have the independent rôle Pears seems to attribute to them. For what could it mean to correlate a name with an object as Pears suggests? Should we think of the correlation as an empirical activity in which both name and object are at our disposal? Within the tractarian philosophy, it seems to me, this is clearly impossible. Firstly, objects are logical notions enabling insight into the pictorial nature of descriptive language. They show themselves in the logical manner, as abstractions over logical space. Cf. chapter 2. They require

---

[49]This question is inspired by a discussion in Pears (1987), page 101.

[50]Cf. Pears (1987), pp. 98-114, for more details, and for its contrast with the interpretations of Ishiguro and McGuinness.

the holism of context, form, and the combination with other objects (2.0251). In all this, objects are independent of being sensed. Secondly, the proper objects of everyday life, with which we are acquainted empirically, are better thought of as complexes of tractarian objects, just like proper names are best seen as abbreviated complexes of logical names. If at all, empirical correlation takes place at the complex level, not at the logical level of objects and names. Thirdly, there is no 'independent nature' of objects as Pears seems to assume. To be is to be essential to description. The context-principle holds for objects just as much as it holds for names: names only have reference in the context of a proposition, objects can only occur in the context of a state of things described. Even if empirical correlation were feasible, tractarian objects could not take the lead, because they are in the same web of dependencies as names are.

Since empirical correlation cannot get language started, the correlation of names and objects that Pears puts forward, can only be of an non-empirical kind; say, a logical, intentional act of some sort. Within the tractarian system such an act should not only be prior to language but also prior to any kind of thought. In the system, language and thought can hardly be separated from one another: a thought is a propositional sign used as projection into logical space, and in its fully analyzed form a thought is a logical picture just as any proposition is. In consequence, the non-empirical correlation of name and object, as a human phenomenon remaining local to an elementary proposition, starts a regress. For how, in turn, can the correlation between name and object be correlated to language in use? After all, the correlation is assumed to be independent of thought and hence of language. Name and object cannot be correlated in a local manner, if establishing the correlation requires independence of all thinking and picturing.

In a tractarian setting, the idea is mistaken that a correlation between name and object needs to be set up locally to get meaningful language started. This includes any attempt from the other end of the spectrum; namely, to start the correlation from the name of the propositional sign rather than the object. Then the idea would be that since names only have reference in the context of a proposition, these contexts somehow fix the referent of a name. In other words, the form of a name, as abstracted from its occurrences in propositional signs, would define its referent. To see that this approach is untenable, too, let logical space be as simple as:

$$(n_0, n_0)_c, (n_1, n_1)_c,$$

$$(o_0, o_0)_c, (o_1, o_1)_c,$$

with the first two states of things used as propositional signs. In this space the referent of, say, n_0 is not fixed by its form $(n_0, n_0)_c$, for the form still allows to project n_0 either onto o_0 or onto o_1.

Although it is tempting to ask how the use of language comes of the ground if sense requires such an intricate system, I think that at a human scale there is no answer forthcoming from the *Tractatus*. Rather than explaining how humans are able to employ the hidden complexities of unanalyzed language, the *Tractatus* posits the ability as given.

4.002a Humans are able to build languages capable of expressing any meaning, without having the slightest idea how each word has meaning or
what this meaning is – just as people speak without knowing how the
individual sounds are produced.

We, humans, do not have to put the complexities of the system in place in
order to get language started; they are given as a whole with no priority among
its elements. In the abstract, the framework of the *Tractatus* does show how
meaningful language is possible. But when using language, the framework is
presupposed. It does not concern how sense is created.

Tractarian logic and sense are little to do with empirical, psychological
processes or with local, small-scale intentionality.[51] In this respect Wittgenstein is a direct heir of Frege's anti-psychologism. Logic takes care of itself
(5.473). It is not in need of a thinking, representing subject; i.e., a totality of
thoughts that does not exists as an object that can be described or referred
to (5.631). Sense, too, is independent of an empirical subject. It can only be
willed metaphysically, so to speak, as long as this misleading notion is understood as bare as possible. In a way, the tractarian system of representation
is inversely related to what is important for solving the problems of life. The
metaphysical will is manifest only in the change of contingencies in logical
space, which occur independent of my will. Similarly, the subject involved in
sense is the metaphysical subject, which is only manifest as a projection used
within logical space. In the end, the metaphysical solipsism in which logic and
semantics are grounded, is reduced to its barest essence, where it coincides
with a logico-semantic realism (5.64).

Pears suggests the question: 'How does the use of language ever come of the
ground if even elementary propositions are as complex as described here?' As
I read the *Tractatus*, humans do not need to know the internals of language to
use or recognize its meaning. For philosophical purposes, logical analysis can
make such specificities of language manifest. In this context, an analysis may
indeed start with an initial correlation of a name-like structure in a propositional sign with what appears to be an object. Initially this correlation will
take the form of defining an expression in terms of a complex, which derives
its sense from the proposition in which it occurs. The correlation can be retained as long as it remains faithful to the form of names and objects as they
emerge from further analysis. If unfaithful, the correlation is broken. But this
kind of analysis is nothing to do with the *creation* of sense, for it presumes
the tractarian system to be available. Analysis can only detail the framework
that makes the given sense and use of propositions manifest.[*End of interlude*]

Now that we have sufficient insight into the semantics of elementary proposition we shift attention to logically complex propositions.

---

[51]Hacker (1984) has a good overview of how this radical position grew out of Wittgenstein's earlier work, which at times seemed to allow for a more human notion of sense.

5

Complex propositional signs

W. lays too great value upon signs.[52]

When comparing specific languages, such as Lao, Polish, or Afrikaans, descriptive propositions appear vastly different from one another. From a tractarian point of view these differences are superficial and will dissolve as soon as logical analysis uncovers their essentials; i.e., a truth-functional complex of elementary propositions. The main topic of the current chapter is to compare the different ways in which truth-functionality occurs in the *Tractatus* – as truth-tables, as graphical signs or as truth-operations, – and to show that they are essentially identical to each other.

5.1   The general form of propositions

In principle, the results of analysis could exhibit a plethora of form-types that we, humans, cannot anticipate; cf. the diversity of formal systems now in use for the analysis of natural language fragments. By contrast, a few decennia before the current diversity developed, early Wittgenstein held that we *are* able to predict all possible propositional forms, and that they should therefore be of one and the same kind.

4.5c That there is a general propositional form is proved through the impossibility of there being propositions whose form we could not have foreseen (that is: constructed). The general form of propositions is: the facts are so and so.

At first reading the form indicated in 4.5c – the facts are so and so – appears to be some kind of pun. But the general form of propositions is that of truth-functions, and 4.5 just phrases their core feature – truth-possibilities – in a loose fashion.

Since a unique general form of propositions does not strictly follow from our ability to foresee all possible instances of form – after all, our anticipation of form would be consistent with there being a finite number of form-types, or with infinitely many form-types generated in a recursive fashion, – 4.5 marks again Wittgenstein's strong tendency to unify all formal aspects into one system. The most important reason for this is systematic: an absolute and unique

---

[52]Heinrich Scholz' abstract of a letter from Frege to Wittgenstein, dated: 9 November 1913, which is now assumed no longer to exist; cf. Floyd (200x), p. 7.

logic is as important for early Wittgenstein as a unique arithmetic is for Kant. Also, there may have been historical reasons why Wittgenstein held a unique general form to be possible. His main sources of logical inspiration, Frege and Russell, were after a single system to provide foundations for all of mathematics, too, and the tractarian system was clearly meant as a more general successor of both, also encompassing the semantics of descriptive language. It is interesting to note that at about the same time Emil Leon Post with his meta-mathematical approach to truth-functionality with no overt appeal to philosophical issues, was after pluriformity, which is mathematically more interesting (Post, 1967).

In the *Tractatus* propositions, as expressions *par excellence*, have form and content. Here we are mainly concerned with the truth-functional form of analyzed propositions. Form is the possibility to express sense, but the details of expressing sense is left to chapter 7. (Chapter 6 is about how complex logical structure fits within the tractarian ontology.)

Although Wittgenstein distinguishes form and content, he is always careful not to separate them completely. Even in case of logical propositions, which come close to being pure form, he stresses that their form is highlighted due to dissolved content.

4.462b In the tautology the conditions for correspondence with the world – the pictorial relations – cancel one another, so that it has no pictorial relation with reality.

The point is that any suggestion of 'platonic', abstract forms with independent existence should be blocked. Analysis unveils the form of a tractarian propositions *in* the sign. Sense does not come with an intermediate entity between sentence and reference, as Frege held. Also, the *Tractatus* does not consider uninterpreted signs, not even in the borderline case of logic where content is annulled. Yet, there are circumstances where one may disregard the sense of a sign and concentrate on its formal aspects (6.126b), as e.g. when transforming signs in analysis. For in spite of the care taken not to posit independent abstract forms, a considerable part of the *Tractatus* is concerned with determining the general form of propositions. The general form is crucial to get at a perfect notation that makes the sense of a proposition manifest. Truth-functionality, in turn, is key to the general form. It is the topic of the main theses 5 and 6. Thesis 5 is about the essence of propositional form.

5 A proposition is a truth-function of elementary propositions.
  (An elementary proposition is a truth-function of itself.)

And in thesis 6 the general form of truth-functions, and so of propositions, is given uniformly in terms of the truth-operation N.

6 The general form of a truth-function is $[\bar{p}, \bar{\xi}, N(\bar{\xi})]$.
  This is the general form of a proposition.

The theses indicate there are at least two ways to present the truth-functional structure of a proposition: 5 presents propositions as truth-functions, and 6 uses so-called truth-operations. Thesis 6.1203 indicates yet another way: the graphical notation. The remainder of this chapter should make clear that what appear to be different approaches are just variants of one and the same kind of truth-functionality.

5.2 The nature of truth-functions

Since we are all familiar with truth-functions, it seems that a formalization of theses 5 and 6 is readily available. But on the current approach we should ask to what extent the truth functions in the *Tractatus* are similar to the truth-functions of modern logic?

In the *Tractatus* truth-functions are mentioned for the first time in 3.3441 when treating of symbols, and they surface again in thesis 5, which identifies propositions with truth-functions. The question is whether what is called 'truth-function' in the *Tractatus* is the same as what is commonly presented as such in more recent logic texts; cf. Chang and Keisler (1990). In recent logic, truth-functions are (set-theoretic) functions that map—often finite—sequences of the truth-values 1 (true) and 0 (false) onto these truth-values:

$$\{0,1\}^n \longrightarrow \{0,1\}.$$

To see why this semantic notion cannot be adopted when formalizing the *Tractatus*, two things should be recalled. Firstly, although harmless in this case, the set-theoretic representation of truth-functions is alien to the *Tractatus*. In this text, sets as a way to turn totalities into objects, have no rôle to play. It even aims to show that an all-embracing philosophy can do without sets. Secondly, to use truth-values as arguments of a function requires the True (1) and the False (0) to be available as objects. These Fregean ideas are explicitly rejected in 4.431, 4.44, 4.441, and elsewhere. Instead the *Tractatus* treats truth-tables as signs, i.e., structures whose projection help determine sense (if any). To be more specific: truth-tables are complex propositional signs, and consequently truth-functions are treated as abstractions – formal concepts – based on such signs.[53]

For logicians the differences pointed out may seem of no importance. They will be interested, say, in the expressive, the deductive or the algorithmic strengths of a logic, and they regard logics that are the same in these respects as identical. But from a philosophical point of view the choice is far-reaching. The Fregean and Russellian approach to logic posits a third realm of abstract objects whose basic properties are intuited as self-evident. Early Wittgenstein was strongly opposed to such a platonistic view, and argued for what may be called an Aristotelean variant in which logic is inherent to the structure of language and the world, where content and form are not to be separated from one another.

As early as in the *Notes on Logic* of 1913 there is a stark formulation against the assumption that there are logical objects: 'The false assumption that propositions are names leads us to believe that there must be logical objects: for the meanings of logical propositions will have to be such things' (Wittgenstein, 1914-16, 107(7)). And later, in 4.431, Wittgenstein again re-

---

[53] A formal concept is shown in a propositional variable consisting of all its instances. The variable indicates the characteristics of the concept that the instances have in common (4.126 ff.).

jects the Fregean idea of truth and falsity as objects.

4.431c2 But Frege's explanation of the truth-concept is incorrect: If 'the True' and 'the False' really were objects and arguments of ¬p etc. then according to Frege's definition, the meaning (*Sinn*) of ¬p would by no means be defined.

Setting aside the question to what extent the True and the False are fit to give Fregean meaning to negated propositions, in the *Tractatus* objects are combined into configurations in logical space. If the True and the False were reified into logical space, necessity becomes part of the realm that is now capturing the contingent. As a consequence, the fundamental insight of reaching the transcendental necessary as limit of the contingent must be given up, and with it the Tractarian philosophy of ethics, language, logic, mathematics, and science, which are based on it. Indeed, rather than as abstract objects, truth and falsity should be understood in terms of the sense of a proposition being realized or not.

In line with the idea that there are no logical objects, 4.44 introduces truth-tables as signs rather than as semantic graphs of truth-functions.

4.44 The sign that arises from correlating the mark 'T' with the truth-possibilities is a propositional sign.

4.441 It is clear that a complex of signs 'F' and 'T' has no object (or complex of objects) corresponding to it; just as there is none corresponding to the horizontal or vertical strokes or to the parentheses.—There are no 'logical objects'.

The same applies of course to all signs that express what the schemata of 'T' and 'F' do.

Examples of 'signs that express the same as the schemes of 'T' and 'F' are the logical constants: ∧, ∨, →, ¬, etc., which are presented as truth-operations. Again, truth-operations concern signs; they transform one or more propositional signs into another propositional sign. In thesis 6 a uniform way is presented to generate truth-functional signs using the truth-operation N of joint negation (5.502b).

Now I will start reconstructing the three kinds of complex signs that can be found in the *Tractatus*: truth-table signs, introduced step by step in theses 4.26-4.45, the graphical method in 6.1203, and truth-operations, which are mainly explained in 5.3 ff.. The reconstruction in this chapter and the next clarifies the formal content of theses 5, 6, and some of their sub-theses, and shows that the three approaches are essentially equivalent. The gist of the discussion is a sign-based variant of truth-functional completeness, which states that each complex of the truth-operation N results in a truth-table sign, and conversely that each truth-table sign is obtained in this way.

5.3 Truth-table signs: a general form of propositions

According to thesis 5, an elementary proposition α can be regarded as a truth-function of itself:

$$\frac{\alpha}{\begin{array}{cc} T & T \\ F & F \end{array}}.$$

Although there is no essential difference between α and the truth-table sign of α, when introducing the general notion of a truth-table sign it is helpful to distinguish between elementary signs and truth-table signs. Elementary signs are a basic ingredient of truth-table signs. Strictly, elementary propositions as truth-functions of themselves are a borderline case of truth-functionality.[54] They are identity-functions, which 'show' the proposition is true if true, else false. Projection into logical space supplements this truth-conditionally empty form, and determines what truth or falsity consists of. Here elementary signs come in, which help showing how the truth or the falsity of a logically complex proposition depends on the truth or falsity of its elementary parts.

For now we abstract from the structure of elementary signs. Like in 4.24, they are introduced as letters p, q, r, s, possibly with sub- or superscripts. The elementary signs in a complex proposition describe states of things, and Wittgenstein observes that the number of truth-possibilities for n elementary signs – i.e., the different ways in which these signs can be marked for truth or falsity, – is given by the binomial coefficient (4.27).

$$K_n = \sum_{k=0}^{n} \binom{n}{k}.$$

As is well known: $K_n = 2^n$. Implicit in this calculation is the assumption that elementary propositions are logically independent of each other, in the sense that one can be true or false while the truth-values of all others are unaltered. This is in line with ontology:

1.21 Something can be the case or not the case while everything else remains the same.

This is in line with proposition 21 that shows elementary propositions to be without logic: they do not imply, contradict or validate each other.

Truth-table signs list all truth-possibilities and mark them with 'T' for truth or with 'F' for falsity to indicate the truth-value of the complex.

Definition 30 *A truth-table sign is a structure of the following form:*

$$\frac{\begin{array}{cccc} & p_1 & \cdots & p_n \\ \hline X^1 & X^1_1 & \cdots & X^1_n \end{array}}{\begin{array}{cccc} \vdots & \vdots & \cdots & \vdots \\ X^{2^n} & X^{2^n}_1 & \cdots & X^{2^n}_n \end{array}}$$

Here, the $p_1, \ldots, p_n$ are distinct elementary signs. The X's vary over the marks 'T' and 'F'. A sequence $X^i_1, \ldots, X^i_n$ ($1 \leq i \leq 2^n$) is called a truth-possibility. The 2^n distinct truth-possibilities for $p_1, \ldots, p_n$ are listed below them. The X^i to the left of a truth-possibility is called a truth-marking. *The truth-possibilities marked 'T' are the* truth-grounds *of a proposition.* □

Since it is inessential in which order the elementary content or the truth-possibilities are presented, truth-table signs can be taken as invariant under

---

[54]P.M.S. Hacker brought this point to my attention.

such ordering. The columns corresponding to elementary signs or the rows of truth-possibilities can be changed ad lib without changing the signs.[55]

The truth-possibilities of a truth-table sign are often presented in the canonical way of table 5. Table 5 gives the basic case on the left-hand side, and the general schema to generate the truth-possibilities for complex signs on the right-hand side.

$$
\begin{array}{c|}
p \\
\hline
T \\
F
\end{array}
\qquad
\begin{array}{cccc}
p_1 & \cdots & p_n & p_{n+1} \\
\hline
X_1^1 & \cdots & X_n^1 & T \\
\vdots & \vdots & \vdots & \vdots \\
X_1^{2^n} & \cdots & X_n^{2^n} & T \\
X_1^1 & \cdots & X_n^1 & F \\
\vdots & \vdots & \vdots & \vdots \\
X_1^{2^n} & \cdots & X_n^{2^n} & F
\end{array}
$$

TABLE 5   Canonical listing of truth-possibilities

If for n propositions the listing of all possible truth-possibilities is fixed canonically, a logically complex sign is fully specified by giving its elementary content and the truth-markings; cf. 4.442c, or also 5.101, 5.5, 5.502. Then, e.g., $(TF)(p)$ is an abbreviation of truth-table sign:

$$
\begin{array}{cc}
& p \\
\hline
T & T \\
F & F
\end{array}
$$

In the sequence (TF) 'T' is the truth-marking of the top-most truth-possibility in the canonical table, etc. Connectives are defined in terms of the canonical table. For instance, the canonical truth-table sign for $p \to q$ is·

---

[55] In more detail, the invariance amounts to the following. Let $\pi(1),\ldots,\pi(n)$ be a permutation of the sequence $1,\ldots,n$, then the following signs are two notations for the same truth-table sign:

$$
\begin{array}{cccc}
 & p_1 & \cdots & p_n \\
\hline
X^1 & X_1^1 & \cdots & X_n^1 \\
\vdots & \vdots & \cdots & \vdots \\
X^{2^n} & X_1^{2^n} & \cdots & X_n^{2^n}
\end{array}
\qquad
\begin{array}{cccc}
 & p_{\pi(1)} & \cdots & p_{\pi(n)} \\
\hline
X^1 & X_{\pi(1)}^1 & \cdots & X_{\pi(n)}^1 \\
\vdots & \vdots & \cdots & \vdots \\
X^{2^n} & X_{\pi(1)}^{2^n} & \cdots & X_{\pi(n)}^{2^n}
\end{array}
$$

And let $\pi'(1),\ldots,\pi'(2^n)$ be a permutation of the sequence $1,\ldots,2^n$, then the following signs are two presentations of the same truth-table sign:

$$
\begin{array}{cccc}
 & p_1 & \cdots & p_n \\
\hline
X^1 & X_1^1 & \cdots & X_n^1 \\
\vdots & \vdots & \cdots & \vdots \\
X^{2^n} & X_1^{2^n} & \cdots & X_n^{2^n}
\end{array}
\qquad
\begin{array}{cccc}
 & p_1 & \cdots & p_n \\
\hline
X^{\pi'(1)} & X_1^{\pi'(1)} & \cdots & X_n^{\pi'(1)} \\
\vdots & \vdots & \cdots & \vdots \\
X^{\pi'(2^n)} & X_1^{\pi'(2^n)} & \cdots & X_n^{\pi'(2^n)}.
\end{array}
$$

	p	q
T	T	T
T	F	T
F	T	F
T	F	F

which in the new notation becomes:

$$(\text{TTFT})(p, q) \;=\; (p \to q) \quad \textit{Def.}$$

In this notation the abbreviated truth-table sign for joint negation of n elementary propositions – which is true if all its elementary propositions are false, else false, – is:

$$(\underbrace{\text{F}\ldots\text{F}}_{2^n - 1}\text{T})(p_1, \ldots, p_n).$$

In 5.502 Wittgenstein abbreviates the sign even further as:

$$\text{N}(p_1, \ldots, p_n),$$

i.e., the sign for the truth-function N, first introduced in 5.5. It should be noticed that this notation is subtly different from the notation $\text{N}`(\bar{\xi})$ of N as a truth-operation with the apostrophe added. With apostrophe, the result of applying the operation to its bases is indicated. Without apostrophe, the notation is ambiguous between indicating the application of the operation to its bases or indicating the result of the application. Context should make clear which use is intended.[56]

Besides the identity functions of elementary propositions, truth-functional signs have two other extremes: tautologies, with all truth-possibilities marked 'T', and contradictions, with all truth-possibilities marked 'F' (4.46). These logical propositions are respectively necessarily true and necessarily false. In the final analysis, they lack a descriptive content that puts restrictions on what is the case: they say nothing. See section 7.4. Contradictions and tautologies play a prominent rôle in the notion of proof, but as long as they are not supplemented with rules they do not form a proof-system by themselves. See chapter 8.

Thesis 5.101 indicates what the sixteen possible truth-functions are for two elementary propositions. For instance, the sign $(\text{TFFF})(p, q)$ is that of p and q, the sign $(\text{TTTF})(p, q)$ that of p or q, and so on. From this we see that per truth-possibility truth-tables specify truth conjunctively, while per truth-marking of the possibilities truth is specified disjunctively. E.g., the truth-table sign:

	p
T	T
F	F

shows that p is true *or* p is false, and it does not show, e.g., that p is true *and* p is false. This apparent triviality is in fact crucial. Tractarian truth-tables are signs. On a modern view signs could allow for different interpretations.

---

[56]Sundholm (1992, p. 61) traces the apostrophe back to the elevated inverted comma notation in Whitehead and Russell (1910, p.245).

But there is no such ambiguity here: the logic of complex signs is determinate and fixed.

5.4 Graphical signs: the logical structure of complexes

In 6.1203 Wittgenstein introduces graphical signs, which are an alternative to truth-tables. The signs are sketched just enough to see what they are. The basic idea is that each proposition ξ has two poles W (true) and F (false):

$$W\xi F$$

In earlier phases, Wittgenstein used the poles 'a' and 'b' instead of 'T' and 'F', perhaps to abstract over the duality of truth-tables. To ensure determinate sense, the *Tractatus* keeps the poles fixed.

The poles of a proposition can be connected with each other using brackets that are themselves assigned to one of the two 'outer' poles of the resulting logically complex proposition. Thesis 6.1203 offers some examples. The simplest example is negation, which reverses the polarity of ξ's poles:

$$
\begin{array}{c}
W \\
\diagdown \\
\text{»}W\xi F\text{«} \\
\diagdown \\
F
\end{array}
$$

Next there is a graphical sign for the conjunction $\xi \wedge \eta$:

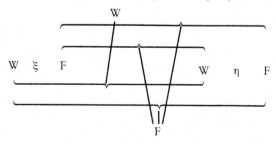

In these signs brackets correspond to the truth-possibilities of truth-tables, while the poles to which the brackets are assigned, correspond to the truth-markings of truth-possibilities. Thus, the sign is conjunctive: only the bracket that connects both truth-poles is connected to the outer truth-pole, all other brackets are connected to the outer falsity-pole. Finally, the most complex example in 6.1203 is the sign for $\neg(\neg q \wedge p)$:

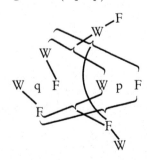

Here we see that for more complex logical structures the perspicuity of a graphical sign may rapidly decline. This is even worse when considering brackets connecting more than two poles. Be this as it may, these are just superficial features of graphical signs that can be remedied easily. To do so I introduce a box variant of the signs that clearly indicates which graphical signs are used to compose a new graphical sign. Figure 5 respectively has the alternative graphical signs for primitives, negations and conjunctions. In the boxes, 'o'

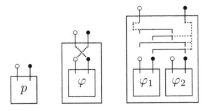

FIGURE 5 Box graphical signs: primitives, negations, conjunctions

is used for the 'T'-pole, and '•' for the 'F'-pole. Figure 6 gives the canonical form of a graphical sign. In this figure the lines that connect the brackets to the outer poles are just indicated, since for the general case the precise assignment can of course not be given.

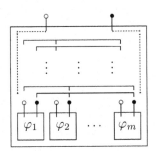

FIGURE 6 General form of graphical sign

The canonical graphical sign makes crystal clear that graphical signs allow for nesting in a straightforward way: complex graphical signs are as suited for forming a new graphical sign as primitive ones are. Figure 7 has the graphical sign for $\neg(\neg q \wedge p)$.

In 6.1203 Wittgenstein observes that besides a truth-functional notation graphical signs are the basis of a proof-system. Its rule – contradictory connections in a sign must be eliminated, – reminds one of tableaux rules. This aspect of the notation is explored further in chapter 8, and especially in the chapters 11 and 12 on infinity.

As to the complexity of signs, from the above explanation one should not conclude that complex nesting is available for graphical signs but not for truth-table signs. To begin with, thesis 5.31 states that truth-table signs can

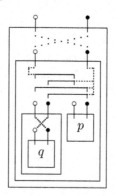

FIGURE 7   Graphical sign for $\neg(\neg q \wedge p)$

be used for this purpose as well.[57]

5.31 The schemata No. 4.31 make sense even when 'p', 'q', 'r', etc. are not elementary propositions.

   And it is easy to see that the propositional sign in No. 4.442 even in case 'p' and 'q' are truth-functions of elementary propositions, expresses a truth-function of elementary propositions.

Still, how to view the complex formulas in a truth-table sign? In the setting of the *Tractatus* I see but one possibility: the formulas are abbreviations of truth-table signs, and so the resulting structures are nested truth-tables. For instance, using the condensed notation of 4.442c one has:

- p is short for: $(TF)(p)$;
- $\neg q$ is short for: $(FT)(q)$;
- $\neg q \wedge p$ is short for: $(TFFF)(\neg q, p)$,
  or in full: $(TFFF)((FT)(q), p)$;
- $\neg(\neg q \wedge p)$ is short for: $(FT)(\neg q \wedge p)$,
  or in full: $(FT)((TFFF)((FT)(q), p))$.

The third and the fourth example are clearly nested. The first example appears to be circular as it seems to initiate the infinite sequence:

$$(TF)(p), (TF)((TF)(p)), (TF)((TF)((TF)(p))), \ldots,$$

and similar observations can be made for the other cases. However, these observations only signal an ambiguous use of p, namely its use as what I have called an elementary sign and its use as a propositional sign. It just so happens that because an elementary proposition is a truth-function of itself (5), it suffices to indicate its elementary sign to specify the propositional sign completely. As a propositional sign p is best thought of as short for: $(TF)(p)$ with '(p)' indicating the elementary sign. In general truth-table signs are logically prior to their abbreviations. The propositional sign p, in particular, abbreviates the sign '$(TF)(p)$', but plays no rôle in specifying it.

---

[57]As to the theses named in 5.31, recall that 4.31 introduces the list of all truth-possibilities for a given number of elementary propositions, and that 4.442 shows the truth-table sign for: if p then q.

Thinking along these lines one sees that the truth-table sign for $\neg(\neg q \wedge p)$, i.e. (FT)((TFFF)((FT)(q), p)), with all abbreviations eliminated is:

$$
\begin{array}{ccc}
 & q & \\
\hline
F & T & \\
T & F & p \\
\hline
T & T & T \\
F & F & T \\
F & T & F \\
F & F & F \\
\hline
F & & T \\
T & & F \\
\end{array}
$$

This structure is very similar to its graphical sign in figure 7, but growing downward instead of upward. In fact, only a little reflection is required to see that graphical signs and nested truth-table signs are essentially the same.

Proposition 31 *Graphical signs and nested truth-table signs are notational variants of each other.*

Proof. That the result holds for elementary propositional signs is clear, so let us consider a complex propositional sign. For a truth-function of complex φ_i $(1 \leq i \leq m)$, a quick look at the canonical graphical sign shows that it is a notational variant of the corresponding truth-table sign with the φ_i temporarily thought of as place-holders. The brackets correspond one-to-one with the truth-possibilities in the truth-table, and the (dashed) links of the brackets with the outer poles correspond one-to-one with the truth-markings of these truth-possibilities. By induction hypothesis, the graphical signs and the nested truth-table signs of the complexes φ_i are notational variants. So substituting the nested truth-table signs for the place-holders, or conversely will result in notational variants of the complexes. □

Although the statement in 5.31b is correct that nested truth-table sign (and hence: graphical signs) are easily seen to express a truth-function, it may be quite hard to see in a particular case *which* truth-function is expressed. Despite the equivalence, graphical signs are a more perspicuous notation in this regard than nested truth-table signs. In a nested truth-table sign, the links between the truth-possibilities of the nested tables are hard to trace. In the graphical sign, especially in its boxed form, they are immediate from the connections of poles using dashed lines.

In the previous section we have seen that truth-table signs give a general form of propositions, which at once makes clear the truth-function that is expressed. As such, truth-table signs are well suited for the end-result of propositional analysis. But they have little or no regard for the form of propositions at intermediate stages of an analysis where truth-functional structure is typically indicated in terms of non-elementary, complex content. In the same spirit, truth-table signs lack rules to determine whether a complex proposition is tautological or contradictory. By contrast, graphical signs provide for the nesting of logical complexes and quickly suggest appropriate proof-rules, but they have the drawback that it may not be immediate which truth-function is at stake. Somehow the gap has to be bridged between the perspicuous truth-table signs on the one hand, and the logically complex structures used for reasoning on the other. The third logical method of the tractatus – truth-operations – provides such a bridge.

5.5   Truth-operations, or rule-based composition

Round about 1913 Wittgenstein must have had the brilliant insight that truth-functionality has two aspects to it:

- It shows the truth-conditions of complex-propositions;
- It determines a rule to transform several truth-functions into a new one.

The first aspect is captured in truth-tables and graphical signs. The second aspect is captured in truth-operations. The essentials of truth-operations are hinted at in the stilted formulation of 5.3 (the main idea is also in 5.234):

> 5.3 All propositions are results of truth-operations on elementary propositions.
>
> The truth-operation is the way in which elementary propositions give rise to a truth-function.
>
> It is of the essence of a truth-operation to yield in the same way as from elementary propositions a truth-function, from truth-functions a new one. Each truth-operation produces from truth-functions of elementary propositions again a truth-function from elementary propositions, a proposition. The result of each truth-operation on the results of truth-operations on elementary propositions is again the result of *a* truth-operation on elementary propositions.
>
> Each proposition is the result of truth-operations on elementary propositions.

According to 5.3c, a truth-operation yields from truth-functions a new one in the same way as a truth-function is obtained from elementary propositions. So, to arrive at more insight in the notion of truth operation one must first answer the question: the same in what sense? I see two possibilities:

1. Formation rule: Just as elementary signs are used to form basic truth-table signs, so truth-table signs can be used to form other complex truth-table signs.
2. Transformation rule: Just as elementary propositional signs, which are truth-functions of themselves, may be transformed into a complex truth-table sign, so complex truth-table signs may be transformed into other complex truth-table signs.

The two options are directly related to the ambiguity noted in the previous section between elementary signs, that help showing the sense of an elementary proposition, and elementary propositional signs, which are truth-functions of themselves. On a reading as formation rule, 5.3c says that the same rule that turns elementary signs into a truth-table sign can be used to turn truth-table signs into a complex truth-table sign. By contrast, the reading as transformation rule considers the sign p to be the same as the propositional sign $(TF)(p)$, i.e. a truth-function of p itself. On this interpretation 5.3c says: just as elementary truth-functions can be transformed into complex truth-functions, so complex truth-functions can be transformed into a complex truth-function. Clearly the differences between the two readings are substantial. In my opinion the transformational reading is preferred. The formation-rule reading implies there are two kinds of truth-operations: one from elementary signs to truth-functions and one from truth-functions to truth-functions. This naturally leads to a kind of nested truth-table signs. On the reading as a transformation rule there is just one kind of truth-operation: from truth-function signs to truth-function signs. Apart from elegance, this interpretation has the virtue of making the perspicuous truth-table signs a basis for both the perfect notation and for a full-blown proof-system. See chapters 8, 9, and 12 for details.

Definition 32 *An operation Ω is given with a rule to transform a given sequence of signs into a sign. The apostrophe (') indicates the result of applying Ω to a sequence of signs.* □

In definition 32 the notion of operation is kept general. Since Wittgenstein presents N, $\neg$, $\wedge$, $\vee$, etc. as *truth*-operations, it is clear however that not just any transformation will do; the rule of such operations should have truth-functional character. Since the *Tractatus* offers no detail, one has to reconstruct the character of the rules oneself (but the space of interpretation is narrow).

Before specifying truth-operations in detail, it is best to consider some examples. Firstly, the truth-operation $\neg$ transforms a truth-table sign τ into a new one by reversing each truth-marking of τ: 'T' becomes 'F', and 'F' becomes 'T'. For instance, the negation $\neg'p$ is the transformation:

$$\begin{array}{cc} & p \\ \hline T & T \\ F & F \end{array} \quad \overset{\neg}{\longmapsto} \quad \begin{array}{cc} & p \\ \hline F & T \\ T & F \end{array}$$

or in condensed notation:

$$(\text{TF})(p) \quad \overset{\neg}{\longmapsto} \quad (\text{FT})(p).$$

The result of applying the truth-operation $\neg$ to a propositional sign p is written as $\neg'p$. In this notation $\neg$ indicates a rule, much like e.g. 'MP' is used to indicate *modus ponens*. So, '$\neg$' does not abbreviate a truth-function. In nuce, the example shows the core of Wittgenstein's perfect notation, namely that in the end redundancies such as the repeated application of a truth-operation are absent. In particular, the double application of $\neg$ to p, i.e., $\neg'\neg'p$ leads to exactly the same sign as the one started from:

$$(\text{TF})(p) \quad \overset{\neg}{\longmapsto} \quad (\text{FT})(p) \quad \overset{\neg}{\longmapsto} \quad (\text{TF})(p) .$$

Indeed, some truth-operations are involutions that eliminate themselves (5.254):

$$\neg'\neg'p = p.$$

Any sequence of negations $\neg' \ldots \neg'\varphi$ will result in either the sign $(\text{TF})(\varphi)$ or the sign $(\text{FT})(\varphi)$; no more, no less. This insight is crucial to interpret 5.43.

5.43 That from a fact p infinitely many *others* would follow, namely $\neg\neg p$, $\neg\neg\neg\neg p$, etc., is even at first sight hard to believe. And it is no less remarkable that the infinite number of logical (mathematical) propositions follow from half a dozen 'basic laws'.

But all logical propositions say the same. That is, nothing.

See section 6.7 for a detailed interpretation. As a second example, consider conjunction. For simple signs such as $\wedge'(p, q)$ one has the transformation:

$$(\text{TF})(p), \quad (\text{TF})(q) \quad \overset{\wedge}{\longmapsto} \quad (\text{TFFF})(p, q).$$

But in more complex cases the subsigns could have overlapping elementary material, and when considering the cartesian product of all truth-possibilities contradictory truth-values may be assigned to it. For example, the conjunctive arguments $\wedge'(p, q)$ and $\wedge'(q, r)$ would then result in contradictory assignments for q that subsequently would have to be eliminated; cf. the second and third row in the following tables.

	p	q			q	r
T	T	T		T	T	T
F	F	T		F	F	T
F	T	F		F	T	F
F	F	F		F	F	F

Instead of using this devious procedure, the elimination of 'closed', inconsistent possibilities can remain implicit if the result of the transformation $\wedge'(\wedge'(p, q), \wedge'(q, r))$ is determined as follows. First, the elementary signs from the two argument tables are collected (without repetition), and for the resulting sequence the canonical table of truth-possibilities is formed. Finally, for each of the truth-possibilities τ in the canonical table, we consider the restricted truth-possibilities $\tau \upharpoonright_{p,q}$ and $\tau \upharpoonright_{q,r}$ and check in the two arguments what their respective truth-marking is. For the operation '$\wedge$' the following rule must be used: if both are marked 'T', τ in the canonical table is marked 'T', else τ in this table is marked 'F'. All in all, the familiar truth-table for $\wedge(\wedge(p, q), \wedge(q, r))$ results. Using condensed notation the transformation is:

$$(\text{TFFF})(p, q), \quad (\text{TFFF})(q, r) \quad \overset{\wedge}{\longmapsto} \quad (\text{TFFFFFFF})(p, q, r).$$

As in case of negation, the beauty of using the conjunctive truth-operation is that nesting operations does not result in nested signs; one remains in the realm of simple truth-table signs. Again, this is a crucial step toward the perfect notation that captures the form of sense without redundancy. In section 5.6 we shall show in detail that signs whose operational specification contains logically 'redundant' material are the same sign as the sign without the redundancy; cf. the idempotence of conjunction: $\wedge'(\varphi, \varphi) = \varphi$.

Now we are in a position to define the notion of truth-operation.

Definition 33 *Let TF be a truth-function. A truth-operation Θ_{TF} is an operation that transforms a sequence of truth-table signs into a truth-table sign according to a rule that is obtained from TF as follows:*

1. *Form the sequence of all elementary signs of the argument truth-tables eliminating repetitions;*
2. *Add the canonical list of truth-possibilities to the elementary signs;*
3. *For each truth-possibility τ in this list check what the truth-marking is in each of the arguments;*
4. *Given these markings, use TF to determine the marking of τ in the resulting truth-table.* □

Definition 33 formalizes the idea that truth-operations use proof-like techniques to compose the subsigns of a complex sign into one. The transformations either result in a logical sign – a tautology or a contradiction, – or in a contingent sign. It remains to be checked that the transformation does not alter the sense of the signs involved. This is done in chapter 8.

The rule of a truth-operation is related to a truth-function. But since the length of the operation's input may vary – in the current setting it could be fixed finite or arbitrary finite, – the truth-functional content of the rule does not need to be finite. In particular, the rule of the truth-operation N puts no restriction on the number of arguments. Regardless of number, a truth-possibility τ in the resulting table is marked 'T' if in all argument tables it evaluates to 'F'. For now we concentrate on fixed finite truth-functions. Then it is immediate that each truth-operation Θ_{TF}, if applied to a finite number of elementary propositions, yields the truth-table sign of the truth-function TF on which the rule of Θ_{TF} is based.

Proposition 34 *For distinct elementary propositions $p_1, \ldots, p_m$:*

$$\Theta_{\text{TF}}{}'(p_1, \ldots, p_m) = \text{TF}(p_1, \ldots, p_m).$$

Proof. Based on the different elementary propositions the canonical table is formed.

p_1	$\cdots$	p_m
T	$\cdots$	T
$\vdots$	$\vdots$	$\vdots$
F	$\cdots$	F

As argument of Θ_{TF}, p_i is short for the truth-table $(\text{TF})(p_i)$, so when input to Θ_{TF}'s rule, the resulting truth-function can only be:

	p_1	$\cdots$	p_m
X^1	T	$\cdots$	T
$\vdots$	$\vdots$	$\vdots$	$\vdots$
X^{2^m}	F	$\cdots$	F

with $X^1, \ldots, X^{2^m}$ the truth-markings that TF assigns to the truth-possibilities. By definition this is the same as $\text{TF}(p_1, \ldots, p_m)$. □

Proposition 34 can be taken as a first indication that the rules are correct, but a reflection on the 'meaning' of truth-operations is left to section 7.7 on the compositionality of sense. From the proposition we have in particular:

$$\neg'(p) = \neg p,$$

$$\wedge'(p_1, p_2) = p_1 \wedge p_2,$$
$$N'(p_1, \ldots, p_m) = N(p_1, \ldots, p_m),$$

and so on. This result sanctions a somewhat sloppy use of the apostrophe ', which comes with truth-operations but not with truth-functions (cf. footnote 56). As said, often the context makes clear which reading is intended.

In the *Tractatus* there is not much discussion on specific truth-operations; the six theses concerning the notion – 5.234, 5.3, 5.32, 5.41, 5.442, 5.54 – all remain abstract. This suggests that at the time of composing the text, Wittgenstein had generalized truth-operations to operations whose rule can be based on any truth-function, or so they are reconstructed in definition 33.

5.6 Composing truth-operations

According to 5.3 truth-operations can be composed with one another; cf. the following formulation:

5.3c3 The result of each truth-operation on the results of truth-operations on elementary propositions is again the result of *a* truth-operation on elementary propositions.

Perhaps the most straightforward way to read this claim is that each complex of truth-operations $\varphi(p_0, \ldots, p_n)$, with all its elementary propositions indicated, results in a truth-table sign of $p_0, \ldots, p_n$. That this is so, is not difficult to see.

Proposition 35 *Let $\Theta'(\varphi_0, \ldots, \varphi_m)$ be a composition of truth-operations involving the elementary propositions $p_0, \ldots, p_n$. The result of this composition is a truth-table sign, i.e., the result of a truth-operation on $p_0, \ldots, p_n$.*

Proof. By induction we may assume that each complex φ_i reduces to a truth-table sign that involves $p_0, \ldots, p_n$ or a subset thereof. Next, the rule of Θ transforms the arguments into a truth-table sign TF of $p_0, \ldots, p_n$. $\square$

Proposition 35 captures the simplest interpretation of 5.3c3. A slightly subtler reading of it holds that truth-operations are closed under composition; that is, truth-operations can be defined in a uniform way as a composition of given ones. The notion of truth-operation in definition 33, when restricted to truth-operations of a fixed finite arity, clearly allows such definitions. In the more general setting of truth-operations there is a slight itch, because they allow an arbitrary finite number of arguments. The definition of composed truth-operations over finite sequences of arbitrary length should somehow fix how the elements of the argument are distributed over the truth-operations used to define the composition. Definition 36 assumes the distribution of the input sequence to be given as part of analysis.

Definition 36 *Let ξ be a sequence of truth-table signs subdivided in sequences: $\xi^1, \ldots, \xi^k$. The composition $\Theta^0(\Theta^1, \ldots, \Theta^k)$ is defined by:*

$$\Theta^0(\Theta^1, \ldots, \Theta^k)'(\xi) := \Theta^{0'}\left(\Theta^{1'}(\xi^1); \ldots; \Theta^{k'}(\xi^k)\right). \qquad \square$$

Some examples of truth-operational compositions are:

$$\begin{aligned}
\vee &= N(N) & def. \\
\rightarrow &= N(-,N) & def. \\
\wedge_n &= N(\underbrace{N,\ldots,N}_{n\ times}) & def.
\end{aligned}$$

So that:

$$\begin{aligned}
\vee`(p_0,\ldots,p_n) &= N(N)`(p_0,\ldots,p_n)) \\
&= N`(N`(p_0,\ldots,p_n))
\end{aligned}$$

And:

$$\begin{aligned}
\rightarrow `(p;q) &= N(-,N)`(p;q) \\
&= N`(p,N`(q))
\end{aligned}$$

And:

$$\begin{aligned}
\wedge_n`(p_0;\ldots;p_{n-1}) &= N(\underbrace{N,\ldots,N}_{n\ times})`(p_0;\ldots;p_{n-1}) \\
&= N`(N`(p_0),\ldots,N`(p_{n-1}))
\end{aligned}$$

In the examples the semi-colon indicates how the elements of a given sequence ξ are distributed over the composed truth-operations. The appropriate distribution must result from analysis.

Proposition 37 *A composition of truth-operations* $\Theta^0(\Theta^1,\ldots,\Theta^k)$ *is itself a truth-operation.*

Proof. Since the $\Theta^{1`}(\xi^1),\ldots,\Theta^{k`}(\xi^k)$ reduce to a sequence of truth-table signs, $\Theta^{0`}\big(\Theta^{1`}(\xi^1),\ldots,\Theta^{k`}(\xi^k)\big)$ reduces to a truth-table sign. So $\Theta^0(\Theta^1,\ldots,\Theta^k)$ is a truth-operation by definition. □

From proposition 37 it is immediate that propositions 35 holds in general: the truth-operations involved in the complex $\Theta'(\varphi_0,\ldots,\varphi_m)$ may themselves be compositions of truth-operations. This makes also clear the rule-based truth-operations allow for a parsimonious notion of propositional form; in essence truth-table signs suffice. For instance, any complex involving the elementary sign p only reduces to one of the four tables:

$$(TT)(p),\ (FT)(p),\ (TF)(p),\ (FF)(p).$$

This means that truth-operations achieve in general what graphical signs only achieve in case of double negation: irrelevant logical complexity is eliminated. In chapter 6, I discuss the philosophical importance of this insight, and how it relates to Wittgenstein's quest for a perfect notation.

5.7 Thesis 6 and truth-operational completeness

That truth-operations transform truth-table signs into a truth-table sign comes with a question: is there a small number of truth-operations that suffice to generate all truth-table signs? Based on an insight of the logician Sheffer, Wittgenstein already knew in 1913 that this is indeed the case. In the *Notes on Logic* he wrote that 'the repeated application of one <u>ab</u>-function suffices to generate all these propositions' (Wittgenstein (1913), 3rdMS, 22; At the time Wittgenstein used the term <u>ab</u>-function for truth-function.) In the *Tractatus*, thesis 5 states that propositions are truth-functions of elementary propositions. In 6 this is sharpened by holding that all truth-functions can be generated from the truth-operation N alone.

6 The general form of a truth-function is $[\bar{p}, \bar{\xi}, N(\bar{\xi})]$.
 This is the general form of a proposition.

The insight is phrased in an ultra-condensed way that is in need of explanation.

5.7.1 The variable $[\bar{p}, \bar{\xi}, N(\bar{\xi})]$

According to 6.001, 6 just states that each proposition results from repeated application of the truth-operation N. But how does the notation capture the idea? To understand the tractarian variable $[\bar{p}, \bar{\xi}, N(\bar{\xi})]$, it must be seen how it shows the rule with which the totality of all truth-functions, thought of as a sequence, is obtained.

The generation starts with the sequence of all elementary propositions, i.e., truth-functions of themselves, which is assumed given.

$$\bar{p} = p_0, \ldots, p_n, \ldots$$

This totality must be assumed, because there is no general form shared by all elementary propositions that could be the basis of a rule. As I have argued in chapter 2, elementary facts are given holistically, and the totality of elementary facts used as signs is no exception. If Wittgenstein had thought that elementary propositions have a general form, he would have specified it as a variable. Notice, by the way, that the variable $\bar{p}$ has a bar. The bar indicates that the sequence concerns all elementary signs (5.501).

Given the case of elementary propositions, the generation proceeds with forming a variable $\bar{\xi}$ that determines a totality of propositions from those already generated. The variable $\bar{\xi}$ is used to yield the sign $N(\bar{\xi})$ by applying the truth-operation N to the truth-functional elements in $\bar{\xi}$. The resulting truth-function is added to the already available propositions, and the generation proceeds with forming a new variable $\bar{\xi}'$ over the new totality etc. So, $\bar{\xi}$ in $[\bar{p}, \bar{\xi}, N(\bar{\xi})]$ does not determine a sequence that is given once and for all. It rather figures conditionally: if $\bar{\xi}$ is a sequence of signs over the signs already given, then $N(\bar{\xi})$ is a sign.[58]

The variables $\bar{\xi}$ play a crucial rôle in showing how the generation of all truth-functional signs proceeds. According to 5.501 it is inessential in which manner the elements of the variable $\bar{\xi}$ are given, but Wittgenstein indicates three ways in which it can be done: (i) by enumeration, (ii) by use of a material function $f(x)$, or (iii) by giving a rule that generates the elements. In the finite case an enumeration suffices. But when generalizing the tractarian system to the infinite case the use of variables must come with a genuine restriction. Then it is impossible to select an arbitrary infinite subsequence of the sequence that is already given. Instead a sequence should be chosen of signs that share a common form. This form is shown in the notation of the variable, which also indicates how the form is to be used in generating the sequence. Totalities cannot be formed arbitrarily, but should be definable in

---

[58]This explanation of the variable is the same as that of Russell, except for the fact that Russell speaks of sets while sequences are intended (Wittgenstein, 1922a, p. 272).

this special way. See chapter 11 and 12.

Ule (2001) objects that $\bar{\xi}$ is formed as a sub*sequence* of propositions already generated; *ibidem*, p. 273, footnote 11. Instead, he claims, the generation should take the form of a well-founded tree. In my opinion Ule's claim is mistaken as the approaches are equivalent. For take any formula φ. On Wittgenstein's approach one can first form the subformulas of φ with N applied to just elementary proposition and add those to the totality; next one generates its subformulas with N applied to signs with at most one application of N, *etc.* After sufficiently many steps φ, thought of as a well-founded tree, is obtained. Conversely each well-founded tree φ gives rise to a sequence of signs that can be used in its generation. As Wittgenstein has assumed, the use of sequences will do.

5.7.2 Truth-operational completeness

That all truth-functions can be generated using the variable of 6 amounts to a variant of so called truth-functional completeness. From propositions 35 and 37 we know that any complex of truth-operations results in a truth-table sign. This holds in particular of the complexes of N. But does the converse hold as well? Can any truth-table sign be seen as obtained from a composition of the truth-operation N? This is what is called truth-operational completeness: the truth-operation N suffices to generate all finite truth-table signs. In this section I shall make the result explicit.

We start with the treatment of disjunction. Write the result of applying the truth-operation $\vee$ in $\vee'(p_1, \ldots, p_m)$ as $\{p_1, \ldots, p_m\}$.

Proposition 38 *For distinct elementary propositions* $p_1, \ldots, p_m$:

$$N'(N'(p_1, \ldots, p_m)) = \{p_1, \ldots, p_m\}.$$

Also, for $q_1, \ldots, q_n$, *not necessarily distinct, and* π *a permutation:*

1. $\{q\} = (TF)(q)$
2. $\{q_1, \ldots, q_n\} = \{q_{\pi(1)}, \ldots, q_{\pi(n)}\}$
3. $\{q_1, \ldots, q_i, \ldots, q_i, \ldots, q_n\} = \{q_1, \ldots, q_i, \ldots, q_n\}$

Proof. The result $N'(p_1, \ldots, p_m)$ is the canonical truth-table sign:

	p_1	$\cdots$	p_m
F	T	$\cdots$	T
F	F	$\cdots$	T
$\vdots$	$\vdots$	$\vdots$	$\vdots$
F	T	$\cdots$	F
T	F	$\cdots$	F

So, the rule of N applied to this one argument results in the negation of this sign, i.e., the sign for disjunction.

	p_1	$\cdots$	p_m
T	T	$\cdots$	T
T	F	$\cdots$	T
$\vdots$	$\vdots$	$\vdots$	$\vdots$
T	T	$\cdots$	F
F	F	$\cdots$	F

To show 1., just observe that $\{q\} = N^{\prime}(N^{\prime}(q)) = (TF)(q)$. And 2. is immediate from the invariance of N under permutation. Finally, 3. is due to the second step in the rule for truth-operations in which different occurrences of the same elementary content are eliminated. □

Next, we consider conjunction. Write the result of applying the truth-operation $\wedge$ in $\wedge^{\prime}(p_1, \ldots, p_m)$ as $[p_1, \ldots, p_m]$.

Proposition 39 *For distinct elementary propositions $p_1, \ldots, p_m$:*

$$N^{\prime}(N^{\prime}(p_1), \ldots, N^{\prime}(p_m)) = [p_1, \ldots, p_m].$$

Also, for $q_1, \ldots, q_n$, not necessarily distinct, and π a permutation:

1. $\qquad\qquad\qquad [q] = (TF)(q)$
2. $\qquad\qquad\qquad [q_1, \ldots, q_n] = [q_{\pi(1)}, \ldots, q_{\pi(n)}]$
3. $[q_1, \ldots, q_i, \ldots, q_i, \ldots, q_n] = [q_1, \ldots, q_i, \ldots, q_n]$

Proof. It is clear that $N^{\prime}(p_i) = (FT)(p_i)$, so written in full the truth-table signs for the arguments are:

	p_1						p_m	
F	T			$\cdots$			F	T
T	F						T	F

Forming the canonical truth-table based on the rule for N results in:

	p_1	$\cdots$	p_m
T	T	$\cdots$	T
F	F	$\cdots$	T
$\vdots$	$\vdots$	$\vdots$	$\vdots$
F	T	$\cdots$	F
F	F	$\cdots$	F

I.e., the sign for conjunction. The three remaining properties are proved as in case of disjunction. □

Now that conjunction and disjunction are defined in terms of N, it is not hard to show that any truth-function can be obtained from this truth-operation.

Proposition 40 *Any truth-table sign σ based on elementary signs*

$$p_1, \ldots, p_m$$

can be obtained by repeated application of N on this elementary basis.

Proof. Let σ be a truth-table sign in which, say, k truth-possibilities are marked 'T'. For each truth-possibility $\tau_j = X_1, \ldots, X_m$ marked 'T', define

$$p_i^{\tau_j} = \begin{cases} N^{\prime}p_i & \text{if } X_i = F \\ p_i & \text{if } X_i = T \end{cases}$$

Set $[\tau_j] = [p_1^{\tau_j}, \ldots, p_m^{\tau_j}]$, and $\{\sigma\} = \{[\tau_1], \ldots, [\tau_k]\}$. Clearly, $\{\sigma\}$ yields the truth-table sign σ, as required.   □

According to proposition 35 each repeated application of N results in a finite truth-table sign and according to proposition 40 each finite truth-table sign can be obtained from repeated application of N. This is how truth-operations help capturing Sheffer's insight that N is truth-functionally complete for finite tractarian signs. The variable that generates all possible combinations of N from the elementary propositions indeed gives the general form of finite truth-functions and therefore, according to the *Tractatus*, of all finite propositions.

Wittgenstein phrases the insight that the truth-operation N generates all possible truth-functional signs as follows:

6.1261 In logic process and result are equivalent. (Hence no surprises.)

The processes of truth-operations and the signs that result from them, are both of truth-functional character and essentially equivalent to each other. Although in a particular case it may be unexpected which specific truth-function results, in any case it is clear it will be an instance of one and the same general form. Hence no surprises.[59]

Truth-table signs, graphical signs and truth-operations are equivalent to each other. The following proposition makes clear how this claim should be taken.

Proposition 41 *Truth-table signs, graphical signs and truth-operations are essentially the same truth-functional notation.*

Proof. That each truth-table sign can be seen as a graphical sign is clear from the canonical graphical sign (proposition 31). Proposition 31 states that graphical signs are nested truth-table signs. If in a nested truth-table sign, truth-operational structures replace the truth-table signs form the bottom up – as in proposition 40, – a nesting of truth-operations results. This nesting reduces to a truth-table sign (proposition 35).   □

Again, it remains to examine that the signs in this cycle have the same sense. This is checked in chapter 8.

5.8 *Interlude*: September–November 1913

In texts on the history of logic, Wittgenstein is often presented as one of the inventors of truth-tables. Nowadays this suggests that he introduced truth-tables semantically, as graphs of Boolean functions, but as we have seen in section 5.2, this reading is incorrect. In the *Tractatus* truth-tables are signs, and as such they are more akin to tableaux – like the analytical tableaux of Smullyan (1968) based on signed formula's $T\varphi$ and $F\psi$, – than to purely semantical notions; cf. also chapter 7, 11, and 12. Tractarian semantics concerns picturing, and although truth is rooted in picturing, in a logically complex sign there is no reference to logical objects such as the True and the False.

The idea of marking propositions with 'T' or 'F', for truth or falsity, is basic to graphical signs, where such markings occur as poles. In the *Tractatus*

---

[59]In section 12.8 this observation is generalized so that it is independent of truth-operational completeness.

graphical signs almost appear as an afterthought; they are mentioned quite late in 6.1203. Sources prior to the *Tractatus* suggest that the invention of truth-table signs, graphical signs and truth-operations have occurred at about the same time; namely in the period from September until November 1913. These sources are:

1. The *Notes on Logic* (Wittgenstein, 1913, Vol.2), which originate from September 1913; and

2. The correspondence with Russell (Wittgenstein, 1995), especially the letters 28-30, 32, which are mainly from November 1913.

In the *Notes on Logic* truth-table signs – called *WF*-schemes, – are mentioned just prior to graphical signs – called *ab*-functions (Wittgenstein, 1913, Summ., 8-9). In these notes Wittgenstein phrased the systematic relation between truth-table signs and graphical signs:

> 'In place of every proposition '*p*', let us write '$^a_b p$'. [...] Let *n* propositions be given. I then call a 'class of poles' of these propositions every class of *n* members, of which each is a pole of one of the *n* propositions, so that one member corresponds to each proposition. I then correlate with each class of poles one of two poles (*a* and *b*).'　　　　　　　　　　　　　(NL, 3rdMS, 21)

This passage first introduces truth-possibilities in terms of the poles of elementary signs. The indicated class of poles should be thought of as being bracketed, and the bracket in turn is marked for truth or falsity using just one of the poles *a* or *b*. Truth-marking of a class of poles, i.e. a truth-possibility, is of course at the heart of truth-table signs. It is exactly in this way that truth-table signs are correlated with graphical signs using the box notation; cf. section 5.4.

In section 5.4, graphical signs were shown to be isomorphic to nested truth-table signs. Such signs capture logical complexity analogous to the way in which formulas as $\psi \wedge \neg\varphi$ do. Within the context of the *Tractatus*, these more elaborate structures give rise to a certain tension, noted earlier, that if each proposition is a truth-function of elementary propositions, nested structures do not show at once which truth-function is intended. If at all, graphical signs are most useful at intermediate stages of analysis, but not so much for its end-result. In the third manuscript of the *Notes on Logic* the tension is presented as an apparent contradiction, and I think the way in which it is resolved contains the germs of the notion of a truth-operation:

> 'It is easy to suppose a contradiction in the fact that on the one hand every possible complex proposition is a simple <u>ab</u>-function of simple propositions, and that on the other hand the repeated application of one <u>ab</u>-function suffices to generate all these propositions. [...] But the <u>ab</u>-functions must be introduced as follows: The function *p|q* is merely a mechanical instrument for constructing all possible symbols of <u>ab</u>-functions. The symbols arising by repeated application of the symbol '|' do <u>not</u> contain the symbol '*p|q*'. We need a rule according to which we can form all symbols of <u>ab</u>-functions, in order to be able to speak of the class of them; and we now speak of them e.g. as those symbols of functions which can be generated by repeated application of the operation '|'. And we say now: For all <u>p</u>'s and <u>q</u>'s, '*p|q*' says something indefinable about the sense of those simple propositions which are contained

in $\underline{p}$ and $\underline{q}$.' (NL, 3rdMS, 22)

In more tractarian terms the quote can be read as follows. Every proposition is essentially a truth-function, shown in a simple, unnested truth-table sign of elementary signs. But although Sheffer's stroke '|' – with '$p|q$' read as: $\neg p$ or $\neg q$, – somehow corresponds to such a truth-function, its repeated application seems to yield a nested structure rather than a simple truth-table sign. According to the above quote, however, '|' should not be taken as part of the sign. Instead it indicates the rule of its corresponding truth-operation that is applied to the truth-functions of its arguments. So, when made explicit, '|' in the simple case of $p|q$ amounts to the transformation:

$$(\text{TF})(p), \ (\text{TF})(q) \quad \overset{|}{\longmapsto} \quad (\text{FTTT})(p,q).$$

And since '|' corresponds to a rule, its application, even if repeated, is no part of the resulting sign. Indeed, Wittgenstein's fascination for rules surely goes a long way.

The quote notes further that Sheffer's stroke as a rule, just like any other truth-operation, is indefinable; it can only show itself in the systematic passing from a few propositional signs to another; cf. section 7.7 for more discussion. The indefinability is, I think, nothing to do with the fact that '|' may be used as a primitive from which all other truth-functions can be defined. Whether or not it is indefinable in this sense depends on which truth-operations are taken as basic; as is well-known, e.g., the pair $\lor$ and $\neg$ may serve this purpose as well (5.42). Such indefinability is relative, not absolute, as the indefinability of rules is.

6

On signs and symbols as facts

From its origin the *Tractatus* kept to the idea that propositional signs take center stage. It is the unifying rôle of signs that contributes to the beauty of the tractarian system, but only after their necessary and contingent aspects are delicately balanced. In this chapter I discuss what the status is of signs and symbols as facts. This should help to understand more clearly that the *Tractatus* offers a descriptive essentialism.

6.1 The need for a perfect notation

Signs are at the heart of the tractarian system. Signs are used to describe the contingencies in logical space; signs, when analyzed, exhibit the form of sense; signs show the necessities of form and logic; signs and their meanings delimit the world.

5.475 The only thing that matters is to generate a system of signs with a specific number of dimensions – with a specific mathematical multiplicity.

However, in a philosophy where signs are facts, the balance between necessary and contingent is easily disrupted and so unjustified relations among them may result. One may be forced to hold, for instance, that the existence of each proposition as a fact comes with infinitely many other facts, as in:

$$p, \quad p \wedge p, \quad p \wedge p \wedge p, \quad p \wedge p \wedge p \wedge p, \quad \ldots$$

Without proper care, logical relations among signs could induce contingent relations of signs as facts. To preclude this, early Wittgenstein aimed for a perfect notation that kept the necessities of logic and the contingencies of description sufficiently close but separated. E.g., on the truth-operational approach the above infinite sequence reduces to the single:

$$p.$$

For early Wittgenstein, it was crucially important to realize a perfect notation with all redundancies eliminated; this not just to attain logical elegance, but to avoid some pernicious philosophical problems. In doing so, he concentrated on the finite case, mostly ignoring the delicacies of generalizing it into the infinite. We do the same here, but see chapter 11 for further discussion.

6.2 Logically elementary and complex signs

In the *Tractatus* there are logically elementary and logically complex signs. The status of elementary signs as facts seems to be without doubt. In section 3.2, I've argued that in the final analysis elementary signs are elementary facts $(n_0, \ldots, n_k)_c$ used as signs. As any fact, elementary signs have form and content.[60] Their sign-content consists in the content of the names of which the sign is formed; i.e., the token aspect of names that allows their form to be instantiated more than once. As to form, although the realization of form c is contingent, the form itself is necessarily one of the forms that are possible in logical space.

Besides elementary signs there are complex signs. Signs can be complex in many different ways, but in the end the main kind of complexity is truth-functional. (For the rest of this chapter 'complex sign' is short for: logically complex sign.) Since complex signs are generated from elementary ones, the above remarks on content and form hold to a large extent for the complex case as well. Yet, there is an important difference. Despite its complexity, all content of a complex propositions is due to its elementary parts. The difference is crucial when considering the status of complex signs as facts.

The point is this. At the heart of early Wittgenstein's philosophy is the idea that there are no logical objects; cf. section 5.2.

4.0312 The possibility of propositions is based on the principle that objects are represented by signs.

My fundamental idea is that the 'logical constants' do not represent anything. That there is no way to represent the *logic* of facts.

Thesis 4.441 clarifies the same idea in relation to truth-functional signs.

4.441 It is clear that a complex of signs 'F' and 'T' has no object (or complex of objects) corresponding to it; just as there is none corresponding to the horizontal or vertical strokes or to the parentheses.—There are no 'logical objects'.

The same applies of course to all signs that express what the schemata of 'T' and 'F' do.

Besides these negative characterizations, 5.4611 has a crisp formulation of how logical signs should be understood in a more positive way.

5.4611 The signs for logical operations are punctuation marks.

Given this much the question is: how do the logical punctuation marks, i.e. marks that indicate logical structure, contribute to a complex sign as a fact? Interestingly, in answering this question it becomes clear that the status of the apparently trivial logical marks strongly underpins interpreting the tractarian ontology as a descriptive essentialism.

---

[60]As has been noted before, in the context of signs the word 'content' is ambiguous. In the *Tractatus*, the content of elementary signs could both refer to the (sign-)content of signs as fact, and to the (sense-)content of the sign due to its projection onto another fact. In the present discussion, sign-content is meant.

6.3   Ontology and logically complex signs

In the literature the beginning of the *Tractatus* is often read as sketching the content and structure of all that exists. This reading still allows for a broad variety in how the basic elements of this ontology should be taken: as matter, as sense-data, as logical categories....[61] But it severely restricts the possibilities for the rôle of logical punctuation marks as part of signs as facts.

On the assumption that theses 1 and 2 and their offspring give the tractarian ontology in full, logical punctuation as part of a realized sign, a fact, must itself be a configuration of objects. This, I think, is impossible. A first objection is that all configuration is contingent, and so in the end unsuitable to show the necessities of logical structure. But perhaps more important is the second objection. The picture theory of meaning holds that sense concerns the projection of the forms of object-configurations onto one another. Now, what rôle can the configurations have that would correspond to logical constants? If such configurations were possible, they would be superfluous. Since there are no logical objects to refer to, only the elementary parts of a complex sign are involved in projection. But the assumption of such partial projection, which would leave out the assumed structure of punctuation, runs counter to the idea that projections are isomorphisms. Meaning is based on isomorphic structure in sign and sense; if the sense of a sign lacks logical objects than so does the sign, i.e., a fact used as a sign. Hence, logical punctuation marks should not be seen as configurations of objects.

If logical constants come without any object, their contribution to a sign can only be purely formal. Then, they provide the form that turns the elementary content of a proposition into a mirror of the independent, possible elementary facts in logical space. This form shows how the truth-conditions of the complex depends on these possibilities.

That logical constants must be purely formal is best seen in tautologies and contradictions, where all content is dissolved.

4.4661   Certainly in the tautology and the contradiction, too, signs are still connected with one another, i.e. they are related to one another, but these relationships are senseless, inessential to the *symbol*.

The insight is stated most clearly in the *Notebooks* (where truth-functions are sometimes called *ab*-functions):

17.12.14.   The signs of *ab*-functions are not material, else they could not vanish.

Still, the presence of purely formal signs-elements does not square well with the idea that the ontology at the beginning of the *Tractatus* concerns all there is, in sign and in sense. In this ontology form and content are inseparable, and accordingly pure forms do not exist. I think it is precisely for this reason that the theses on ontology should be taken in a wider context that also includes theses on propositional structure. As we shall see, there is more to ontology than theses 1 and 2 and their offspring indicate.

---

[61] Again, I refer to Stokhof (2002), chapter 3, for an overview.

6.4   Signs, symbols, and descriptive essentialism

The status of logical punctuation in signs as facts comes with the question of how to account for pure form in ontological terms? I see two possible answers. The first is to stick to the interpretation of 1-2.063 as giving the ontology of all there is, and to observe that it fails to account for all aspects of signs, especially the formal ones indicating logical structure. Since this failure would be a major omission in a philosophy that wants to give logic its proper status, I argue for a second more favorable heuristics. This interpretation weakens the ontological contribution of 1-2.063 in holding that the theses rather present a form of descriptive essentialism concerning elementary description, which is supplemented later in the *Tractatus* to allow for aspects of signs that may be purely logical or even inessential from a semantic-logical point of view.

3.34 Propositions have essential and accidental features.

Accidental are those features that result from the particular way in which the sign is produced. Essential are those which alone enable a proposition to express its sense.

On this reading, 1-2.063 indicate the essentials of the world as a framework for elementary picturing to be possible, but leaves out of scope those aspects of signs that contribute to logically complex picturing, and those aspects that clothe the essential structure of a sign for human use (4.002). Only later in the text the inessential aspects of signs are regarded.

4.011 At first sight the proposition – when for instance it is printed on paper, – does not seem to be a picture of the reality it is about. But also musical notation at first sight does not seem to be a picture of music, and our phonetic (alphabetic) notation not a picture of our speech.

And yet these notations prove to be pictures, even in the ordinary sense, of that which they represent.

But in a more logical notation the inessential aspects of signs are nearly absent.

4.012 It is clear that we experience a proposition of form '*aRb*' as a picture. Here the sign is clearly a simile of what is signified.

To elucidate the phenomenon that the sense of some signs is more apparent than that of others, Wittgenstein introduces the distinction between symbols and signs.

3.31 Each part of a proposition that characterizes its sense I call an expression (a symbol).

(A proposition is itself an expression.)

Expression is all that is essential for the sense of a proposition, what propositions may have in common.

An expression typifies a form and a content.

3.32 The sign is what can be perceived of a symbol.

Humans require signs that can be perceived and used. But the aspects that make signs available in this way are inessential, and often cloak the symbol that the sign expresses.

4.002d Language disguises thought. In fact so that from the outward form of
the clothing the form of the clothed thought cannot be derived; since the
outward form of clothing is designed for completely different purposes
than to reveal the body's form.

To arrive at the symbol in a sign there is the heuristics of analysis, in which
determining the appropriate definitions of apparent simples is combined with
pragmatics.

3.24d The contraction of a symbol for a complex into a simple symbol can be
expressed by a definition.

3.326 To see the symbol in a sign one must pay attention to meaningful use.

In analysis, a sign or some of its parts may be meaningless and useless.

3.328 If a sign is *not used*, it is senseless. That is the point of Occam's maxim.
(When all appears as if a sign has sense, then it does have sense.)

However, such frills that may help to make up a sign's clothing are inessential
from a logical or semantic point of view. Some forms of nonsense consist of
such frills only, but the symbol of signs that have sense – and I'll concentrate
on those in the following, – represents no inessentials.

In its barest appearance logically complex signs have just two ingredients.
Firstly, its has elementary signs.

3.2 Propositions can express a thought so that elements in the propositional
sign correspond with the objects of the thought.

3.201 These elements I call 'simple signs', and the proposition 'completely
analyzed'.

Secondly, it has logical punctuation marks, which may include brackets.[62]

5.461 Significant is the apparently unimportant fact that the logical pseudo-
relations, such as $\vee$ and $\supset$, need brackets; unlike real relations.
Indeed, the use of brackets with these apparent ursigns already indicates that they are not the real ursigns. For after all no one will believe
that the brackets have independent meaning.

In short, each fully analyzed sign exhibits the main ideas of the *Tractatus* in
a prototypical way. The sign shows the logical and semantic essentials of the
world, and once the symbolic turn has been taken these essentials are all there
is.

NB 22.1.15. My *whole* task is to elucidate the essence of propositions.
That is to give the essence of all facts of which the proposition *is* a
picture.
Indicate the essence of all being.
(And here 'to be' does not mean: to exist – for than it would make
no sense.)

The notebook-entry stresses that a new view on ontology is required; especially
the parenthetical remark states explicitly that in the *Tractatus* ontology is not

---

[62]As we now know, brackets are inessential for finite signs. One could resort to Polish
notation.

on existence *per se*. The tractarian ontology is rather a descriptive essentialism in which to be is to be essential to description.

Essential to a symbol are its elementary parts embedded in a purely formal logical structure. There may be inessential aspects of a sign that help to make it available for human use, and some signs may even lack an essence. But neither the logical sign-elements nor the inessential aspects of a sign should be analyzed as contingent complexes of tractarian objects. The perceptible aspects of logical sign-elements, in particular, are inessential, without any object. Elementary facts used as signs figure in showing what is essential to a sign's sense, in the symbol that it expresses. Logical punctuation is part of the symbol as well, but strictly speaking there are no pure logical forms. Logical structure is always a formal *extension* of elementary form and content.

6.5   The coherence of signs

Symbols indicate the semantic and logical essentials of signs. At this point we know that not all elements of logical punctuation in a symbol can be analyzed away, but it is still unclear how they relate to the other parts of a sign and to one another. Even if we restrict ourselves to signs as the end-result of analysis – i.e., logical complexes of elementary propositions, – the insight that logical structure is purely formal 'raises the question how the coherence of a proposition comes about' (4.221b). For elementary signs one can hold that its coherence is due to the form of their names that enables them to immediately relate to each other (4.22). But this cannot be upheld for the complex case, as the sign-elements indicating logical structure are no names, no objects used as names.

The question that 4.221b poses is answered, I think, at a later point in the *Tractatus* where Wittgenstein observes that logical punctuation marks put a restriction on the elements with which they combine.

5.515   It must show itself in our symbols that that which '∨', '∧', etc. connects with each other must be propositions.

   And this is indeed the case, for the symbol 'p' and 'q' itself already presupposes '∨', '¬', etc. If the sign 'p' in '$p \vee q$' does not stand for a complex sign, it cannot have sense all by itself; but then also the signs '$p \vee p$', '$p \wedge p$', etc., which are equisignificant with p, cannot have sense. If however '$p \vee p$' does not have sense, then '$p \vee q$' cannot have sense either.

Thesis 5.515 makes plain that the restriction is mutual: logical connectives only combine propositions (5.515a), but also conversely, elementary signs must have the complexity of a configuration to have sense at all (5.515b2), and must already possess the ability to combine with connectives to express a logically complex sense (5.515b1). In other words, just as each object has a form, and so presupposes other objects as a range of possible, immediate combinations, so elementary signs and logical punctuation marks have forms that enable them to relate immediately with each other. The elementary signs and the logical punctuation marks differ, however, in that elementary signs have content, namely the content of the names of which they are formed, while logical punctuation marks are formal additions to the content of the elementary parts. Whatever appears to give a logical mark content is inessential, is not part of the symbol in which the mark occurs. There are no logical objects, neither in sense nor in signs.

6.5.1 The coherence of truth-table signs

It is worth to observe that 5.515a not only allows the simple logical structure of truth-table signs, but also the nested logical structures of, for instance: '$\neg((p \wedge q) \vee \neg r)$'. In case of nesting, each logical part – here: '$\neg r$', '$p \wedge q$', and '$(p \wedge q) \vee \neg r$', – is again a proposition, and thus must have a form that relates immediately with other logical sign-elements to form new complexes. Be this as it may, in a complex the rôles of logical sign-elements, such as '$\neg$', '$\wedge$', '$\vee$', can differ vastly from one another, and it is hard to see how to account for the differences in terms of purely formal, contentless additions to propositions. Fortunately, there is no need to do so. As we have seen in chapter 5, it is sufficient to understand the coherence of logical complexity in terms of the single, unnested general form of truth-table signs. Nesting is accounted for by means of truth-operations, which are absent in the end-result of logical analysis, and thus inessential to ontology.

The logical structure of truth-table signs is non-nested. The form of its logical marks enables an immediate relationship with the forms of its elementary parts. But this is the end of it: the resulting combination is not involved in any further combination, and so lacks the form that would be required for such nesting.

6.5.2 Ontology and truth-operations

In a strict sense, truth-operations show logical structure indirectly. They presuppose their arguments to be truth-table signs, and indicate how signs are transformed into other such signs. Therefore, the sign-elements for truth-operations have a quite different status than the marks partaking in truth-table signs as facts. The sign-elements for truth-operations rather indicate a rule of transformation. In this regard they are analogous to the marks for proof-rules in a proof; e.g., to an 'MP' – for: *Modus Ponens* – that is put in a proof's sideline.

6.1264 A proposition that has sense states something, and its proof shows that this so; in logic every proposition is the form of a proof.

Every proposition of logic is a modus ponens represented in signs. (And one cannot express modus ponens by means of a proposition.)

As a consequence, it is less urgent to understand the status of truth-operation marks in a sign as fact. The occurrence of an operation does not characterize the sense of a proposition (5.25); they are semantically inessential. Truth-operations, in particular, indicate a process to arrive at a sign. As such they occur at the intermediate stages of sign-analysis, and help to make the logical structure of a sign humanly accessible. But one could hold, as I do, that only the end-result of the transformation, i.e. a truth-table sign, is a fully analyzed logical sign. And only for this end-result the question of coherence is pertinent. To indicate a transformation is largely for human convenience, but when it comes to the essentials of logic and sense, simple truth-table signs will do.

6.6 *Interlude*: the pragmatics of sign and sense

We are in a good position to return to a claim of chapter 4: that the *Tractatus* and its system lacks an absolute distinction between pictures, including language, and that what pictures are about. I have argued that the use of language comes with a bi-partitioning of states of things, where one half is used as elementary sign, while the other half concerns their content. The bi-partitioning was presented as non-absolute, and restricted only by the projections that are possible given the logical space of states of things. By contrast, one may now have the impression that the distinction between pictures and their senses should be absolute after all, because logical punctuation marks, which seem inherently linguistic, are elements of a sign that cannot be eliminated. I think this impression is illusive.

3.1431 The essence of a propositional sign becomes very clear if we imagine it composed, instead out of written signs, out of spacial objects (for instance, tables, chairs, books).

Then the mutual spatial position of these things expresses the sense of the proposition.

It is obvious that the *Tractatus* retained an early insight of Wittgenstein that goes back to the *Notes on Logic* (1913): the fact that this inkpot is on this table may express that I sit in this chair. And 3.1431 leaves no doubt that the same options are still at hand in the construction of logically complex pictures. In the final analysis, each state of things can form the symbolic essence of an elementary proposition, and so each state of things should have a form that allows to be extended into a logical complex structure. Cf. also section 7.3 on the nature of logical picturing. Therefore, even in the context of logically complex propositions, what constitutes a sign and what its sense depends on which part of the system is used in logical projection to which end.

6.7 Perfect notation: the economy of showing sense

In the introduction to this chapter I observed that the unifying rôle of signs contributes to the beauty of the *Tractatus*, but that this rôle requires a delicate balance between a sign's necessary and contingent aspects. The marks showing the necessities of logic, in particular, may quickly have an undesired influence on the contingencies of signs as facts. For instance, if the sign φ is true, do all signs of propositions that follow logically from it also have to be facts? Such a mixture of contingency and necessity is undesirable. It is therefore crucial to the tractarian system to provide for a proper understanding of all aspects of signs, the contingent as well as the necessary. To do so, requires a perfect notation that captures its essence, its symbol; nothing more, nothing less.

Elementary propositions The combination and separation of the contingent and the necessary can already be found in elementary signs. At this level, a contingent configuration $(n_0, \ldots, n_k)_c$ is used to describe states of things, where description is projection of content to mirror form. This contingent configuration is necessarily one of the possibilities in logical space. Further, form itself cannot be content, and thus cannot be described. Both the form c of a sign and the necessity of its possibility shows itself in how the names $n_0, \ldots, n_k$ are configured. This means that the contingent and the necessary aspects of an elementary sign are in balance. Although they co-occur in the elementary sign, contingent description cannot describe the necessities of form. Both aspects of the elementary sign remain as it were each in its own realm.

Complex propositions As we have seen, propositions are complex in different ways. But with a view to the status of signs as facts all inessentials should be disregarded, and so only the complexity of symbols remain.

Like elementary propositions, logically complex propositions require a delicate balance between the contingent and the necessary. In general, signs show necessities, but they must do so without forging unguarded bridges towards the contingent; especially the necessities shown in the logical structure of a sign must have no effect on the contingencies of the sign as a fact. This means that a proper notation must be found to do justice to these mutually exclusive aspects of a sign. But how? In this regard, tractarian signs sometimes remind one of rabit-ducks: now their contingent aspect is used for description, then their necessary aspect shows a logical form.

To see more clearly what is at stake, let us consider the logical operation of negation. In the classical logic of the *Tractatus*, a proposition p has

$$\underbrace{\neg \ldots \neg}_{2n} p$$

among its logical consequences. The sign p being a fact suggests therefore that its realization and use implies the existence of infinitely many other facts $\neg \neg p$, $\neg \neg \neg \neg p$, $\ldots$. This would be unjustified; not because p does indeed have infinitely many consequences, but because the totality of facts would change solely due to logic, to result in infinitely many signs all with the same sense (5.43).

Wittgenstein avoids the embarrassment by holding that logical operations have no reference. On the one hand there are no logical objects, on the other hand logical punctuation marks are contentless formal additions to elementary form and content. As we have seen, the idea resulted in the notion of symbol, which in its barest form is a truth-table sign.

By holding that logical punctuation marks are formal add-ons to the elementary parts of a sign, Wittgenstein had already found a proper balance between the contingent and the necessary aspects of complex signs. In his descriptive essentialism, logical complexity has no effect on the content of a sign as a fact. Truth-tables are the only complexes with ontological import. By contrast, the marks of truth-operations indicate a rule to transform truth-table signs. These marks occur at intermediate stages of analysis only. As such they are inessential to ontology, and are absent in the resulting truth-table sign.

Since symbols must be restricted to what is essential to the sense of a sign, a perfect notation is called for that keeps the indication of logical form minimal.

4.0621b That a sign has an occurrence of negation is as yet no feature of its sense $(\neg\neg p = p)$.

Section 9.4 shows in detail how truth-operations help realize the perfect notation, with its economy of the sign showing the form of sense.

6.8 *Interlude*: the genesis of a perfect notation

Round about 1913, Wittgenstein thought that graphical signs offered the logical minimality required if propositional signs are facts. His idea was that the poles of a graphical sign are 'transitive'; that is, a complex pole -W-F-W- reduces to -W-, and a pole -F-W-F- to -F-. Thus, transitivity should ensure that the sign

$$-W-F-W-p-F-W-F-,$$

for $\neg\neg p$, is identical to the sign

$$-W-p-F-,$$

for p, and so the apparent infinity of negations would reduce to either p or to $\neg p$. See Wittgenstein (1914), p. 114-5.

At a later stage, Wittgenstein's fascination for graphical signs waned. In the *Tractatus* they only surface in 6.1203. The reason for his reduced interest may have been that transitivity of poles shows the equivalence of some signs, but it is quite unclear how the idea generalizes to more complex signs, and so the embarrassment remains that logical structure appears to affect facts. Indeed, the idea of transitivity seems limited to the case of negation, but other connectives have similar effects. For instance, p has $p \wedge \ldots \wedge p$ and $p \vee \ldots \vee p$ as logical consequences, and a quick look at the corresponding graphical signs shows that here the transformation required is more involved than just transitivity. Figure 8 shows the stages in case of conjunction.

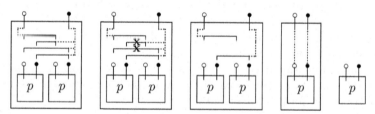

FIGURE 8 Transforming a conjunction

At the left-hand side there is the graphical sign for $p \wedge p$. The first step in its transformation is to check for inconsistent truth-markings – using '$\mathbf{x}$', – and next to eliminate them. Due to the uniformity in the resulting truth-markings – either all true or all false, – the elementary parts p can be identified. And only in the graphical sign resulting from this identification, we arrive at a stage where something like transitivity of poles applies; now from -W-W- to -W-, and from -F-F- to -F-. Finally, the sign p remains. The transformation for such propositions as $p \vee p\, p \to p$ or $p|p$ are different but similar.

The introduction of truth-operations offered a general solution to the problem of separating the consequences of a proposition from the number of facts there are. The idea surfaced in 1913,[63] but Wittgenstein may have realized its full impact only at a later stage. Truth-operations as general rules are first mentioned in August 1916.

NB 17.8.16. An operation is the transition from one term to the next one in a series of forms.

To recall how truth-operations solve the issue, consider the sequence:

$$p, \neg`p, \neg`\neg`p, \neg`\neg`\neg`p, \neg`\neg`\neg`\neg`p, \ldots$$

Since $\neg$ is a truth-operation, the sequence is not truly infinite. In section 5.5 we have seen that the result of the operations is either $(TF)(p)$ or $(FT)(p)$. In fact, truth-operations as in definition 33 give a solution for all logical forms, not just for negation. Proposition 35 shows that any complex of truth-operations based on elementary proposition $p_0, \ldots, p_n$ results in a truth-table sign for $p_0, \ldots, p_n$. In other words, truth-operations ensure that the result of analysis has an optimal complexity relative to the given elementary basis.

NL, 3rdMS, 20 The very possibility of Frege's explanation of 'not-p' and 'if p then q', from which it follows that 'not-not-p' denotes the same as p, makes it probable that there is some method of designation in which 'not-not-p' corresponds to the same symbol as 'p'. But if this method of designation suffices for logic, it must be the right one.

One may object that on this approach, too, signs still have marks of truth-operations, and with it some kind of infinity. But this view is misleading. In the ontology of descriptive essentialism there is nothing corresponding to them. Rather, their operation is transcendental, and shown outside the realm of what can be said or described (5.24). Logically only the result of their application exists, namely: truth-table signs, in which logical structure is a purely formal add-on to elementary signs. The next chapter shows that after a simple adaption to capture symbols, truth-table signs lack notational redundancy, as required.

4.1213 Now we also understand our feeling that we have a proper logical conception once everything in our sign-language is all right.

---

[63]See the quote from the *Notes on Logic* (3rdMS, 22) in section 5.8

7

Complex propositions

How is it possible for descriptive language to have sense? On a tractarian approach, the analysis of language makes clear that this question can be answered in two stages. First one elucidates the sense of elementary propositions, and next the sense of logically complex ones.

The sense of elementary propositions is already discussed in chapter 4. Chapter 5 started to answer the question for logically complex propositions with its study of logically complex signs that exhibit the general form of sense. It remains to complete this study by detailing the full sense of logical complexes.

The present chapter begins with a sketch of Frege's view on propositions that Wittgenstein took as point of departure. Next, I introduce an approach to Wittgenstein's notion of sense, but argue that it has to be refined to do justice to senseless logical propositions. The insights are used to comment on the extent in which the system is contextual and compositional.

7.1 Frege on propositions

Despite considerable differences, it is fair to say that Wittgenstein's notion of sense is indebted to Frege, and detailing the tractarian concept of sense is best done against the backdrop of the great works of his predecessor.

7.1.1 Sense and reference

Frege held that each expression has both a sense (*Sinn*) and a reference (*Bedeutung*).[64] For propositions the sense is the thought it expresses, and its reference is a truth-value. The sense of an expression is close to what one would normally call 'its meaning'; it is the informative, objective content of an expression that can be shared among people (Frege (1892b), footnote 5). More in particular Frege presents the sense of an expression as a way in which its referent is given.

To capture the sense and reference of propositions, Frege introduced truth-values as objects and extended the notion of function accordingly. On Frege's view, propositions are a particular kind of name that refer to the truth-value the True or the truth-value the False. And the predicate of a propositional expression refers to a concept or a relation, which are functions from objects to truth-values. Without going into too much detail of what their appropriate

---

[64]This section is based on Frege (1892b).

analysis would be, one might say that (5a,b) differ in the way in which they present the reference of the predicates, i.e., a concept. Similarly, (5a,c) differ in the way in which they present the reference of the subject, i.e., an object.

(5) a. M. Ciccone wears a red jacket.
   b. M. Ciccone wears a jacket with the colour of a rose.
   c. The author of 'The English Roses' wears a red jacket.

The differences are of sense, not of reference; the changes affect the thoughts that the propositions express, but not their truth-value.

On Frege's view, when inquiring into the truth of a proposition one is drawn from its sense, a thought, to its reference, a truth-value. In case of the propositions (5a-c) the senses of the subject term all present the object Madonna as referent, and the senses of the predicate all present the concept about her jacket that the predicates refer to. The thought determines how these referents combine to yield a truth-value, and in judging the thought we claim the propositions to be true. Whether the judgment is correct or not, the propositions all name the same truth-value: they are equivalent.

Frege offers an analogous but more general analysis for propositions that have logical structure. In section 2.1.1 we have already seen how quantifiers are treated, so it remains to give an indication of truth-functional structure. Using Frege's notation, the conditional (6a) is written as (6b).

(6) a. If God manifests itself, the square-root of three is one.
   b. $\models$ The square-root of three is one
      $\quad\quad$ God manifests itself
   c. Antecedent: God manifests itself
   d. Consequent: The square-root of three is one
   e. Function: If ( ), then [ ]

The proposition is false only if the antecedent is the True, and the consequent is not the True. The formulation 'not the True' is chosen instead of 'the False', because Frege allowed other objects than truth-values to be argument to the conditional

$$\models \begin{array}{l} \zeta \\ \xi \end{array} .$$

However, using the so-called 'content stroke' he circumvented having to deal with more than two truth-values. The context stroke before ξ in:

$$-\xi$$

yields the True if ξ is the True and yields the False otherwise. His notation of, e.g., the conditional is considered to merge content strokes with the notation for conditionals proper. With the content strokes made explicit, the conditional reads:

$$\models \begin{array}{l} -\zeta \\ -\xi \end{array} .$$

For the cases considered here the content stroke is not needed: the referents of the antecedent and the consequent proposition are composed with the referent of the conditional expression, and since propositions refer to truth-values the result will be a truth-value. Cf. Frege (1879), §5, or Frege (1893), §29, 30.

Apart from its sense and its reference a propositional expression can be asserted. The content of a sentence – its sense, – is a thought, and a sentence names a truth-value. The thought is that the conditions obtain for the reference to be the True. Asserting the sentence, it is presented as naming the True. Cf. Frege (1893), §32.

7.1.2 Compositionality and contextuality

Although Frege did not state it explicitly, it is often assumed that he adhered to a principle of compositionality:

> **Compositionality** The reference of an expression is a function of the reference of its parts and the way in which they are combined. The sense of an expression is also obtained in a compositional way.

As we shall see, the principle does not hold in general for the tractarian system. Wittgenstein accepted it only for the sense of logical complexes.

Apart from compositionality, Frege assumed the principle of contextuality. For different types of function and objects Frege indicates what kind of sense and what kind of reference they have. This might suggest that the functions and objects have meaning independent of any context, but this reading would be mistaken:

> **Contextuality** One should ask for the meaning of words within a propositional context, not in isolation.

The principal of contextuality is stressed at the beginning of Frege's *Grundlagen*, Frege (1884), 23. It receives much less emphasis in the classic trilogy: *Funktion und Begriff*, *Über Sinn und Bedeutung*, *Über Begriff und Gegenstand*, which is published about seven years later. Dummett (1973) argues that the principle is indeed absent in Frege's later work (Dummett, 1973, p. 7). By contrast, Hacker shows convincingly that Frege continued to adhere to it (Hacker, 1979).

7.2 Proposition and sense

From an abstract point of view, the work of Frege and early Wittgenstein may appear quite similar, but zooming in to the details important differences come to the fore. The similarity mainly consists in the prominent rôle that is given to the sense of a proposition, which for Frege and for Wittgenstein consists in the thought that the truth-conditions of the proposition obtain. Also, both held that if a proposition is used to make a statement, the sense is claimed to be true, and hence the truth-conditions of the proposition are claimed to obtain. Wittgenstein puts it thus:

4.022 A proposition *shows* its sense.

A proposition *shows* how things relate *if* its true. And it *states that* they do so relate.

From this point onward, differences between Frege and Wittgenstein start to emerge. Other than Frege, Wittgenstein does not grant all expressions a sense and a reference. Contingent propositions have sense but no reference, and names have reference but no sense. Wittgenstein's fundamental idea that logical constants do not refer, leads to substantial deviations, too, as does his idea that sign and sense can hardly be separated from one another. In fact, Frege must have been aware of the dissimilarities. It made him comment that Wittgenstein 'lays too great value upon signs'. See chapter 5, motto.

Early Wittgenstein held sense to be prior to the facts, but other than Frege he leaves no doubt how sense and facts are related. To this end he first elucidates how the truth-value of elementary propositions is based in contingent configuration of objects, and next how the sense of logically complex propositions depends on elementary sense. Before giving details, let us first recall the main theses on sense.

The leading idea is that the sense of a proposition is a picture, a logical model of reality.

4.021 A proposition is a picture of reality: for I know the situation that it represents if I understand the proposition. And I understand the proposition without its sense being explained to me.

A picture show its sense, which is a situation in logical space (2.11, 2.202, 2.221).[65]

4.031 In a proposition a situation is composed as it were by way of experiment.
One could say without more ado: instead of, this proposition has such or such a sense, this proposition represents such or such a situation.

To clarify sense it is therefore necessary to come to grips with situations.[66] A first, too naive way of looking at situations would be to see them as the totality of states of things involved in the projection of an analyzed proposition, a truth-table sign. For the current discussion we keep a frame Λ, π fixed.

Definition 42 *Let Λ be logical space, and let π be one of its projections in use. The combination $\mathcal{F} \equiv \Lambda, \pi$ is called a frame.* □

The projection π bi-partitions Λ in elementary propositions and their projections (definition 13, p. 56). The states of things used as signs are the basis from which all possible logical structures are formed, in the way thesis 6 specifies.

Definition 43 (Tentative) *Let $(X^1, \ldots, X^{2^n})(p_1, \ldots, p_n)$ be a complex proposition. A naive rendering of its sense in Λ, π is:*

$$\pi[(X^1, \ldots, X^{2^n})(p_1, \ldots, p_n)] = \pi[p_1], \ldots, \pi[p_n] \ .$$ □

Only a little reflection is required to see that this tentative definition will not do: within the frame it gives all truth-table signs based on $p_1, \ldots, p_n$ the same

---

[65] I follow Pears and McGuinness in using 'situation' as translation of 'Sachlage'. Wittgenstein's use of 'Sachlage' is ambiguous. In the first three theses in which it is used – 2.0121, 2.0122, 2.014, – it is synonymous with 'Sachverhalt': state of things, a configuration of objects. In the remaining theses it means: situation. Situation, too, is somewhat ambiguous: it may correspond to the disjunction of truth-possibilities in a sign, or, more determinate, to just one of its disjuncts. Here, I use the first, most general reading.

[66] In 2.1, one reads 'We make ourselves pictures of facts'. The use of 'facts' instead of 'situations' is unexpected. Sense must be independent of facts (4.061), because facts are realized states of things, while sense requires purely logical possibility. Perhaps 2.1 should be read in the context of 2.06 that uses the older terminology of positive and negative facts, where negative facts are not realized either.

sense! For example, both (TF)(p) and its negation (FT)(p) get the sense $\pi[p]$. By contrast, they should have opposite sense (4.0621c). In the *Tractatus* it is already stated at an early stage that although projection is important, it does not constitute sense entirely.

3.13 A proposition includes all that the projection includes; but not what is projected.

Therefore, the possibility of what is projected but not what is projected itself.

A proposition, therefore, does not yet contain its sense, but does contain the possibility of expressing it.

('The content of a proposition' means the content of a proposition that has sense.)

A proposition contains the form of its sense, but not its content.

In 3.13d 'content' is used to indicate sense-content: the totality of states of things the proposition is about, i.e., what I tentatively suggested to be its full sense.

Thesis 3.13 presents sense as an interaction between a propositional sign and reality. The form of sense is *in* the propositional sign, which shows the forms of the states of things described. The content of sense are the projected states of things, which are part of reality but not of the proposition (3.13c). So, projection presents the content of sense from a logical perspective. To get at the logical perspective, recall that in the *Tractatus* sense and situation are near synonyms, and that situations are stated to be logical complex.

4.032a A proposition is a picture of a situation only in so far as it has logical structure.

The question is: how much logical structure may a situation have? One would expect situations to be contingent, but in 5.525 Wittgenstein stretches the notion to its limit to allow certain and impossible situations.

5.525 It is incorrect to paraphrase the proposition '$(\exists x).fx$' – like Russell does, – as 'fx is possible'.

The certainty, possibility or impossibility of a situation is not expressed by a proposition, but through this, that an expression is a tautology, a proposition with sense, or a contradiction.

The precedent to which we are constantly inclined to appeal must already reside in the symbol itself.

Below I shall argue that this use of 'situation' is too broad; it is inconsistent with the idea that logical propositions have no sense. But it does indicate that the sense of a proposition, the situation it describes, is the states of things it is projected onto but shown, as it were, in the logical shadows of the propositional sign. Consider, for instance, an elementary proposition (TF)(p) projected via π onto the state of things β. The sense of p shows β as realized – that is, how things relate *if* it were true, – and p is used to state that β is realized. If β is indeed realized, p's statement is correct and so p true, but else p's statement is incorrect and p false. Figure 4 on page 62, with α as p, gives the idea in visualized form.

The next simplest example is the negation $\neg p$ or $(FT)(p)$ of p. Again, π projects $\neg p$ onto the content β of its sense. But now, the form of $\neg p$'s sense shows β as not realized and $\neg p$ is used to state that β is not realized. Thus, if β is realized $\neg p$ is false, but else it is true. This is how I understand 4.0621c.

4.0621c The propositions 'p' and '$\neg p$' have opposite sense, but one and the same reality corresponds with them.[67]

Figure 9 visualizes the idea for the negation of an elementary propositions. As for an elementary proposition, the projection π turns the state of things α into a logical picture of β. But the sense of $\neg\alpha$ ($\neg p$ in the above) shows β as not realized. This is indicated by the double circle around β. So, if β is realized (β°), a statement made with $\neg\alpha$ is false. But a statement made with $\neg\alpha$ is true, if β is not realized ($\beta^\bullet$).[68]

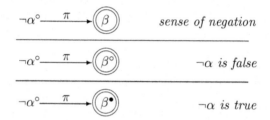

FIGURE 9   Negation

In much the same way the sense of the conjunction $(TFFF)(p, q)$ shows both states of things $\pi[p]$ and $\pi[q]$ as realized, while the sense of the disjunction $(TTTF)(p, q)$ shows either $\pi[p]$ as realized or $\pi[q]$ as realized. In logic, at the borderlines of sense, states of things are either left unconstrained – in case of tautology, – or they disappear completely – in case of contradiction. In the next interlude I consider to what extent a picture can be logical, but first I give a general definition of sense. The definition will be refined later with a view to logical propositions.

The sense of a proposition φ shows alternative realization patterns in logical space; i.e., what would be the case if φ were true. The realization patterns are determined by φ's truth-grounds as follows.

Definition 44 *Let $(X^1, \ldots, X^{2^n})(p_0, \ldots, p_n)$ be a propositional sign, and let $Y^i_1, \ldots, Y^i_n$ be the truth-possibility of the truth-ground X^i (so $X^i = T$). In the frame Λ, π, the realization pattern $\rho[X^i]$ of X^i is defined by:*

---

[67]On the current treatment of sense the verb 'to say' in 4.0621a does not mean: to state, but rather: to describe; cf. 4.0621 in full.

4.0621 But that the signs 'p' and '$\neg p$' *can* say the same, is important. For it shows that the sign '$\neg$' has nothing corresponding to it in reality.

That a sign has an occurrence of negation is not enough to characterize its sense ($\neg\neg p = p$).

The propositions 'p' and '$\neg p$' have opposite sense, but one and the same reality corresponds with them.

[68]For a discussion of the ontological status of '$\neg$' (or of '$\wedge$' and '$\vee$' in the examples to come), see chapter 6.

- $\rho[X^i]$ shows $\pi[p_j]$ as realized if in its truth-possibility $Y_j^i = T$;
- $\rho[X^i]$ shows $\pi[p_j]$ as not realized if in its truth-possibility $Y_j^i = F$.

In other words, $\rho[X^i]$ is the conjunction of (non-)realizations that the $Y_1^i, \ldots, Y_n^i$ ascribe to $\pi[p_1], \ldots, \pi[p_n]$. □

The basis for considering only T-markings is 4.022b: a proposition *shows* how things are *if* its true. Now, it is a small step to get from realization patterns to sense.

Definition 45 *Let φ be a proposition and let $(X^1, \ldots, X^{2^n})(p_1, \ldots, p_n)$ be its truth-table sign. The sense $\sigma[\varphi]$ of φ in the frame Λ, π has form and content:*

- *The form of $\sigma[\varphi]$ is the disjunction of realization patterns $\rho[X^i]$ for each of φ's truth-grounds.*
- *The content of $\sigma[\varphi]$ is the states of things $\pi[p_1], \ldots, \pi[p_n]$.* □

Definition 45 captures what Black (1964), 217, calls the same-level interpretation of sense. On this interpretation truth-tables are not seen as condensed meta-statements about elementary propositions; cf. the Tarskian:

'$p \wedge q$' is true, if and only if: 'p' is true and 'q' is true.

According to the *Tractatus*, meta-phrases such as "p' is true' are pseudo-propositions that attempt to say what can only be shown, namely the sense of p (cf. 5.542). Instead, the sense of a logically complex proposition remains at the same level as the sense of elementary propositions: it shows possible ways in which the content of its elementary propositions can be realized or not. In this regard, the core of the same-level reading is stated in 2.11, be it in a highly condensed way (cf. also 4.2).

2.11 A picture presents a situation in logical space, the existence and non-existence of states of things.

To make the same observation in a more logical jargon: truth-table signs are disjunctive normal-forms of elementary propositions that help making the sense of a proposition manifest.

For tractarian sense to be possible at all, it is pertinent that the non-pictorial part of logical space lacks overt logical complexity; it should consist of independent, positive states of things. Although the independent realizability of states of things is at the hart of logic, all overt marking of logical structure resides in the sign. Only in this way the situation a proposition presents can be a disjunction of contradictory perspectives onto one and the same reality. This will often be the case, for if a truth-table sign has more than one truth-ground its realization-patterns are inconsistent with each other. In particular, non-pictorial logical space does not have the form of a Boolean algebra (as has been suggested by Ishiguro and is formalized by Lokhorst). This would have been queer anyway, for why hold that logical marks do not refer if such structure is overtly available in reality?

7.3 *Interlude*: logical pictures

In the literature, Wittgenstein's use of the term 'picture' has been found misleading. For one, Hintikka and Hintikka never tire of stressing how ill-conceived they think the term 'picturing' is. See Hintikka (1986), chapter 4. But what's in a name? That which Wittgenstein calls 'a picture' called by any other name would depict as well.[69]

In the *Tractatus* Wittgenstein uses the word 'picture', especially 'logical picture', to capture determinate sense, and this use is quite removed from the common meaning of 'picturing'. Normally pictures are similar to what they depict, not strictly isomorphic, as determinate sense requires. However, a departure from everyday language is quite common in philosophy, and is hardly a problem as soon as a proper interpretation is found. In fact, analogous observations can be made with regard to Wittgenstein's use of the words 'world', 'fact', 'situation', 'object', 'name', *et cetera*.

Visser (1999) explains in detail how Wittgenstein use of the term 'picture' puts his work in the continental tradition of Hertz and Boltzmann. Hertz asked the Kantian question 'How is an *a priori* science of nature possible?', and used the term 'picture' to develop a logico-mathematical view on scientific models and explanation. Cf. also Janik and Toulmin (1973), chapter 5. Early Wittgenstein broadened Hertz' question to language in general: 'How is an *a priori* notion of sense possible?'. As we have just seen, his answer consists in sense as an *a priori*, logical model, a picture of possible ways the world may be.

Apart form its philosophical use, I think there is a natural, quite mundane

---

[69] According to Hintikka and Hintikka, the picture theory of language 'is often thought of as an original and mysterious creation of Wittgenstein's. In reality it can be considered (subject to a number of qualifications) being little more than a dramatization of certain Fregean ideas which Tarski later build into his truth-conditions of atomic propositions' (*ibidem*, 92). Indeed, Hintikka and Hintikka ignore Wittgenstein's ties with the tradition of Hertz and Boltzmann, and put him squarely into that of Frege as it was continued by the Vienna Circle, Tarski, among others (*ibidem*, 87).

I hope to have made plain already that the tractarian system can hardly be called Fregean. Frege's influence on Wittgenstein's philosophy of language is restricted to its broad outlines, but at a detailed level their use of basic concepts such as 'function', 'object', 'sense', 'reference', 'truth', 'composition' is different to the extent of being incompatible.

To reduce the gap between the work of Frege and Wittgenstein, Hintikka and Hintikka assume Frege was familiar with the idea of truth-conditions of elementary propositions (*ibidem*, 95). Hintikka and Hintikka do not give sources for this assumption, and I have found none in the works of Frege. It rather seems that Frege himself would have disagreed: 'Am I one of those who will understand your book? Without your assistance, hardly. What you write about atomic facts, facts, and states of affairs would never have occurred to me, although possibly I come close to your opinion at one place in my essay' (Dreben and Floyd, 200x, 44). Frege's reference is to *Der Gedanke* (1918), which was unavailable to Wittgenstein while writing the *Tractatus*.

In my opinion it is a major accomplishment of Wittgenstein to have devised a full-fledged intensional system that does justice to the similarities and differences between logical and contingent propositions. It may well be that Wittgenstein's work helped Tarski to formulate his theory of truth. But even if Tarski's theory of truth were essentially Fregean, it would still be misleading to suggest, as Hintikka and Hintikka do, that the tractarian system (subject to a number of scarcely indicated qualifications) is basically Tarskian.

I find it quite surprising that in line with the extensional tradition of the Vienna Circle, Hintikka and Hintikka pay little attention to the sense of elementary propositions, and only note the modal approach of the *Tractatus* in passing (*ibidem*, 97). Given their keen knowledge of intensional logic, I would rather have expected them to grant Wittgenstein his innovation, and to clarify its ties with the tableau method and with disjunctive normal forms, to which Hintikka made such important contributions.

level at which the logical notion of picturing makes sense. For elementary propositions this is hardly contested, especially when it is reminded that a model just captures what is essential semantically. In this regard most aspects of a picture, and in particular those that make it but similar to what it depicts, can be abstracted from as inessential. In this way, even an isomorphism may result.

For negation in a similar manner, say, a model of a thief, a cook, his wive and her lover may be used to show what did *not* happen. The negation is unique in that it indicates that the objects in the model do not relate as indicated, but the negation is 'weak' in that one does not learn what, if anything, is really the case.

Finally, when more than one model is available, they can be used in a conjunctive or a disjunctive manner. For example, let one model show Lee driving a car hitting Spike's shop, and let another model show Tony riding a bicycle that passes Lee's car. The models can be used to show that Lee drove a car hitting Spike's shop *and* that Tony rode a bicycle that passed Lee's car. But they can be used, too, to show that Lee drove a car hitting Spike's shop *or* that Tony rode a bicycle that passed Lee's car. In the logical use of such models there must be a means to mark their logical combination; e.g., an indication of the connective intended (negation, conjunction, disjunction), or a sign attached to the models marking them as conjuncts or alternatives.

Wittgenstein's notion of sense abstracts from the practice of using concrete models in a logical way, and turns it in his imagery of propositions as most general, logical pictures. The figures 4 and 9, on pages 62 and 118 respectively, visualize the idea for elementary propositions and negations. But the visualization can be extended to include conjunctions and disjunctions. Figure 10 visualizes the idea for conjunctions.

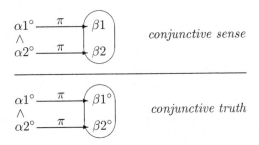

FIGURE 10 Conjunctive proposition

Much as before the projection π turns the conjuncts α_1 and α_2 into logical pictures of β_1 and β_2. Projection π is in use, so α_1 and α_2 are realized, as '∘' indicates. The states of things β_1 and β_2 are encircled to mark that the proposition $\alpha_1 \wedge \alpha_2$ shows both β_1 and β_2 as realized. If β_1 and β_2 are both realized, as in the lower part of figure 10, than the statement of $\alpha_1 \wedge \alpha_2$ is true, else it is false. Figure 11 visualizes the idea for disjunctions. Now the states of things β_1 and β_2 are in a dashed box to indicate that the proposition $\alpha_1 \vee \alpha_2$ shows either β_1 or β_2 as realized. So if, say, β_1 is realized, as in the

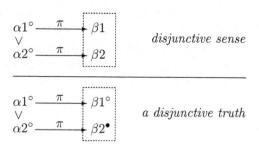

FIGURE 11 Disjunctive proposition

lower part of figure 11, than the statement of $\alpha_1 \vee \alpha_2$ is true. Only if both disjuncts are not realized the statement of $\alpha_1 \vee \alpha_2$ is false.

In complex cases the intensional interpretation of truth-tables is surely more convenient as a representation of sense than the visualizations given here. But the aim was to make clear there are mundane ways of logical picturing that make sense; it is quite natural to speak of 'picturing' even if logical complexity is involved. [*End of interlude*]

7.4 Logical propositions

Relative to projection, definition 45 specifies the sense of a logically complex propositions. On closer scrutiny one could argue that the definition must be refined: logical propositions are not captured correctly. Consider for instance $(TT)(p)$ and $(FF)(p)$. Per tentative definition their sense-content is $\pi[p]$, but as to the form of sense: that of $(TT)(p)$ shows $\pi[p]$ as realized or not realized, while that of $(FF)(p)$ offers no realization pattern for $\pi[p]$. Perhaps this is in line with 5.525 and with the imagery concerning propositions of 4.463.

4.463 Its truth-conditions determine the space that a proposition leaves to the facts.

(A proposition, a picture, a model is in the negative sense like a solid body that restricts the freedom of movement of the others; in the positive sense it is like a space limited by a solid substance in which a body has its place.)

The tautology leaves reality the entire – infinite – logical space; the contradiction fills the entire logical space and leaves reality none of its points. Thus neither of them can determine reality in any way.

By contrast, I think definition 45 is too liberal, and that logical propositions require a more radical interpretation of sense. Definition 45 gives different logical proposition different sense-content; e.g., $(TT)(p)$ has $\pi[p]$ as its content, and $(TT)(q)$ $\pi[q]$. But logical propositions are senseless – they are 'about' nothing (6.124), – and should therefore lack any content.

4.461a-c Propositions show what they say, the tautology and the contradiction show that they say nothing.

The tautology has no truth-conditions for it is unconditionally true; and the contradiction is true on no condition.

Tautology and contradiction are senseless.

Content is reserved for propositions with sense, and the apparent content of a logical proposition should somehow vanish.

3.13d 'The content of a proposition' means the content of a proposition that has sense.

The interpretation must not be so radical that logical propositions become of a different category as descriptive propositions. Logical proposition *are* propositions and may occur as such as part of descriptive ones. Therefore, the division between logical and descriptive propositions should not be absolute.

6.112 The proper explanation of logical propositions must give them a unique position among all propositions.

Refining definition 45 to allow for the senseless extremes of logic is a subtle matter. The point is that on the one hand logical propositions lack content, but on the other hand all signs should reach out to reality, even the logical ones whose content is dissolved in the final analysis.

6.124 Logical propositions describe the skeleton of the world, or rather they exhibit it. They are 'about' nothing. They presuppose that names have reference and elementary propositions sense: and this is their connection with the world. It is clear that it must indicate something about the world that certain connections of symbols – that essentially have a determinate character – are tautologies. This is the decisive point. We have said that much about the symbols that we use is inessential and much is not. In logic only the latter has expression: that is, in logic it is not *we* ourselves that express with signs what we would like to, but in logic the essence of essentially necessary signs expresses itself: If we know the logical syntax of a sign-language, then all logical propositions have already been given at once.

How to reconcile the apparently conflicting claims concerning a connection with reality and the lack of content? In the previous chapter the logical sign-elements – such as 'not', 'and', 'or', – are presented as formal additions to elementary propositions, and 6.124 states that in a logical proposition the semantic contribution of elementary propositions is presupposed unaltered. Still, the formal elements of logic connect elementary propositions so that in a tautology or contradiction their contribution is no part of the symbol.

4.4661 Clearly even in the tautology and the contradiction signs are still connected with each other, that is, they are related to one another, but these relationships are without sense, inessential for the *symbol*.

Thesis 4.4661 indicates that the apparent conflict can be solved using the distinction between sign and symbol. Thinking along these lines the truth-table sign of a proposition is a sign, and it should be left as-is to keep its elementary parts and their connection with the world in place. But to get at the symbol of this sign a further analysis is required to captures the sign's contingent nucleus or, if such a nucleus is absent, its logicality. Since logical propositions can be part of descriptive ones, the reduction should be general, for all propositions.

4.465 The logical product of a tautology and a proposition says the same thing as the proposition. The product is therefore identical with the proposition. For one cannot alter the essence of a symbol without altering its sense.

To determine the contingent nucleus of a proposition, its logical parts have to be eliminated. If in the process no content remains, the proposition is logical; else the sense of the proposition is defined on the basis of the resulting symbol. For example, to determine the symbol of $(p \vee \neg p) \wedge q$ its logical part $(p \vee \neg p)$ must be removed, so that its sense is that of q. Similarly, the process of determining symbols must show $(p \vee \neg p)$ to have no sense at all (cf. 5.513b2). Notice that *occurrences* of an elementary proposition may have to be eliminated, not elementary propositions *simpliciter*. Elementary propositions that occur in a logical part may still contribute to the meaning of the proposition if they also have occurrences in a contingent part; e.g., $p \wedge (p \vee \neg p)$.

   As it happens, the rule that dissolves the logical parts of a truth-table signs is quite elegant.

Definition 46 *Reduction rule: if the truth-table sign of a proposition φ has form: $(Y_1, \ldots, Y_{2^n}, Y_1, \ldots, Y_{2^n})(p_1, \ldots, p_{n+1})$, it reduces to the form: $(Y_1, \ldots, Y_{2^n})(p_1, \ldots, p_n)$. Written more schematically the rule is:*

$$\frac{(Y_1, \ldots, Y_{2^n}, Y_1, \ldots, Y_{2^n})(p_1, \ldots, p_{n+1})}{(Y_1, \ldots, Y_{2^n})(p_1, \ldots, p_n)} \text{ RR}$$

The symbol $\Sigma[\varphi]$ of φ is the structure that remains if all possible reductions have been effected. □

Recall that the notation used in definition 46 is based on so-called canonical truth-table signs. So if a sign has form

$$(Y_1, \ldots, Y_{2^n}, Y_1, \ldots, Y_{2^n})(p_1, \ldots, p_{n+1}),$$

this means that in the canonical sign there are two copies of

$$(Y_1, \ldots, Y_{2^n})(p_1, \ldots, p_n),$$

but one with in addition p_{n+1} everywhere marked with 'T' and the other in addition p_{n+1} everywhere marked with 'F'. So in effect the sign is the conjunction of $(Y_1, \ldots, Y_{2^n})(p_1, \ldots, p_n)$ with the tautology $(TT)(p_{n+1})$, or dually, the disjunction of $(Y_1, \ldots, Y_{2^n})(p_1, \ldots, p_n)$ with the contradiction $(FF)(p_{n+1})$, and as far as sense is concerned the latter can be eliminated. In yet other words, the rule is basically one of resolution. Wittgenstein was aware of this rule. On 2.9.16 he wrote in his notebook: $(p.q) \vee (p.\neg q) = p$. See MS103.

   Some examples may clarify how the reduction rule works. Consider the truth-table sign at the left-hand side in table 6. As it stands the reduction rule does not seem to apply, because the relevant symmetry in the truth-markings is lacking. Recall however that truth-tables signs are invariant under permutation, and such permutation may be required before the rule can be used. Indeed, the permuted sign in the middle has symmetry. Therefore, the sign is reduced to the sign for q at the right-hand side. The truth-table on the left-hand side gives the symbol of $(p \vee \neg p) \wedge q$ or of $(p \wedge \neg p) \vee q$, and so the

	p	q	
T	T	T	
T	F	T	
F	T	F	
F	F	F	

	q	p	
T	T	T	
F	F	T	
T	T	F	
F	F	F	

	q	
T	T	
F	F	

TABLE 6   From sign to symbol

sense of this sign is that of q.

It remains to discuss the logical extremes. In applying the reduction rule, a logical proposition of just one elementary proposition may result. For a single p, only $(TT)(p)$ and $(FF)(p)$ satisfy the premiss pattern of the reduction rule. The first reduces to $(T)()$ and the second to $(F)()$. As required, neither has any specific content. The logical rôle of these empty forms is in line with viewing truth-table signs as disjunctive normal forms. In a way, the empty $(T)()$ has truth-grounds but the corresponding realization patterns are empty. Thus, $(T)()$ can be seen as a disjunction of empty conjuncts. An empty conjunct is true iff all its terms are true, and so it is natural to hold empty conjuncts to be true vacuously. Then, the disjunction of vacuous truths is true as well; it leaves all states of things as either realized or not realized without constraint. The empty $(F)()$ signals the absence of truth-grounds, and can thus be seen as an empty disjunction. An empty disjunction is true if some of its terms is true. In other words, it is always false. It fills out logical space completely to leave a proposition no states of things.

Again, as 4.464 puts it, there are two ways for logical propositions to be senseless.

4.463c The tautology leaves reality the entire – infinite – logical space; the contradiction fills the entire logical space and leaves reality none of its points. Thus neither of them can determine reality in any way.

The emptiness of logical propositions moves the perspective as it were to unconditional projection in logical space: $(T)()$ marks any truth-possibility with 'T', and $(F)()$ any truth-possibility with 'F'. In more detail the idea becomes:

Definition 47 *Let Λ, π be a frame, and let φ be a proposition. The symbol $\Sigma[\varphi]$ expressed by φ is either its contingent nucleus based, say, on the elementary content $p_1, \ldots, p_n$ ($n \geq 1$). Or, if φ is logical, the symbol is one of the empty $(T)()$ or $(F)()$.*

Relative to the projection π, the sense of the contingent nucleus $\Sigma[\varphi]$ restricts logical space to the situation involving $\pi[p_1], \ldots, \pi[p_n]$, as determined in definition 45.

Logical propositions are senseless: the symbol $(T)()$ is a disjunction of empty conjunctions that shows each $\pi[q]$ as either realized or not realized; the symbol $(F)()$ is an empty disjunction that shows no state of things $\pi[q]$ as realized. □

For exegetical reasons, I have presented $\Sigma[\varphi]$ as obtained in two steps: first the truth-table sign is used as given explicitly in the *Tractatus*; next the symbol of the truth-table sign is shown, to make the symbolic aspect of the tractarian philosophy explicit. Definition 48 below gives a one step approach that is in line with the compositionality of logically complex propositions.

To summarize, the standard truth-table signs for tautology and contradiction suggest that projections $\pi[p]$ figure in their sense, but according to 4.461 it does not: logical propositions are senseless and have no content.

To show sense one has to exploit the division between signs and symbols more carefully. The sign of a fully analyzed proposition is a truth-table sign based on its elementary parts. This sign safeguards the presuppositions of reference that are in force for any proposition, but it is the symbol of the sign that shows a proposition's sense or senselessness.

Only those elementary propositions that occur in the contingent nucleus of a proposition contribute to its sense, those occurring in logical parts do not. In the symbol, the logical occurrences are eliminated. Thus, the sense of a proposition – the truth-conditions that it expresses, – are manifest as soon as its contingent nucleus is shown. If the symbol of a proposition is empty it is logical, it is 'about' nothing (6.124).

Now that we have made precise what the sense of a proposition is, it remains to consider – after a short interlude – to what extent the tractarian system adheres to the Fregean principles of contextuality and compositionality.

7.5   *Interlude*: toward signs and symbols in MS102

In the notebooks, especially in MS102, Wittgenstein is outspoken that something like the distinction between sign and symbol is required for a proper treatment of logical propositions. The distinction in an explicit form seems to be of later date.

In December 1914 Wittgenstein makes an observation that later grew into the philosophy of logical proposition in 4.46 and its offspring; cf. e.g. 4.4661 and 4.465 quoted above.

MS102, 12.12.14 p.TAUT $= p$ i.e. TAUT. says nothing!

A few months later the distinction between sign and symbol starts to surface.

MS102, 25.5.15 One may enlarge each proposition, without changing its sense, using 'and' together with ANY tautology or with the negation of a contradiction.

And 'without changing its sense' means: without changing the ES-SENTIALS of the sign itself. For one cannot change the SIGN without changing its sense.

Slightly later Wittgenstein elaborates:

MS102, 10.6.15 'p.$q \vee \neg q$' DOES NOT dependent ON 'q'!!

Entire propositions disappear!

The mere insight that 'p.$q \vee \neg q$' is independent of 'q' while it obviously contains the sign 'q', shows us that signs of the form $\eta \vee \neg \eta$ seemingly but only SEEMINGLY exist.

The reason is of course that the connection '$p \vee \neg p$' is possible externally but it lacks the conditions that gives such a complex something to say and turns it into a proposition.

The entry is in line with early Wittgenstein's descriptive essentialism: only that exists which contributes to description. [*End of interlude*]

7.6 Contextuality

The tractarian system embodies a radical form of Frege's contextuality: only propositions have sense and only names have reference, but names only have reference in the context of a proposition. In the *Tractatus* the sharp distinction between either having sense or having reference is most crucial at the elementary level. Here, the principle of contextuality eliminates any suggestion that names or other sub-propositional expressions have sense independent of (elementary) propositions.

According to Wittgenstein, sense concerns the description of contingencies. Since contingency is understood as possible realization of structure, structure is essential to sense. Names lack semantic structure, and so do not have sense.[70] In general, expressions do have semantic structure – they are configurations of names, – but still they lack an independent sense. At best they contribute uniformly to the sense of all propositions in which they occur. See chapter 4.

The principle of contextuality does not extend, or so I think, to logically complex propositions. There is no such requirements as: only in the context of a truth-function does a proposition have sense. It is rather the other way around: truth-functionality presupposes its parts to have sense. As we shall see shortly, in the tractarian system adhering to contextuality and adhering to compositionality is mutually exclusive.

7.7 Compositionality

It is often assumed that Frege's philosophy of language is based on the principle of compositionality:

> The reference of an expression is a function of the reference of its parts and the way in which they are combined. The sense of an expression is also obtained in this compositional way.

The question is: to what extent can the tractarian system be called compositional? Without doubt the answer is quite different from that for the Fregean system. Not only are their notions of functions different, but their views on sense and reference differs as well.

7.7.1 Functionality

Let us first recall the concept of function in the *Tractatus*. In 3.318 Wittgenstein remarks explicitly that propositions are functional.

3.318 I regard the proposition – like Frege and Russell, – as a function of the expressions it contains.

---

[70]Although the tractarian philosophy restricts reference to names, it is more refined than the Augustinian image of meaning as merely consisting of names having reference. The world is not a totality of things (1.1), and names only have reference in the context of a proposition. Read in this way, both the *Tractatus* and the *Philosopische Untersuchungen* begin with a critique of meaning as just reference of names.

The question is: how should this be taken? For one, Frege's most well-known notion of function – discussed in section 2.1.1, – is quite different from Russell's, who took propositional functions as basic and derived other notions of functions from it. See Hylton (2005), chapter 7, for an overview. In my opinion, Wittgenstein's view on functionality is closer to Russell's than to Frege's, but strictly speaking it is different from both.

At the elementary level, the point is that a proposition is a configuration of names $(n_0, \ldots, n_k)_c$ in immediate combination, but none of the names need to be a relation or a function taking the others as arguments. For example, $(n_0, \ldots, n_k)_c$ does not have to be of form $n_0(n_1, \ldots, n_k)$ with, say, n_0 acting as a functional relation taking the objects $n_1, \ldots, n_{k-1}$ as arguments to yield n_k as their unique functional value. At the logically complex level, logical constants have no reference, in particular they do not refer to functions. This means the functionality alluded to in 3.318 must be rather general and independent of specific functional expressions. It either concerns the uniqueness of a combination of names – argued for in section 2.5, – or the uniqueness of a truth-table sign that results from applying a truth-operation. This is 'like Frege and Russell' in the abstract, but not quite like Frege and Russell in its details.

Since on Wittgenstein's view a sign does not have to have a main functional expression, his approach to compositionality must be interpreted accordingly.

7.7.2 Compositionality of sense

In the *Tractatus*, the only form of compositionality that is stated to hold is that of logically complex propositions.

5.2341a The sense of a truth-function of p is a function of p's sense.

At first 5.2341 may appear to be just about the elementary case, since just one elementary proposition p is mentioned, but it is clear the general case is intended. That only the sense of logically complex propositions is stated to be compositional is deliberate. The sense of elementary propositions cannot be compositional. Elementary propositions are the smallest units of sense, so there are no senses to compose at the sub-elementary level. If Wittgenstein had thought the sense of elementary propositions were compositional, he would have presented a variable indicating the forms from which elementary sense would be composed. But there is no such variable, and there is no such composition. In the next interlude I discuss some changes Wittgenstein went through before taking this position.

The sense of a logically complex proposition is compositional. To see this, consider a proposition ψ that has $\varphi_1, \ldots, \varphi_n$ as its immediate sub-propositions. Given the material developed up until now, it is not difficult to see how the symbol $\Sigma[\psi]$ of ψ is a function of the symbols $\Sigma[\varphi_1], \ldots, \Sigma[\varphi_n]$.

Definition 48 *Let ψ be the complex: $\Omega`(\varphi_1, \ldots, \varphi_n)$. In a frame Λ, π, the sense of ψ has form and content. The form of $\Sigma[\psi]$ is the form of*

$$\Sigma[\Omega`(\Sigma[\varphi_1], \ldots, \Sigma[\varphi_n])],$$

i.e., the symbol that results from the truth-table sign

$$\Omega`(\Sigma[\varphi_1], \ldots, \Sigma[\varphi_n]).$$

The content of $\Sigma[\psi]$ is the content ψ's contingent nucleus, if it has one. Cf. definition 47. □

Of course, $\Sigma[\varphi_i]$ could be either $(T)()$ or $(F)()$, but the tautology and the contradiction can be used in the rule of Ω as usual. See 4.465. In chapter 9, we shall see that $\Sigma[\varphi_i]$ is equivalent with φ_i: using one or the other, the result is essentially the same. Indeed, Wittgenstein has indicated sufficiently how to realize his finite perfect notation.

7.7.3 Comparison with Frege

Though 5.2431 states that sense is compositional, Wittgenstein's approach to compositionality is quite different from Frege's. Frege held that the sense of a proposition is a function of the sense of the proposition's main function as applied to the senses of its arguments. So the sense of an expression $F(o_0, \ldots, o_n)$ would be obtained from the sense of F, the senses of the o_i and the way in which they are combined. In a tractarian composition $\Omega(\varphi_0, \ldots, \varphi_m)$ the operation Ω also appears to be the main function, much as in case of a Fregean expression, but it is not. Operations indicate a transformation of signs, and are no part of the form or content of sense themselves.

5.25 The occurrence of an operation does not characterize the sense of a proposition.
   For an operation does not express anything, only its result does, and this depends on the bases of the operation.
   (Operation and function must not be confused with one another.)

Operations concern a non-material process to obtain a truth-table sign from given truth-table signs. In the resulting sign the operation is absent, but if the proposition is contingent its sign does show its sense. Yet, the sense of a complex is a *function* of the senses of its parts, in that the end-result of the transformation is unique. This notion of compositionality holds in general, for all logical operations.
   In the above an important aspect of sense is taken for granted:

5.233 The operation can appear just in those circumstances in which a proposition results from another one in a logically meaningful way. So in the circumstance where the logical construction of a proposition begins.

The crux of 5.233 is 'in a logically meaningful way', indicating that operations somehow contribute to sense after all. The observation seems to be in conflict with the claim in 5.25 that the occurrence of an operation does not characterize sense. The theses are reconciled up to a certain point as soon as it is noted that 5.25 is about the lack of reference of an operation. In this regard an operation does not leave a material trace. But the operation does have a logical influence on sense in that its rule makes the truth-conditions of its end-result manifest, and with that the sense of the proposition analyzed. In the next chapter, on truth and logical consequence, I will reflect on the 'logical meaning' of truth-operations in the context of Wittgenstein's ideas on rules of inference.

7.7.4 The creativity of language

The compositionality of logically complex propositions also has a bearing on the creativity of language. In section 7.2 and 7.4 we have seen how tractarian sense is independent of facts. The independence of facts is essential to the creativity of description.

4.03a A proposition must use old expressions to communicate a new sense.

The creativity of language is rooted in its general form. Each proposition is a logical picture (4.03c), a truth-table of elementary signs. Since the sense of elementary signs is shown via projection, and since the logical structure of all descriptive propositions is uniform, we understand the new sense without further explanation (4.02). In this way descriptive language is creative.

The creativity of language stops at the elementary level, the rock-bottom of sense. For what is there to create in a unique and absolute logical space? Since the sense of elementary propositions is non-compositional, there is no way in which their sense can be understood in terms of the senses of composing parts. At best, creativity of elementary sense can be had heuristically, based on our assumptions concerning how certain sub-propositional expression contribute uniformly to elementary sense. Be this as it may, the *Tractatus* has little concern for such epistemological processes. The book rather indicates that an extreme form of contextuality cannot be reconciled with compositionality. Elementary propositions adhere to contextuality but are non-compositional, logically complex proposition do not adhere to contextuality but are compositional.[71]

In Wittgenstein's early philosophy the determinateness of sense and the holism that comes with it is much more important than a satisfactory description of the dynamics of linguistic creativity. But even if the sense of a proposition were 'composed' of what is contributed to it by its expressions, one may wonder what the fine-grained combinatorics of names and objects would contribute to understanding the creativity of language? Human creativity is more likely to be understood at the level of logically complex sense. From a tractarian point of view, most phrases that we use, such as proper names, should be thought of as logically complex and so compositional in terms of analysis and the structure of sense. Also, recall 4.002: 'Humans are able to build languages capable of expressing any meaning, without having the slightest idea how each word has meaning or what this meaning is – just as people speak without knowing how the individual sounds are produced.'

7.8 *Interlude*: elementary compositionality lost

In the early sources there are some traces indicating that at times Wittgenstein did consider elementary propositions to be compositional. But when

---

[71] According to Hacker (1979, 232): 'It would be quite misguided [...] to interpret the principle [of contextuality, JDO] as assigning primacy to sentence meaning, just because this would render unintelligible our ability to construct and understand new sentences.' By contrast, I think the *Tractatus* does hold this radical position. Creativity can only be rendered intelligible for those aspects of language that have a general form. Logically complex propositions have a general form (6), but elementary propositions do not. See sections 7.7.2 and 7.8.

completing the *Prototractatus* he must have come to the insight that compositionality at this level cannot be assumed. Indeed, McGuinness (2001), pp. 267-268, argues that the theses on the application on logic (5.55 and its offspring), which state that any specific, *a priori* assumptions on elementary propositions must be shunned, have 'few early parallels', and are presumably added in the final stages of writing.

About all philosophy in the pre-tractatus writings takes place at the propositional level. Sub-propositional elements are mostly considered to see how an entire structure can be picture of another one. There are a few exceptions. Perhaps, most explicit is *Notes on Logic*, First MS Wittgenstein (1913), 123.

NoL, First MS. Indefinables are of two sorts: names, and forms. Propositions cannot consists of names alone; they cannot be classes of names. A name can not only occur in two different propositions, but can occur in the same way in both.

The quote suggests that elementary propositions are composed of forms taking names as arguments. If this is read correctly, the idea is clearly abandoned at later stages. In the *Tractatus* forms are *of* names and enable elementary structure as immediate combination of names. In this way, instead of classes structures result.

There are other quotes suggesting a compositional view on elementary propositions.

MS102, 3.11.14 Knowledge of the representing relation CAN only be founded on knowledge of the constituents of the states of things.

MS102, 7.6.15 The logical form of a proposition must already be given with the forms of its constituents.

The quotes are insufficient to assume Wittgenstein still held such a position in the *Tractatus*. Compositionality of logically complex sense is based on the general form of complex propositions. Similarly, to hold that the sense of elementary propositions is compositional, requires to give the general form of elementary propositions. As the following quotes make plain, this proved to be a dead-end. At first there is a promising observation.

MS103, 16.4.16 EACH simple propositions can be viewed as having form $\varphi(x)$. That is why all simple propositions should be given this form.

Assuming ALL simple propositions were given to me: Then the question would simply arise which propositions could be formed from them. And that are ALL propositions, and THIS is how they are bounded.

Here 'elementary proposition' is called: simple proposition. The idea is that if each elementary proposition is complex, it can always be viewed as consisting of a material property φ, i.e. a complex of names, and a name x *simpliciter*. But this specification of elementary form is too unspecific to single out their totality in a sensible way. It is no wonder that in the entry of the next day doubt starts to creep in.

MS103, 17.4.16 In general the above definition can only be a sign-rule that is nothing to do with the sense of the sign.

But is such a rule possible?

The definition is only possible if it is not a proposition itself.

For a proposition can not be about all propositions, but a definition can.

The topic is addressed again a few months later (with 'elementary proposition' also called: atomic function).

MS103, 21.11.16 The general form of propositions must be given, because all possible propositional forms must be a priori. [...]

Now we need to clarify the notion of atomic function and the notion 'and so on...'. [...]

Without this clarification we are stuck with the ursigns and could not go 'on'.

MS103, 23.11.16 What makes operations possible? In general, a notion of structural similarity. If e.g. I consider the elementary propositions, they should share something; else I could not speak collectively of them all as the 'elementary propositions'.

But then it should be possible to obtain them from each other as the result of operations.

Then if two elementary propositions truly share something what an elementary proposition and a complex proposition do not share, this shared feature should somehow be expressible in a general way.

With these open remarks, Wittgenstein's considerations on the general form of elementary propositions come to an end. I surmise he abandoned the idea that elementary proposition have a general form, and with that the assumption that their sense is compositional. In the *Tractatus*, 5.55 clearly states that we cannot indicate the different forms names may have prior to analysis, and so the same holds with regard to the general form of elementary propositions. Accordingly, the *Tractatus* lacks a variable that generates the elementary propositions; their totality is assumed given without specifying a common form. See section 2.3. Analysis may show that some kinds of elementary proposition are compositional, perhaps even all are, but there is no ground to assume this beforehand.   [*End of interlude*]

This finishes the discussion on complex propositions. We have indicated how a proper understanding of symbols comes with a perfect notation that captures sense without redundancy. It remains to see whether the transformations involved in the notation preserve sense. Since truth-conditions are at the heart of sense, this requires a study of logical consequence and equivalence.

8

Truth and logical consequence

Sense is prior to truth. This principle is so basic to the tractarian philosophy that even the structure of the text highlights it. The book starts with presenting the logical space in which, slightly later, pictures and truth-table signs project sense by indicating possible ways the world may be. Next, truth comes to the fore.

In the previous chapter I have clarified the sense of a proposition. Now we turn our attention to truth and the related notion of logical consequence. This allows us to reflect, among other things, on how the sign-transformation of truth-operations affects sense, and how Wittgenstein's sign-based notion of logical consequence relates to the now more common semantical one.

8.1 World, reality, logical space

The truth or falsity of a proposition is dependent on its projection onto reality. So before being able to concern ourselves with falsity and truth, we must first come to grips with the closely related notions of world, reality and logical space. The *Tractatus* introduces the world first, as the totality of states of things that exists (1, 2.06). Next comes logical space (1.13), and 1.13 and 2.013 make clear that it consists of the totality of states of things as logical possibilities. Reality is introduced last in 2.06. It is the totality of the states of things that exist (also called: positive elementary facts) and the states of things that do not exist (also called: negative elementary facts).

From a systematic point of view a slightly different arrangement is as plausible. Reflecting on how it is possible for language to have sense, we turn to the world, the totality of all states of things that exists.

1. The world is all that is the case

2. What is the case, a fact, is the existence of states of things.

Due to logical structure, especially that of negation, one sees that the world must be part of a more encompassing reality of existing and non-existing states of things. But in reality the states of things that are realized vary contingently, so, finally, from reality logical space is abstracted as the unique realm of logical possibility and sense. To have all notions in one definition, we recapitulate the earlier definition of logical space (definition 1) and of world (definition 10) here.

Definition 49 *Let Λ be the* logical space *of all states of things α. A reality* *is captured as a pair* Λ, R *with* R *indicating which states of things are realized* – α *is in* R, – *and which are not* – α *is not in* R. *A world is the part of a reality that is realized. We write respectively* r *and* w – *possibly with indices,* – *to indicate realities and worlds in* Λ. $\square$

As is well-known, Wittgenstein is not fully consistent in his use of 'world' and 'reality'. Sometimes the two are identified: total reality is the world (2.063). Although it is best to distinguish the two, as in definition 49, relative to logical space they are inseparable: the world (as realized part) and reality (as also including the unrealized part) are given at once.

It is time for truth.

8.2   Truth

Quite near the beginning of the *Tractatus*, in 2.22-2.225, it is made clear that picture and reality are related to each other via the picture's sense. I quote the theses in full.

2.22 A picture depicts what it depicts, independent of its truth or falsity, by means of its pictorial form.

2.221 What a picture depicts is its sense.

2.222 The agreement or disagreement of its sense with reality constitutes its truth or falsity.

2.223 To see whether a picture is true or false we must compare it with reality.

2.224 From the picture alone one cannot see whether it is true or false.

2.225 A picture that is true a priori does not exists.

Later, in 4.01, it is stated that propositions, too, are pictures of reality.

4.01 A proposition is a picture of reality.
      A proposition is a model of reality as we imagine it.

As with pictures in general, only to the extent that a proposition has sense it is internally related with reality, and so true or false (4.05, 4.06). Since definition 47 has detailed what the symbol of a proposition is and definition 49 what the tractarian view on reality is, it remains to clarify how a proposition can be true or false. To this end worlds and truth-possibilities must be compared with each other. This is how a world gives rise to a truth-possibility.

Definition 50 *Let* Λ, π *be a frame with* π *a projection in use. Let* Λ, π, R *be a reality in this frame with world* $w \equiv \Lambda, R$. *The truth-possibility* $\tau_w[\overline{p}]$ *for the elementary signs* $\overline{p} \equiv p_0, \ldots, p_n$ *in the frame is defined by:*

- $\tau_w[\overline{p}]$ *marks* p_i *with 'T', if* w *realizes* $\pi[p_i]$;
- $\tau_w[\overline{p}]$ *marks* p_i *with 'F', if* w *does not realize* $\pi[p_i]$. $\square$

The truth of an analyzed proposition, i.e., a truth-table sign, is now determined as follows.

Definition 51 *Let* Λ, π, R *be a frame with* π *a projection in use and world* $w \equiv \Lambda, R$, *and let* τ *be a truth-table sign for* $\overline{p} \equiv p_0, \ldots, p_n$. *We define:*

- $w \Vdash \tau$, *if* τ *marks the truth-possibility* $\tau_w(\overline{p})$ *with 'T';*
- $w \nVdash \tau$, *if* τ *marks the truth-possibility* $\tau_w(\overline{p})$ *with 'F'.* $\square$

As special cases one has that if τ is a tautology, τ is true in w; and if τ is a contradiction, τ is false in w; (T)() marks any truth-possibility as true, and (F)() any truth-possibility as false.

Definition 51 is much like the familiar truth-definition for propositional logic. Yet there is a difference. Since propositions do not refer to truth-objects, their value is not defined directly but is based on sense and description. Why this apparent detour via sense? Isn't it equivalent to the now common extensional truth-definition? Not quite. An extensional definition may erroneously suggest that all propositions have sense. But logical propositions are senseless, and the commonalities and differences between contingent and logical propositions should be highlighted at the level of sense. The question is rather whether the extensional means that Wittgenstein uses are perfect enough to represent sense. See chapter 9.

8.3 How overt logical structure disappears

Logically complex propositions are truth-table signs that are projected onto non-pictorial reality. The projection is based on the elementary content of the sign, and its logical structure is used to show its truth-conditions (4.431b), i.e., the possible ways in which the content is realized *if* the proposition is true. But in contingent, non-pictorial reality there is nothing the logical constants could refer to: *all* logical complexity is overt in the propositional sign and in the sign only (4.0312, 4.441). This is clear already in the simplest instance. In the sign (TF)(p), p provides the structure of the sign's content. Given a projection π, the truth-ground 'T' of the sign indicates that the content $\pi[p]$ is a state of things shown as realized. But the logical structure of the sign, its structure of 'T's and 'F's, disappears; none of it has reference in reality.

On contemplating the sense of a proposition its overt logical complexity *must* vanish. For firstly, in non-pictorial reality the concept of truth does not apply; truth and falsity are propositional notions. Secondly, as we have seen in the previous chapter, the situation that is projected from a sign may involve a disjunction of realization-patterns that are logically incompatible. To enable such mutually inconsistent possibilities, the depicted part of logical space must lack overt truth-functional structure itself, and just consist of elementary states of things. Its 'logic' is that of independent realizability, based in the combinatorial possibilities of objects; nothing more.

8.4 Truth and truth-operational signs

Up until now, we have considered the truth and falsity of truth-table signs. In terms of it, the truth and falsity of truth-operational signs such as $N'(\varphi_1, \ldots, \varphi_n)$ can be defined.

Definition 52 *Let Λ, π be a frame with logical space Λ and π a projection in use. Let φ be a truth-operational formula. For any world w in Λ, π we set:*

$$w \Vdash \varphi \Longleftrightarrow w \Vdash \tau[\varphi].$$

Here, $\tau[\varphi]$ is the truth-table sign φ reduces to. □

Definition 52 is well-defined. Also, truth for truth-operational propositions can be understood recursively. As we have seen in chapter 5 each truth-operational

φ reduces to a truth-table sign. On the basis of this definition, the familiar truth-clauses are derived rather than stipulated.

Proposition 53 *For any frame* Λ, π, *and any world* w $(= \Lambda, \pi, R)$:

$$w \Vdash p \iff w \text{ realises } \pi[p],$$

$$w \Vdash N'(\varphi_1, \ldots, \varphi_n) \iff w \not\Vdash \varphi_1 \text{ and} \ldots \text{and } w \not\Vdash \varphi_n.$$

Proof. The case for an elementary proposition $(TF)(p)$ is clear. So let Λ, π be a frame; let w be the world Λ, π, R, and let $N'(\varphi_1, \ldots, \varphi_n)$ be a formula based on the elementary propositions $\overline{p} = p_1, \ldots, p_k$. Assume $w \Vdash N'(\varphi_1, \ldots, \varphi_n)$. By definition:

$$w \Vdash \tau[N'(\varphi_1, \ldots, \varphi_n)],$$

so $\tau_w(\overline{p})$ is a truth-ground in $\tau[N'(\varphi_1, \ldots, \varphi_n)]$. According to the rule of N that transforms $\tau[\varphi_1] \ldots \tau[\varphi_n]$ into $\tau[N'(\varphi_1, \ldots, \varphi_n)]$, this means that $\tau_w(\overline{p})$ marks the $\tau[\varphi_i]$ with 'F', when appropriately restricted to the elementary signs p_j occurring in it. Thus, $w \not\Vdash \tau[\varphi_i]$, and $w \not\Vdash \varphi_i$ by definition. The converse is strictly similar. □

The result sheds light on the philosophical status of the truth-operation N. Wittgenstein stresses that logical constants do not refer, but this does not imply that logical constants have no effect on meaning at all. In particular, N's rule does affect the sense of a proposition whose sign it transforms. This is clear already for propositions of minimal logical complexity: p and $N'(p)$ have the same content but opposite sense (4.0621c). The one way to see whether the rule of N changes sense as required, is to check the truth-conditions of the resulting sign. Proposition 53 shows them to be in order.

8.5   *Interlude:* The asymmetry of truth and falsity

Before turning to logical consequence, I would like to pay attention to an observation of Wittgenstein concerning an asymmetry between truth and falsity, which makes one wonder where the asymmetry is to be located. The observation is in 4.061.

4.061 Ignoring that a proposition has a sense independent of the facts, makes one susceptible to the belief that true and false are equal relations between a sign and what its signifies.

Then one could say, e.g., that 'p' signifies in a true way, what '$\neg p$' does in a false way, etc.

The suggestion in 4.061 of allowing sense to be dependent on facts refers, I think, to an earlier phase in Wittgenstein's thinking where he contemplated to allow for positive and negative facts in combination with different modes of representation. See Wittgenstein (1914-16), October 30th, 1914. Russell advocated the idea, too, in his Lectures on the Philosophy of Logical Atomism (Russell, 1918, 211 ff.). And though Frege acknowledged that propositions have sense independent of the facts, in his philosophy all specifics of a proposition are lost in the structureless True and False, which from this perspective appear to be alike. The tractarian system, too, is hardly different in this regard. Logical space can be understood as the realm of all possible contingent combination, which as such is indifferent to the distinction between truth and falsity; both are on a par.

Thesis 4.062 suggests to cash out the equality by giving each proposition a new, opposite meaning, but argues that this is bound to fail.

4.062 Can we not make ourselves understood with false propositions just as we have done up till now with true ones? Just as long as one knows that they are intended to be false? No! For a proposition is true if what we say with it is the case; and when we mean '¬p' with 'p', and things are so as we intend it, then on the new reading 'p' is true and not false.

On the new reading, 'p' means: ¬p. However, the point of 4.062 is not about sense or reference, it is about making statements or judgments, where truth and falsity may well be asymmetrical. With a statement a claim is made that certain conditions obtain, and here the truth of the claim is leading. This is where the asymmetry should be located: one cannot get round the information one truthfully claims to convey.   [*End of interlude*]

8.6   Logical consequence: main theses

The theses in the range of 5 open with the truth-functional character of propositions, and next 5.11-5.143 continue with the notion of logical consequence. In the *Tractatus* logical consequence is a sign-based notion, but since the signs are truth-tables there is a clear connection with semantical consequence. The basic idea is in 5.11.

5.11 If the truth-grounds that are common to a number of propositions are jointly the truth-grounds of a certain proposition, then we say that the truth of this proposition follows from the truth of those propositions.

The truth-grounds of a propositions are defined in 5.101: it are the truth-possibilities marked with 'T'. That a proposition follows from a single other one, as in 5.12 or 5.121, is of course a special case of 5.11.[72]

Logical consequence is set apart from material relations: there is no configuration of objects indicating it. Whether the logical relation obtains or not depends solely on the form of signs.

5.13 That the truth of one proposition follows from the truth of other propositions, we can see from the structure of the propositions.

In other words, the relationship of logical consequence is internal to the form of signs (5.131), and and as such independent of empirical contingencies.

5.133 All deductions are made *a priori*.

Also, there are no 'laws of inference' that would justify a deduction independent of the given signs.

5.132 If p follows from q, from q I can infer p; deduce p from q.
      The nature of the inference can be gathered only from both propositions.

---

[72]In the theses on logical consequence, Wittgenstein often uses p and q, etc., which vary over elementary propositions (4.24). It is clear however that logical consequence for propositions of arbitrary logical complexity is intended.

> Only they themselves can justify the inference. 'Laws of inference'
> that would justify the inference – as for Frege and Russell, – are without
> sense, and would be superfluous.[73]

The interlude traces which views of Frege and Russell Wittgenstein is opposing
to. Next, I give a more detailed version of tractarian consequence.

8.6.1 *Interlude*: Frege and Russell on laws of inference

It is not quite clear which views of Frege and Russell Wittgenstein had in
mind, when he states in 5.132 that they justify a logical consequence on the
basis of 'laws of inference'. In trying to get more clarity in this regard, it is
most important to answer the question: can such a position be attributed to
Frege or Russell at all? The matter is discussed extensively in Proops (2001a),
on which the following is based.

Frege The hardest part is to see which ideas of Frege Wittgenstein is referring
to. Black (1964), XLVI, gives no sources, and the reconstruction in Proops
(2001a) is not entirely convincing. Proops gives a reading based on ideas in
Frege's *Die Grundlagen der Arithmetik* (Frege (1884), §17) that are reflected
in *Über die Grundlagen der Geometrie* (Frege, 1906). Together with Frege
(1903b), also named *Über die Grundlagen der Geometrie*, Frege (1906) is his
polemic against Hilbert's formalism and Korselt's defence of it.

It is part of Hilbert's formalist approach to show that axioms are independent of each other. Frege (1906) indicates how this can be done in a
non-formalist manner. His idea is to use the deductive closure of a group
of thoughts Γ. In particular, a thought γ is dependent on Γ if it is either in Γ
or if it can be deduced from thoughts that have already been deduced from
Γ. See Frege (1906), III (*Schluß*), pp. 423-4. Proops uses this passage to show
that Frege held logical consequence to be justified on the basis of 'laws of
inference'. This is a bit surprising, since earlier in the article – published as
Frege (1906), II, – inference is discussed explicitly in a highly relevant way.
Here, Frege contrasts his notion of inference with an attempt to make sense
of the formal variant.

> An inference [...] is a judgment passed according to logical laws on the ba-
> sis of judgments passed already before. Each premise is a specific thought
> acknowledged to be true, and in the concluding judgment likewise a specific
> thought is acknowledged to be true.
>
> What is a formal inference? One could say: in a certain sense each infer-
> ence is formal in that it follows a general law of inference; but in another
> sense no inference is formal in that the premises and the conclusion have
> their thought-content which occurs in this particular connection only in this
> inference. (*Ibidem*, 387)

According to Frege inference is no formal matter: it concerns judging thoughts,
senses, but the concluding thought in an inference is reached on ground of a
logical law (*logischen Gesetze*) that justifies its inference. This, I think, is the

---

[73]The human phrasing of 5.132a is misleading, unless it is seen as an application of the
structural notion of logical consequence in 5.11. Similarly, given the non-axiomatic nature
of the tractarian system (6.127), the notion of inference in 6.126 should rather be seen as
derived than as primitive. See Sundholm (2009) for the historical context of the notions.

position that Wittgenstein objects to in 5.132.

Note in passing that in 6.126 Wittgenstein seems to hold a position that is akin to Frege's: logical propositions are derived from logical propositions using certain operations. Doesn't this contradict 5.132? Not necessarily: 5.132 can be taken to give the primitive notion of logical consequence from which the notion of logical inference in 6.126 can be derived. Also, even within its restricted area, 6.126 holds that inference is a formal matter that could be bypassed using more direct checks for logicality.

Russell That Russell took logical laws to justify inference is beyond doubt. Based on a suggestion of Moore to order propositions as more or less necessary in terms of the number of propositions they strictly imply – i.e., imply without being equivalent to it, – Russell observed that to make the idea non-trivial a notion of implication is required that is not truth-functional. Truth-functionally $\bot \to \top$ is valid. So, truth-functional implication would classify all falsities as least necessary and all truths as most necessary. But a false premise or a true consequent seems insufficient ground to license a logical consequence. To eliminate this inflatory aspect of truth-functional implication, Russell introduced the notion of deducibility: ψ is deducible from φ if and only if the implication $\varphi \to \psi$ can be deduced from a few intuitively self-evident axioms. An example of such an axiom is: $(p \to q) \to (\neg q \to \neg p)$. On Russell's view, it is the deduction based on such laws that justifies the inference.

Wittgenstein was against laws of inference as justifying logical consequence, but why? It must have been for the same reason why he did not want to separate sign and sense: once the propositions in a logical consequence are parted from the law of inference, they cannot be glued back together again. After separation, can't they obey the laws of another kind of logic just as well? Absolute logical necessity can only be had if logical consequence is rooted without intermediary in propositional form. All justification for a logical consequence should be internal to its propositions (5.132). [*End of interlude*]

8.7 Detailing logical consequence

In 5.122 Wittgenstein characterizes logical consequence in terms of sense. Like Frege he holds consequence to be non-formal; it concerns propositions, not just senseless inscriptions. Yet, tractarian inference is nearly formal: logical consequence requires abstracting from the content of the relevant propositions to leave an ineffable comparison of their logical forms. To elucidate the notion we consider it in the context of truth-table signs, the signs in terms of which truth-grounds are defined.

Comparing truth-table signs with regard to logical consequence is not always straightforward. For example, in table 7 are the truth-grounds of the sign at the right-hand side included in that at the left-hand side, or is it rather the other way around? To see what is at stake, one should realize that the form of a proposition has two aspects to it: (i) a descriptive aspect given by its elementary parts, and (ii) a logical aspect given by its truth-markings. To see logical consequence, the descriptive content should be factored out, so to

	p
T	T
F	F

	p	q
T	T	T
T	F	T
T	T	F
F	F	F

TABLE 7 Comparing truth-grounds

speak, by making it identical in all propositions involved. For instance, since q in table 7 does not figure in the sign for p, the inclusion of truth-grounds cannot be read of the signs without more ado. Instead, the truth-possibilities of both premise and conclusion should be on a par. This process, which is left implicit in the *Tractatus*, I call 'standardization'. It is a refinement of 5.132, which states that in a consequence both premises and conclusion should be considered. The idea is simply to determine all elementary parts occurring in premises and conclusion, and to ensure that all propositions are standardized to this totality. In this way the descriptive aspect of the signs is identified to enable the comparison of their logical forms. For the example at hand we standardize the sign $(\mathrm{TF})(p)$ to include q as a logical part; cf. table 8.

	p	q
T	T	T
F	F	T
T	T	F
F	F	F

	p	q
T	T	T
T	F	T
T	T	F
F	F	F

TABLE 8 $p \vee q$ follows from p

That the truth-grounds of p, i.e. $p \wedge (q \vee \neg q)$, are included in those of $p \vee q$ is now immediate from the logical form of the signs alone. A similar standardization is required in case the conclusion contains less material than the premises; cf. that p follows from $p \wedge q$ in table 9.

	p	q
T	T	T
F	F	T
F	T	F
F	F	F

	p	q
T	T	T
F	F	T
T	T	F
F	F	F

TABLE 9 p follows from $p \wedge q$

It is safest to assume that any logical consequence requires standardization, perhaps only vacuously. It is achieved by adding as logical parts all elementary propositions on which premises and conclusion are based.[74] If the elementary propositions are $\bar{p} = p_1, \ldots, p_k$, I use the notation $\chi \wedge \top(\bar{p})$ for this purpose.

---

[74]The example in Black (1964), 241, is standardized as well, but Black does not name standardization for logical consequence in general.

Definition 54 *Let the truth-table signs* $\tau_1, \ldots, \tau_n, \tau_{n+1}$ *be based on* $\overline{p} = p_1, \ldots, p_k$. *We say that* τ_{n+1} *is a logical consequence of* $\tau_1, \ldots, \tau_n$, *written as:*

$$\tau_1, \ldots, \tau_n \vdash \tau_{n+1},$$

if and only if: the truth-grounds common to all $\tau_1 \wedge \top(\overline{p}), \ldots, \tau_n \wedge \top(\overline{p})$ *are included in the truth-grounds of* $\tau_{n+1} \wedge \top(\overline{p})$. □

Standardization is harmless. It has no effect on the senses of premises and conclusion; the symbol of a proposition and that of its standardized variant is the same:

$$\Sigma[\tau] = \Sigma[\tau \wedge \top(\overline{p})].$$

It appears to be a drawback of definition 54 that it is restricted to truth-table signs. But a notion of consequence can be introduced for truth-operational signs in terms of it, as follows.

Definition 55 *Let* $\varphi_1, \ldots, \varphi_n, \psi$ *be truth-operational signs. We say that* ψ *is a logical consequence of* $\varphi_1, \ldots, \varphi_n$, *notation:*

$$\varphi_1, \ldots, \varphi_n \vdash \psi,$$

if and only if: $\tau[\varphi_1], \ldots, \tau[\varphi_n] \vdash \tau[\psi]$. *Here,* $\tau[\chi]$ *indicates the truth-table sign that results from applying all truth-operations in* χ. □

Clearly, definition 55 is correct. The $\tau[\varphi_1], \ldots, \tau[\varphi_n], \tau[\psi]$ are indeed truth-table signs. But the definition is complete, too, in the sense that no logical consequence of truth-table signs is omitted. Due to truth-operational completeness, for each truth-table sign τ there is a truth-operational sign χ such that $\tau[\chi] = \tau$. See section 5.7.2 for details. And so for each logical consequence $\tau_1, \ldots, \tau_n \vdash \tau_{n+1}$, there are truth-operational formulas $\chi_1, \ldots, \chi_n, \chi_{n+1}$ such that:

$$\chi_1, \ldots, \chi_n \vdash \chi_{n+1} \iff \tau_1, \ldots, \tau_n \vdash \tau_{n+1}.$$

It remains to check whether the notion given here has the properties Wittgenstein observes it to have.

8.8   Some properties of logical consequence

To begin with we have sufficient insight to understand Wittgenstein's claim concerning the relative descriptive power of propositions. The descriptive power of a proposition is its ability to restrict the situation in which it is true. The stronger the descriptive power of a proposition, the fewer truth-grounds it has.

5.14 If one proposition follows from another, then the latter says more than the former, the former less than the latter.

Strictly speaking the claim in 5.14 is incorrect: the premise does not have to say more than the conclusion, it may say as much. See 5.141, discussed shortly. But after this small amendment the observation does hold.

Proof. (5.14) Assume that $\tau \vdash \tau'$. Then the truth-grounds of τ are included in those of τ'. To show this, both τ and τ' need to be standardized, but adding logical parts does not change the descriptive power of a proposition.

After standardization, τ discerns at most as many truth-grounds as τ' by assumption. In other word, the descriptive power of τ is at least as high as that of τ', τ says as much as τ' or more. □

Next there is 5.141, which ensures that propositions with the same descriptive power are identical.

5.141 If p follows from q and q from p, then they are one and the same proposition.

There is a slight hitch in that the truth-table sign τ of 'p' and the truth-table sign τ' of 'q' could differ in their logical parts. But 5.141 is not just about signs, it concerns symbols. In particular, 'one and the same proposition' should be read as: expressing the same symbol. Once read in this way, 5.141 is seen to hold.

Proof. (*5.141*) If τ and τ' follow from one another, after standardization they have the same truth-grounds. Proposition 65, p. 154, shows they express the same symbol, are one and the same proposition. □

As to the remainder of 5.14, it is immediate from definition 54 that a tautology follows from any proposition (5.142):

$$\vdash \top(\overline{p}) \ .$$

And it is as immediate that a contradiction has any proposition as a consequence (5.143):

$$\bot(\overline{p}) \vdash \ .$$

Since the sense of a proposition shows its truth-conditions, observations could be made with regard to sense and consequence that are similar to those of descriptive power and consequence, but the way Wittgenstein phrases them is misleading.

5.122 If p follows from q, the sense of 'p' is contained in that of 'q'.

At first reading it seems that in 5.122 'to contain' (*enthalten*) cannot be taken literally, and surely not in line with 5.14. See also Black (1964), p. 242. If p follows from q, the sense of p, its truth-grounds, according to 5.14 rather includes that of q. E.g., it is queer to hold that the sense of $p \vee q$ is contained in that of elementary p *and* in that of elementary q, given that p and q are independent. But perhaps 5.122 should be read in the context of 5.124:

5.124 A proposition affirms every proposition that follows from it.

If 'the sense of φ' is temporarily taken to mean: the totality of propositions that φ affirms, then as 5.122 has it the sense of φ is contained in that of ψ as soon as φ follows from ψ. Assume φ to follow from ψ: $\psi \vdash \varphi$. If χ is 'in' the sense of φ, φ affirms χ: $\varphi \vdash \chi$. Since $\psi \vdash \varphi$, ψ affirms χ, too: $\psi \vdash \chi$, and thus χ is also 'in' the sense of ψ. Therefore, the sense of φ is contained in that of ψ.

I leave it to the reader to apply the insights given to the cases mentioned in 6.1201, and continue to compare sign-based logical consequence with semantic consequence.

8.9 Semantical consequence

In the *Tractatus* the basic notion of logical consequence is the sign-based one, but within the system a semantic notion of consequence can be defined as follows.

Definition 56 *For all frames* Λ, π*: proposition* ψ *is a* semantical consequence *of the propositions* $\varphi_1, \ldots, \varphi_n$ – *notation:* $\varphi_1, \ldots, \varphi_n \models \psi$, – *if and only if for all worlds* $w \equiv \Lambda, \pi, R$*: if* $w \Vdash \varphi_1$ *and*. . . *and* $w \Vdash \varphi_n$, *then* $w \Vdash \psi$. $\square$

Does semantic consequence play a rôle in the tractarian system? It seems to have little to recommend itself from a tractarian perspective. Firstly, Wittgenstein held that consequence is based on a formal similarity between the signs of premises and conclusion, and such similarity is absent on a semantic approach. For instance, what would the formal similarity be in a semantic consequence such as:

$$\neg(\varphi \vee \psi) \models \varphi \leftrightarrow \psi \; ?$$

If a formal foothold is absent, one may well feel invited to introduce laws of inference for its justification, i.e., precisely the position Wittgenstein objects to in 5.132. Secondly, semantic consequence seems to fare best in a formalist setting, where senseless syntactic inscriptions may receive different interpretations. By contrast, in the tractarian system all signs are interpreted. The closest one gets to syntax is by abstracting the formal aspects from the interpreted signs. Therefore, in introducing consequence it seems sufficient to restrict attention to the forms of interpreted signs; especially if it concerns truth-tables signs, whose notion of logical consequence is a near variant of semantical consequence.

Actually, semantic consequence for truth-table signs can be treated as a manner of speech. It is not hard to see that for truth-table signs, logical and semantical consequence are equivalent. To this end, we first need some information on how worlds and truth-possibilities relate.

A truth-possibility for elementary signs $p_0, \ldots, p_n$ determines a finite segment of a world. Since in the arguments below only $p_0, \ldots, p_n$ matter, we fix a world for the given truth-possibility by setting the possibly infinite remainder *ad lib* in a uniform way.[75]

Definition 57 *Let* τ *be a truth-table sign, and let* $\mathbf{t}$ *be a truth-possibility for the elementary signs* $\overline{p} \equiv p_0, \ldots, p_n$ *in* τ*. In the frame* Λ, π*, the world* $w_{\mathbf{t}}$ *is defined by:*

- $w_{\mathbf{t}} \Vdash p_i$, *if* $\mathbf{t}$ *marks* p_i *with 'T';*
- $w_{\mathbf{t}} \nVdash q$, *if* $\mathbf{t}$ *does not mark* q *with 'T'.* $\square$

The last clause ensures propositions p_j that $\mathbf{t}$ marks with 'F' *and* propositions q that do not occur in $\overline{p}$ to be false in $w_{\mathbf{t}}$. A truth-possibility $\mathbf{t}$ only constrains that part of the world which concerns $\overline{p}$. By contrast, a world determines a truth-possibility for any sequence $\overline{p}$.

Definition 58 *Let* $\overline{p}$ *a sequence of elementary propositions and* w *be a world in frame* Λ, π*. The truth-possibility* $\mathbf{t}_w$ *is defined by:*

---

[75]See chapter 11 and 12 for a discussion on infinite logical space.

- $\mathbf{t}_w$ *marks* p_i *with* 'T', *if* $w \Vdash p_i$;
- $\mathbf{t}_w$ *marks* p_i *with* 'F', *if* $w \nVdash p_i$; □

Proposition 59 gives the main relationships between the concepts.

Proposition 59 *Let* $\overline{p} \equiv p_0, \ldots, p_n$ *be a sequence of elementary propositions in frame* Λ, π. *For a truth-possibility* $\mathbf{t}$ *one has:*

$$\mathbf{t}(p_i) = \mathbf{t}_{w_\mathbf{t}}(p_i),$$

And for a world w *in the frame one has:*

$$w \Vdash p_i \iff w_{\mathbf{t}_w} \Vdash p_i.$$ □

Now we are in a position to characterize semantical consequence in terms of truth-table signs.

Proposition 60 *Let* $\tau_1, \ldots, \tau_n, \tau_{n+1}$ *be truth-table signs. One has:*

$$\tau_1, \ldots, \tau_n \vdash \tau_{n+1} \text{ if and only if } \tau_1, \ldots, \tau_n \models \tau_{n+1}.$$

That is, the logic of truth-table signs is complete.

Proof. Let the truth-table signs $\tau_1, \ldots, \tau_n, \tau_{n+1}$ be based on the elementary propositions $\overline{p} = p_1, \ldots, p_k$.

First let Λ, π be any frame, and assume $\tau_1, \ldots, \tau_n \models \tau_{n+1}$ in that frame. In case $\tau_1 \wedge \top(\overline{p}), \ldots, \tau_n \wedge \top(\overline{p})$ have a truth-ground $\tau[i]$ in common – and so are consistent,[76] – one has for $w_{\tau[i]}$ in Λ, π:

$$w_{\tau[i]} \Vdash \tau_1 \text{ and } \ldots \text{ and } w_{\tau[i]} \Vdash \tau_n.$$

By assumption, $w_{\tau[i]} \Vdash \tau_{n+1}$, too, and thus $\tau_{w_{\tau[i]}}$ is a truth-ground of $\tau_{n+1} \wedge \top(\overline{p})$. But $\tau_{w_{\tau[i]}} = \tau[i]$, according to proposition 59. Therefore, $\tau_1, \ldots, \tau_n \vdash \tau_{n+1}$.

Conversely, assume $\tau_1, \ldots, \tau_n \vdash \tau_{n+1}$. Let w be any world in Λ, π so that: $w \Vdash \tau_1$ and $w \Vdash \tau_n$. Then, τ_w is a truth-ground common to $\tau_1 \wedge \top(\overline{p}), \ldots, \tau_n \wedge \top(\overline{p})$. By assumption, τ_w is also a truth-ground for $\tau_{n+1} \wedge \top(\overline{p})$. So, $w_{\tau_w} \Vdash \tau_{n+1}$. Hence $w \Vdash \tau_{n+1}$ by proposition 59, and therefore $\tau_1, \ldots, \tau_n \models \tau_{n+1}$, as required. □

The near semantic nature of truth-table signs makes their completeness for semantical consequence almost trivial. Perhaps this is also why completeness was no issue for Wittgenstein: fully analyzed signs must say it all! Given the general form of propositions as truth-table signs, the distinction between being logical – true in all worlds or false in all worlds, – can only correspond to having the form of a tautology or a contradiction.

6.124i ... if we know the logical syntax of any sign language, then all the propositions of logic are already given.

---

[76] In familiar completeness proofs establishing that a consistent group of propositions has a model requires effort. That this step appears trivial here is due to the fact that in a finite setting truth-table signs list all possible models, and so it can be checked whether the propositions in the group have one in common. In this regard, reasoning with truth-table signs is similar to reasoning with the systematic tableaux used in the completeness proofs for analytical tableaux. Cf. Smullyan (1968). Also, in case of truth-operational formula's, which are considered shortly, inconsistent options – i.e., the analogues of closed paths in proof trees, – are eliminated implicitly when applying the truth-operation N. See section 11.3.3.

But Wittgenstein did consider the semantical notion of being logical as opposed the syntactical one.

6.1232a Logical general validity we could call essential as opposed to accidental general validity.

6.126a-c Whether a proposition belongs to logic can be calculated by calculating the logical properties of the *symbol*.

And this is what we do when we 'prove' a logical proposition. For without troubling ourselves about sense and meaning, we form the logical proposition by mere *rules that deal with signs*

See also section 11.4.3. 'Being given' in the system, as 6.124 puts it, does not have to be the same as being decidable, just like 'to have sense' is different from being able to specify what sense consists of. Such human considerations do not seem to be at the center of Wittgenstein's problem space. Still, 6.126 does indicate that restricting his attention to the finite part, he did underestimate the complexity of the full system. See section 11.4 for further discussion.

Semantical consequence is of interest anyway. It is quite natural to broaden logical consequence for truth-table signs to other notations, such as the truth-operational one. But to comply with the tractarian philosophy, this should be done so that the extension becomes a matter-of-speech. Correctness and completeness should be shown via a reduction to the most basic notion of consequence among truth-table signs. In this regard, proposition 61 is the missing link to extend correctness and completeness from truth-table signs to the truth-operational part of the tractarian system.

Proposition 61 *For any truth-operational signs* $\varphi_1, \ldots, \varphi_n, \psi$:

$$\varphi_1, \ldots, \varphi_n \vdash \psi \quad \Leftrightarrow \quad \varphi_1, \ldots, \varphi_n \models \psi.$$

Proof. Let $\varphi_1, \ldots, \varphi_n, \psi$ be truth-operational signs. Then:

$$\varphi_1, \ldots, \varphi_n \vdash \psi$$
$$\Leftrightarrow \quad \tau[\varphi_1], \ldots, \tau[\varphi_n] \vdash \tau[\psi]$$
$$\Leftrightarrow \quad \tau[\varphi_1], \ldots, \tau[\varphi_n] \models \tau[\psi]$$
$$\Leftrightarrow \quad \varphi_1, \ldots, \varphi_n \models \psi$$

The first equivalence holds by definition 55, the second by proposition 60, and the third by definition 52. □

This result may be seen as a generic analysis showing that it is a safe manner-of-speech to show a logical consequence via a semantic consequence using truth-operational signs. Checking the relevant proofs one sees that the justification for the semantic consequence is based solely in the forms of truth-table signs. Therefore to show, e.g. that:

$$N'(\varphi_0, \ldots, \varphi_n) \vdash N'(\varphi_i)$$

it is sufficient to observe that:

$$N'(\varphi_0, \ldots, \varphi_n) \models N'(\varphi_i).$$

which is immediate.

At this point we should notice, too, that the deduction system presented respects Wittgenstein's non-axiomatic position. The axiomatic approach comes with the assumption of a hierarchy of propositions where some are 'self-evidently' more basic than others. Instead Wittgenstein holds that all propositions are on a par (6.4). The current system complies with his egalitarian view; it allows determining of any proposition whether it is logical, contingent or contradictory.

The core of the tractarian system is now in place. Before speculating on an infinite variant, we continue to look at some main features of the system, such as its perfect notation.

9

Perfect notation

At the center of Wittgenstein's symbolic turn is the idea of a perfect notation, which captures the determinate sense of a proposition, especially its form, without redundancy. The notation is based on propositions as truth-functional complexes, but in such a way that the notation captures their symbols.

Since it concerns sense, the tractarian notion of a perfect notation is a bit puzzling. In the *Tractatus* truth-functional signs represent sense. However, truth-functions are extensional: their value just depends on the truth-values of the arguments, not on the way in which the arguments are given. But then: are the invariances under substitution that are typical of extensionality desirable when representing sense? If not, how does the system circumvent them? Is the system flawed in this regard?

Also, notation can be perfect in varying degrees. This chapter begins with charting the kind of perfection we are after. Next it is shown how this perfect notation is realized in the finite part of the system. This will clarify to what extent the tractarian system can be called extensional, both at the elementary and at the logically complex level. I will make clear that there are no substitutional side-effects of using an extensional system to represent sense. Chapter 12 investigates to what degree infinitary symbols can be perfect.

9.1   Degrees of perfection

The *Tractatus* aims to delimit our descriptive capabilities by showing what can be said with perfect clarity. Thus the text hints at a perfect notation that makes the symbol of a proposition, its sense, entirely manifest. Since the sense of a proposition concerns its truth-conditions, a perfect notation makes the logical relationships between propositions apparent. It should leave no doubt which seemingly different signs cloak one and the same proposition, or which propositions are logical consequences of one another. Symbols also safeguard a delicate balance between the contingent and the necessary aspects of signs as facts (section 6.7).

As is to be expected, the requirements for a perfect notation are even more stringent than those for the conceptual notations of Frege and Russell (3.325). Inquiring in more detail what a perfect notation may amount to, one must distinguish between the elementary and the logically complex case. In the next section, I shall argue that elementary propositions are indeed perfect. With regard to logical complexity, however, a few degrees of freedom come to the fore, not all of which can be traced back to the *Tractatus*:

1. A perfect notation should lack logical parts. See 4.46 and its offspring, or the earlier sources named in section 7.5.

2. A perfect notation should lack inessential repetitions of negations, disjunctions or conjunctions. See e.g. 4.0621, 5.254, 5.43, or 5.512.

3. A perfect notation should be composed of logically independent parts.

As to the first degree, in a perfect notation such signs as: $p \wedge (q \vee \neg q)$ or $(p \wedge q) \vee (p \wedge \neg q)$ have the symbol p. Similarly, the symbol of logical signs such as $p \leftrightarrow p$ should not depend on any elementary propositions. As to the second degree, in a perfect notation signs like: $\neg\neg p$, $p \wedge p$, $p \vee p \vee p$, or $(\neg\neg p \vee p) \wedge p \wedge \neg\neg p$ again have the symbol p. Wittgenstein does not seem to have noticed the third degree of perfection. To see what it amounts to, assume e.g. that φ implies ψ. Then, the sign $\varphi \vee \psi$ should have the symbol of its equivalent ψ. In the same vein, call an elementary proposition or the negation thereof a *literal*. If there are sequences of conjunctions of literals $\theta_0, \ldots, \theta_n$ such that each literal $(\neg)p$ in θ_i also occurs in θ_{i+1}, then the sign $\bigvee_{0 \leq i \leq n} \theta_i$ has the symbol of θ_0. See Quine (1962, part I), for further discussion of the finite case.

The question is: which degree of perfection is required for tractarian symbols? If forced to choose, the first degree is without doubt mandatory. As long as a sign has logical parts, it offers a distorted view on its sense. Logical parts are about nothing, so a symbol must shun any suggestion that its logical parts contribute to its content. The remaining two degrees are surely desirable but could in general be difficult to realize.

Wittgenstein introduced the distinction between truth-operations and truth-table signs to realize finite symbols. As defined in section 7.4, finite symbols have all three degrees of perfection. The reduction rule eliminates logical parts (first degree), and the resulting truth-table sign is in effect a disjunctive normal form of logically incompatible conjuncts (second and third degree). By contrast, for the infinite symbols to be discussed in chapter 12 the situation is much more complex. They still lack logical parts (first degree), but the infinitary disjunctive normal forms proposed as symbol may still contain identical conjuncts (partial second degree), or may contain conjuncts that imply each other (no third degree). Be this as it may, I will argue that in the context of an appropriate notion of symbol identity, a perfect notation of first degree suffices. This will also dispel the illusion that symbols may trivialize logic using graphical means. Logic should take care of itself, and the identity of its symbols must be logical. Any suggestion that logical notions can be reduced to empirical ones should be shunned.

From a modern perspective, a perfect notation is a regulative ideal at best. A perfect notation enables propositions to wear their symbols, and thus their logical relationships upon their sleeves. For this reason, it should concern the limit of human experience: in general logic is too complex to allow for symbols that can be inspected in an empirically feasible manner. It is open to debate whether Wittgenstein aimed for a perfect notation of human measure. It is fair to say that Wittgenstein grossly underestimated the complexity of infinitary logic, and with that the complexity of its symbols. Yet, logical complexity and human feasibility did not seem to have his prime interest. The unique logic that is inherent in our everyday languages is of untainted necessity, and should take care of itself (5.473). If so, for Wittgenstein too, symbols are much like a regulative ideal. One only attains full clarity in the metaphysical subject, a limit of the world, my world. See section 11.4 for discussion.

9.2 Elementary propositions

A perfect notation shows the determinate sense of a proposition. The core of such a notation consists of elementary propositions. We therefore have to clarify first to what extent the tractarian notation for elementary propositions is perfect.

In the tractarian system, the sense of a proposition is its analyzed sign used in projection. So the discussion will have to be based on a frame Λ, π that bi-partitions logical space Λ in a totality of states of things used as signs and a totality of states of things the signs are about. That the notation for elementary propositions is perfect, is mainly due to projection being an isomorphism between the two parts of logical space, for such holistic projection makes:

1. Naming determinate, in that each name has a referent;
2. Naming unique, in that no two names have the same referent;
3. An elementary sign's form unique, in that this form is invariant under the appropriate substitutions.

That naming is determinate is a direct consequence of descriptive completeness; Wittgenstein presupposes that meaningful, everyday language is referentially adequate (4.26; cf. also Frege (1892b), 46). It remains to discuss the unicity of naming and of form, which is best done in the wider context of identity and substitution.

9.2.1 Identity

Wittgenstein's approach to identity ensures that each name has a unique reference, and so contributes in a unique way.

Wittgenstein's position is different from Russell's theory of identity, adapted from Leibniz, which states that two objects are the same if they share all properties of the appropriate type (5.5302). By contrast Wittgenstein held that different tractarian objects may instantiate the same logical form. Two such instantiations share all internal properties – the possibilities to combine with other objects, – even if they differ in content. In the *Tractatus*, identity as a sub-propositional relation is eliminated to make way for the approach to identity that shows itself in naming and reference (5.53). More in particular a projection π in logical space Λ ensures different names n, m to refer to different objects:

$$\pi[n] \neq \pi[m].$$

Wittgenstein's approach to identity is directly related to the perfect notation of elementary propositions. Clearly, a notation for propositions that contains two names referring to one and the same object is less perfect than one in which each name refers uniquely. On the given treatment of identity this aspect of the notation is as perfect as can be.

In fact, not only different names have different referents, the isomorphic nature of the projection π implies that also different expressions make different contributions to sense. Consider for example two expressions ε and ε' that have the same form; they occur in the same contexts $\alpha(-)$. If ε and ε' are different from each other, then for any projection π and for all appropriate elementary material functions $\alpha(-)$:

$$\pi[\alpha(\varepsilon)] \neq \pi[\alpha(\varepsilon')].$$

Mutatis mutandis the same holds for material functions $\alpha(-)$ and elementary propositions α. Thus, relative to projection, the *Tractatus* embodies a radical form of rigidity (cf. Kripke (1980)). Sense is prior to truth, and no sense is allowed to vary with the contingencies of how worlds may be.

As a consequence of this approach, the principles of identity that are typical of the extensional point of view and the substitutions that come with it are 'blocked'. The familiar principle of identity for propositions holds vacuously, and can even be strengthened to an equivalence.

IND For all Λ, π: $\pi[\varepsilon] = \pi[\varepsilon']$ iff $\Lambda, \pi \models \forall\varphi[\varphi(\varepsilon) \leftrightarrow \varphi(\varepsilon')]$.

Here, $\Lambda, \pi \models \forall\alpha[\alpha(\varepsilon) \leftrightarrow \alpha(\varepsilon')]$ means truth in all worlds:

$$\Lambda, \pi, R \Vdash \forall\alpha[\alpha(\varepsilon) \leftrightarrow \alpha(\varepsilon')],$$

for all R. To see that IND holds: if the left-hand side obtains, ε and ε' are strictly the same, and so the right-hand side is trivially true. But also conversely: if the right-hand side of IND holds for all propositions, it holds in particular for all elementary ones. But due to their independence, elementary propositions are only guaranteed to have the same truth-value in all worlds, if just one elementary proposition is involved. The right-hand side can only be true if ε and ε' are one and the same expression.

Along similar lines, one sees e.g. the following principle to hold for material functions:

IND^m For all Λ, π: $\pi[\alpha(-)] = \pi[\alpha'(-)]$ iff $\Lambda, \pi \models \forall\varepsilon[\alpha(\varepsilon) \leftrightarrow \alpha'(\varepsilon')]$.

Again, the identity and equivalence can only hold if the material functions are strictly identical to one another.

The above discussion makes clear that elementary tractarian sense is so refined that each change in an elementary symbol induces a change in its sense. I think it is apt to call such fine-grained representation 'hyper-intensional'. Proposition 65 below shows the same to hold for contingent nuclei in general.

At this point one may have the impression that the system bans identity entirely. This would be mistaken: identities have their use in case of unanalyzed propositions. In analysis Wittgenstein employs the identity sign '=' to indicate that apparently different, unanalyzed expressions make the same contribution to sense. For example, E = E' indicates that the analysis of E and E' results in the same expression:

$$\begin{aligned} \varepsilon &= \text{E} \quad \textit{Def.} \\ \varepsilon &= \text{E'} \quad \textit{Def.} \end{aligned}$$

And then it trivially holds that:

IND^{an} $\qquad\qquad\qquad E = E' \rightarrow \forall\varphi[\varphi(E) \leftrightarrow \varphi(E')].$

To summarize, we have seen that unique naming ensures that the structure of an elementary proposition is fully determinate. It remains to show that especially the form of the sign, i.e., the way in which its names are configured, is essential to sense. To do so, we need to show that the sense of elementary propositions is invariant under substitution.

9.2.2 Substitution

With a view to perfect notation, besides uniqueness of content, there should be uniqueness of form. Uniqueness of form requires invariance under a suitable kind of transformation. Here, the relevant notion is a holistic substitution which simultaneously exchanges names for each other that have the same rôle, with 'rôle' as explained in chapter 2. In other words, the relevant substitutions are automorphism on elementary signs.[77]

Given the previous discussion on projection, object-form and the like, it is clear that substitution cannot be a local matter; it should involve logical space as a whole. To see this, consider a name n having different occurrences, as in $(n, n)_c$ or as in $(n, k)_{c'}$ and $(p, n)_{c''}$. In both cases it will not do to change just one occurrence of n, e.g., for m so that one respectively gets $(m, n)_c$, or $(n, k)_{c'}$ and $(p, m)_{c''}$. The reason is that in this case different names n and m result that refer to the same object. Instead, substitution should involve all names at once.

To arrive at the appropriate notion of substitution, recall that a projection in use partitions logical space in the totalities of signs and the depicted. Consequently, projection induces a partition of names and the objects they refer to. Names and referents are even one-one related.

Proposition 17 *In the context of projection, each object is either a name or a referent, not both. Names $\widehat{n}$ are one-to-one related to objects $\widehat{o}$ via the map:* $\widehat{\pi}$: $n, F(n) \mapsto \pi(n), F(\pi(n))$. $\qquad\qquad\qquad\Box$

Substitution should be defined for all names in all pictures so that names with the same form, the same combinatorial possibilities, can be substituted simultaneously for one another.

Definition 62 *Let Λ, π be a frame. A substitution σ is an automorphism σ on the totality of names $\mathcal{N}$ that preserves form:*

$$F(n) \overset{\sim}{=} F(\sigma(n));\qquad\qquad\qquad\Box$$

It is not difficult to see that an automorphism on the totality of names is equivalent to having an automorphism on the totality of pictures $\mathcal{P}$. That is, σ is a bijection from the names occurring in $\mathcal{P}$ onto itself, so that for any configuration:

$$(n_0, \ldots, n_k)_c \text{ if and only if } (\sigma(n_0), \ldots, \sigma(n_k))_c.$$

---

[77]I have profited from an e-mail correspondence on this topic with Albert Visser.

On the holistic approach substitution is simultaneous, it replaces all names at once; for each elementary proposition $\alpha = (n_0, \ldots, n_k)_c$:

$$\sigma\alpha = [\sigma(n_0)/n_0 \ldots \sigma(n_k)/n_k]\alpha = (\sigma(n_0), \ldots, \sigma(n_k))_c.$$

An example should make clear what this quite abstract notion of substitution amounts to. Consider the following collection of pictures:

$$(n)_c, (n, m)_{c'}, (n, m, n)_{c''}, (m)_c, (m, n)_{c'}, (m, n, m)_{c''}.$$

In this collection the names n and m have the same form. The map:

$$n \mapsto m, m \mapsto n$$

is a substitution. The idea is that in projection the rôles the different names play, can be interchanged for one another. More in general the insight is:

If π is a projection, so is the composition $\pi \circ \sigma$ of π with σ.

To continue the example, in the frame Λ, π the name n refers to $\pi[n]$ and the name m to $\pi[m]$. But in the frame $\Lambda, \pi \circ \sigma$ the rôles of n and m are interchanged: the name n refers to $\pi \circ \sigma[n]$ and the name m to $\pi \circ \sigma[m]$:

$$\pi \circ \sigma[n] = \pi[\sigma(n)] = \pi[m]$$
$$\pi \circ \sigma[m] = \pi[\sigma(m)] = \pi[n].$$

What happens to the sense of a sign under substitution? Elementary sense was defined as follows:

Definition 19 In frame, Λ, π, an *elementary proposition* $\pi[\alpha]$ is the elementary sign α together with the projection of α via π. The form of $\pi[\alpha]$ is the form $(n_0, \ldots, n_k)_c$ of α itself, and the *content* of its sense $(\pi(n_0), \ldots, \pi(n_k))_c$ is shown as realized. □

Thus it is clear that the proposition $\pi[\sigma\alpha]$ has the sense $\pi[\alpha]$ had prior to substitution: $\pi[\sigma\alpha]$ and $\pi[\alpha]$ have the same form, and after substitution the content of $\pi[\sigma\alpha]$ is that of $\pi[\alpha]$. Since $\pi[\sigma\alpha] = \pi \circ \sigma[\alpha]$, there is another way to make the same point. Instead of considering the substitution $\sigma\alpha$ in the context of the projection π, one could as well consider the sign α in the context of projection $\pi \circ \sigma$. But whether names are changed or their referents, the sense of a proposition in the one frame is identical with the sense of its substitute in the other.[78]

Substitution maps elementary propositions onto each other, without changing sense. Again, Wittgenstein's treatment of identity ensures that the hyperintensionality of elementary sense is safeguarded. Substitution, too, is vacuously extensional: the antecedent of the principle

$$\bigwedge_{0 \leq i \leq k}[n_i =_r \sigma(n_i)] \longrightarrow [(n_0, \ldots, n_k)_c \longleftrightarrow (\sigma(n_0), \ldots, \sigma(n_k))_c]$$

with $=_r$ a non-tractarian notation for identity of reference, only holds if σ is the identity mapping $n \mapsto n$.

---

[78]Substitution highlights the true identity of names and object. Cf also: 'Identity is the very Devil and *immensely important*; *very* much more so than I thought. It hangs – like everything else – directly together with the most fundamental questions, especially with the questions concerning the occurrence of the SAME argument in different places of a function.' (Wittgenstein, 1995, 45).

In sum, the notation for elementary propositions is perfect: elementary sense is shown without redundancy. The factors that enabled perfectness were:

1. Each name has reference;

2. The referent of each name is unique to it;

3. Elementary sense is shown using a sign's unique form.

We now continue with discussing the perfect notation for logically complex propositions.

9.3 Logically complex propositions

In section 7.4, a method is introduced to determine the symbol of a proposition. Here, I will argue that the distinction between signs and symbols is crucial in arriving at the perfect notation for logically complex propositions. We begin with establishing that equisignificant propositions – i.e., propositions that have the same symbol, – are equivalent. This is as it should be: the sense of a proposition shows its truth-conditions – how things relate *if* it is true (4.022), – so propositions are equisignificant if their truth-conditions are identical. The proof of this result makes clear that the contingent nucleus of a proposition is as hyper-intensional as elementary propositions are: any change in its elementary parts gives a change in sense. Further confirmation of this insight comes from the discussion on propositional substitution. It highlights that substitution requires proposition that differ, if at all, in parts that are inessential to sense. Thus, the tractarian notation of all propositions is perfect: it uses truth-functional signs to shows the essentials of sense, and it does so in a way that the substitutions allowed leave sense unaffected.

9.3.1 Equisignificance and equivalence

The *Tractatus* strongly suggests that in spite of the plurality in signs and symbols, truth-table signs can be singled out as the canonical signs, and the symbols to which they reduced as the canonical symbols. If propositions reduces to the same symbol, they are called *equisignificant*.

Definition 63 *The propositions φ and ψ are equisignificant – notation: $\varphi = \psi$, – if and only if they reduce to one and the same symbol: $\Sigma[\varphi] = \Sigma[\psi]$.* $\square$

To indicate equisignificance the identity sign is used. But according to 6.23 this sign is superfluous, for whether two propositions are essentially the same or not must be shown by their propositional signs and by these signs alone. Examples of equisignificant propositions are:

$$p \quad = \quad \neg\neg p$$
$$p \leftrightarrow q \quad = \quad (p \wedge q) \vee (\neg p \wedge \neg q)$$
$$p \quad = \quad p \wedge (q \vee \neg q)$$
$$p \wedge \neg p \quad = \quad q \wedge \neg q$$
$$p \vee \neg p \quad = \quad q \vee \neg q$$

Here are some examples of propositions that are not equisignificant:

$$p \neq q$$
$$p \wedge q \neq p \vee q$$
$$p \wedge \neg p \neq p \vee \neg p$$

Observe that we have included logical propositions among the identities. Although logical propositions are senseless, they may still occur in propositions that do have sense. And when considered on their own, it should be clear that tautologies and contradictions are senseless in different ways. As we shall see, this is precisely what definition 63 achieves. E.g., '$p = p \wedge (q \vee \neg q)$' shows that the presence of a logical proposition has no effect on the equisignificance of contingent propositions. Also, if logical propositions are of the same kind, they are equisignificant – for instance: $p \wedge \neg p = q \wedge \neg q$, – but if not they are senseless in different ways – for example: $p \wedge \neg p \neq p \vee \neg p$. Indeed, the notion of a symbol in definition 47 is correct.

Proposition 64 *Each proposition is equivalent with its symbol:*

$$\models \varphi \leftrightarrow \Sigma[\varphi]$$

Proof. The symbol $\Sigma[\varphi]$ results from removing all its logical parts from the truth-tables sign $\tau[\varphi]$. Say these parts are $p_{k+1}, \ldots, p_m$, and indicate their logicality with $\top(p_{k+1}, \ldots, p_m)$. Then:

$$\models \varphi \leftrightarrow \Sigma[\varphi] \wedge \top(p_{k+1}, \ldots, p_m) \leftrightarrow \Sigma[\varphi],$$

as required. □

For example, in case φ is a contradiction based on $p_0, \ldots, p_n$ one has:

$$\models \varphi \leftrightarrow (F)() \wedge \top(p_0, \ldots, p_n) \leftrightarrow (F)().$$

Equisignificant propositions must be equivalent to each other. The sense of a proposition shows its truth-conditions (4.022), so if the truth-conditions of propositions are identical they are equisignificant. Proposition 65 shows that also in this regard symbols are as intended.

Proposition 65 *The propositions φ and ψ are equisignificant if and only if φ and ψ are equivalent:*

$$\Sigma[\varphi] = \Sigma[\psi] \text{ iff } \models \varphi \leftrightarrow \psi.$$

Proof. First suppose that φ and ψ are equisignificant: $\Sigma[\varphi] = \Sigma[\psi]$. Then, equivalence is immediate from proposition 64:

$$\models \varphi \leftrightarrow \Sigma[\varphi] = \Sigma[\psi] \leftrightarrow \psi.$$

Next suppose that φ and ψ are equivalent: $\models \varphi \leftrightarrow \psi$. If either φ or ψ is logical, then $\Sigma[\varphi] = (T)() = \Sigma[\psi]$ or $\Sigma[\varphi] = (F)() = \Sigma[\psi]$, and we are done. So suppose that both φ and ψ are contingent. Since in general $\models \chi \leftrightarrow \Sigma[\chi]$, $\models \varphi \leftrightarrow \psi$ implies: $\models \Sigma[\varphi] \leftrightarrow \Sigma[\psi]$. It therefore suffices to proof that $\Sigma[\varphi]$ and $\Sigma[\psi]$ are based on the same elementary propositions – say $p_0, \ldots, p_k$, – for then by equivalence the symbols would be strictly the same.

To arrive at a contradiction, suppose that $\Sigma[\varphi]$ and $\Sigma[\psi]$ are based on different elementary propositions:

$$\Sigma[\varphi] \text{ is based on } p_0, \ldots, p_k, p_{k+1}, \ldots, p_m,$$

$\Sigma[\psi]$ is based on $p_0, \ldots, p_k, q_{k+1}, \ldots, q_n$,

where possibly $m = k$ or $n = k$ but not both, and where $p_{k+1}, \ldots, p_m$ and $q_{k+1}, \ldots, q_n$ are disjoint. We show that on this assumption $\Sigma[\varphi]$ and $\Sigma[\psi]$ are unequivalent, *quod non*.

Consider $\Sigma[\varphi]$, based on $p_0, \ldots, p_k, p_{k+1}, \ldots, p_m$. Call $Y'_{k+1}, \ldots, Y'_m$ a negation of the truth-possibility $Y_{k+1}, \ldots, Y_m$ if for some i with $k + 1 \leq i \leq m$ Y'_i is 'T' and Y_i is 'F' or conversely. By contingency, there must be a truth-possibility $\mathbf{t}_\varphi \equiv Y_0, \ldots, Y_k, Y_{k+1}, \ldots, Y_m$ that makes $\Sigma[\varphi]$ true, while one of its negations $\mathbf{f}_\varphi$, which combines $Y_0, \ldots, Y_k$ with a negation of $Y_{k+1}, \ldots, Y_m$, makes $\Sigma[\varphi]$ false. For if all such pairs of truth-possibilities were to yield the same truth-value, $p_{k+1}, \ldots, p_m$ could be assigned any truth-value. That is: $p_{k+1}, \ldots, p_m$ would occur logically in $\Sigma[\varphi]$ and could have been removed by the reduction rule. This contradicts $\Sigma[\varphi]$ being a contingent nucleus.

Using the same reasoning, there must be a truth-possibility

$$\mathbf{f}_\psi \equiv Y_0, \ldots, Y_k, Z_{k+1}, \ldots, Z_n$$

that makes $\Sigma[\psi]$ false and a truth-possibility $\mathbf{t}_\psi$, combining $Y_0, \ldots, Y_k$ with a negation of $Z_{k+1}, \ldots, Z_n$, that makes $\Sigma[\psi]$ true.

By assumption the $q_{k+1}, \ldots, q_n$ do not occur in $\Sigma[\varphi]$, so its truth-value does not depend on them, and *mutatis mutandis* the same for $p_{k+1}, \ldots, p_m$ and $\Sigma[\psi]$. Consider a world w that is consistent with the truth-possibility:

$$Y_0, \ldots, Y_k, Y_{k+1}, \ldots, Y_m, Z_{k+1}, \ldots, Z_n.$$

In w, $\Sigma[\varphi]$ has the same truth-value as in $w_{\mathbf{t}_\varphi}$ and so: $w \Vdash \Sigma[\varphi]$. Further, $\Sigma[\psi]$ has the same truth-value in w as it has in $w_{\mathbf{f}_\psi}$ and so: $w \nVdash \Sigma[\psi]$. Therefore, $\Sigma[\varphi]$ and $\Sigma[\psi]$ are unequivalent; a contradiction.

We conclude that $\Sigma[\varphi]$ and $\Sigma[\psi]$ are based on the same elementary propositions, and so by equivalence: $\Sigma[\varphi] = \Sigma[\psi]$ □

Proposition 65 has some immediate consequences.

Proposition 66 *Proposition φ is a tautology if and only if $\Sigma[\varphi] = (\mathrm{T})()$; φ is a contradiction if and only if $\Sigma[\varphi] = (\mathrm{F})()$; φ is contingent if and only if $\Sigma[\varphi]$ is a contingent nucleus.* □

Propositions may express the same symbol but still differ in their truth-table signs. Proposition 64 makes clear that in this case the differences are logical.

Proposition 67 *If φ and ψ are equisignificant, they differ at most in logical parts.* □

Proof. If φ and ψ are equisignificant, they may still be based on different elementary propositions. However, from the proof of proposition 65 we learn that $\Sigma[\varphi]$ and $\Sigma[\psi]$ must be based on the same elementary propositions (possibly none). That is:

$$\models \varphi \leftrightarrow \Sigma[\varphi] \wedge \top(p_{k+1}, \ldots, p_m)$$
$$\models \psi \leftrightarrow \Sigma[\psi] \wedge \top(q_{k+1}, \ldots, q_n)$$

with $p_{k+1}, \ldots, p_m$ in φ but not in ψ, and $q_{k+1}, \ldots, q_n$ in ψ but not in φ. Since $\Sigma[\varphi] = \Sigma[\psi]$, φ and ψ differ at most in the logical parts $\top(p_{k+1}, \ldots, p_m)$ and $\top(q_{k+1}, \ldots, q_n)$. □

Along the same vein, proposition 65 can be seen to give a kind of interpolation.

Proposition 68 *If* $\vdash \varphi \leftrightarrow \psi$, *there is a proposition* σ, *which only depends on the elementary propositions that* φ *and* ψ *share, so that* $\vdash \varphi \leftrightarrow \sigma \leftrightarrow \psi$. $\qquad\square$

This would establish the observation of Göran Sundholm (*private communication*) that 5.135 is related to interpolation.

5.135 In no way the existence of a situation can be inferred from the existence of a situation entirely different from it.

See section 12.10 for further discussion.

To summarize, the contingent nucleus of a proposition, if it has any, is hyper-intensional: the sense it represents changes as soon as its elementary content is changed. Despite the extensional means used, symbols are a perfect notation for sense. It remains to be checked that propositional substitutions comply with this insight; they should change senseless logical parts or forms at best.

9.3.2 Propositional substitution

In the *Tractatus* there is not much detail on how the distinction between sign and symbol applies to propositions. It seems as if Wittgenstein was mainly interested in showing whether a proposition has sense or is senseless, which is fully determined by its truth-table sign. But identity of sense does figure in whether or not a proposition can be substituted for another one.

3.341 So what is essential to a proposition is what all propositions that express the same sense have in common.

And likewise, in general the essence of a symbol is what all symbols that can serve the same purpose have in common.

In definition 63, the identity sign is used to indicate equisignificance. This is deliberate, for equisignificant propositions can be substituted for each other.

6.23 When two expressions are joined by the identity sign, this means that they can be substituted for one another. Whether this is indeed the case, must show itself in the expressions.

It characterizes the logical form of two expressions that they can be substituted for each other.

It remains to check how substitution leaves the representation of sense intact. This, however, is an immediate consequence of proposition 65 and its insight that equisignificant propositions represent sense in essentially the same way.

Proposition 69 *Let* φ, ψ, χ *be propositions. The rule SUB holds:*

$$\frac{\varphi = \psi}{\chi(\varphi) = \chi(\psi)} \; \text{SUB}$$

Proof. If neither φ nor ψ occurs in χ there is nothing to proof. So assume that φ occurs in proposition χ. If χ is elementary, we are done. But if χ is obtained from elementary propositions by repeated application of the truth-operation N, χ has form: $\text{N}'(\chi_1, \ldots, \chi_i, \varphi, \chi_{i+1}(\varphi), \ldots, \chi_n(\varphi))$ with $\chi_1, \ldots, \chi_i, \chi_{i+1}(\varphi), \ldots, \chi_n)(\varphi)$ propositions and all relevant occurrences of φ indicated.

Now if $\varphi = \psi$, then $\Sigma[\varphi] = \Sigma[\psi]$, and according to proposition 65 φ and ψ are equivalent to each other. By induction hypothesis: $\chi_j(\varphi) = \chi_j(\psi)$, so these propositions are equivalent too. Consequently, the following are equivalent:

$$\mathrm{N}'(\chi_1, \ldots, \chi_i, \varphi, \chi_{i+1}(\varphi), \ldots, \chi_n(\varphi))$$
$$\mathrm{N}'(\chi_1, \ldots, \chi_i, \psi, \chi_{i+1}(\psi), \ldots, \chi_n(\psi)),$$

i.e.: $\chi(\varphi)$ and $\chi(\psi)$ are equivalent. So, again by proposition 65, they are equisignificant: $\chi(\varphi) = \chi(\psi)$. □

Substituting equisignificant propositions ensures the representation of sense is unaltered. This is in line with symbols capturing the essence of sense. In a perfect notation of sense just the contingent nucleus counts.[79]

A similar observation can be made from a slightly different perspective. It is a fundamental thought of the *Tractatus* that logical constants do not refer. Therefore, equivalent truth-operations can be substituted for each other without affecting sense. This insight can be detailed as a consequence of proposition 69.

Definition 70 *Call truth-operations Ω_1 and Ω_2 equivalent – notation: $\Omega_1 \equiv \Omega_2$, – iff for all truth-table signs $\tau_0, \ldots, \tau_n$:*

$$\Omega_1(\tau_0, \ldots, \tau_n) \equiv \Omega_2(\tau_0, \ldots, \tau_n).$$

On the same arguments, the truth-operations yield the same result. □

Proposition 71 *Let Ω_1 and Ω_2 be truth-operations; let $\varphi(\Omega_1)$ be a propositional sign in which Ω_1 occurs; let $\varphi(\Omega_2)$ be the result of replacing an occurrence of Ω_1 in φ by Ω_2. Then:*

$$\frac{\Omega_1 \equiv \Omega_2}{\varphi(\Omega_1) = \varphi(\Omega_2)} \, \mathrm{SUB}^{to}$$

Proof. Let $\Omega_1(\tau_0, \ldots, \tau_n)$ be the occurrence of Ω_1 to be substituted in φ. Since $\Omega_1 \equiv \Omega_2$, $\Omega_1(\tau_0, \ldots, \tau_n) = \Omega_2(\tau_0, \ldots, \tau_n)$. Therefore:

$$\varphi(\Omega_1(\tau_0, \ldots, \tau_n)) = \varphi(\Omega_2(\tau_0, \ldots, \tau_n)),$$

and thus $\varphi(\Omega_1) = \varphi(\Omega_2)$. □

This end our discussion on substitution. We finish the chapter with showing that the symbol of a proposition is indeed unique. Algebraically, it can be seen as a canonical representative of an equivalence class.

9.4 *Reflection*: Symbols and Lindenbaum algebra's

A most characteristic feature of the tractarian system is its economy of means. This is particularly true for analyzed propositions: truth-table signs are based on a perfect notation of elementary signs that are combined logically without redundancy. Symbols obtained from truth-table signs have the same effect as working with the elements of a Lindenbaum algebra; each symbol can be

---

[79]It should be interesting to see to what extent Wittgenstein's symbolic approach to sense can be used in non-tractarian settings, such as the semantics of perception reports or of propositional attitudes. Then, the problem of logical omniscience does not seem to arise, because in the intensional context only the contingent nucleus should contribute semantically. In particular, all logical parts would be eliminated from such contexts, which in many cases, but perhaps not all, seems as required. Cf. 5.1362.

seen as a canonical element of a class of equivalent propositions. Lindenbaum algebra's are introduced here for the sake of comparison. They are alien to the classless philosophy of the *Tractatus*, but the comparison is instructive nonetheless.

9.4.1 Lindenbaum algebra

Each frame Λ, π gives a totality of propositions, which could be considered for equivalence:

$$\models \varphi \leftrightarrow \psi \text{ iff for all } R: \Lambda, \pi, R \Vdash \varphi \leftrightarrow \psi.$$

In a Lindenbaum algebra logical connectives are defined in terms of classes of formulas that are equivalent to each other. The domain of the algebra – in a frame that is left implicit, – consists of equivalence classes of propositions φ:

$$[\![\varphi]\!] := \{\psi \mid \models \varphi \leftrightarrow \psi\}.$$

This class is independent of its representative φ. For if $\chi \in [\![\varphi]\!]$, then $\models \varphi \leftrightarrow \chi$, and $[\![\varphi]\!] = [\![\chi]\!]$. So χ would represent the class $[\![\varphi]\!]$ as well as φ does. Here, Lindenbaum algebra's have the single truth-operation N of joint denial. As an operation on equivalence classes, N is defined as follows:

$$N([\![\varphi_1]\!], \dots, [\![\varphi_n]\!]) := [\![N'(\varphi_1, \dots, \varphi_n)]\!].$$

In this definition n is any natural number; the number of argument places of N is not fixed. The resulting Lindenbaum algebra is written:

$$\langle \{[\![\psi]\!] \mid \psi \text{ a proposition}\}, N \rangle,$$

with $\{[\![\psi]\!] \mid \psi$ a proposition$\}$ the domain of the algebra and N its operation. Notice that equivalence classes abstract from logical redundancy. For instance, all double negations $N'^{2n}(p)$ are in $[\![p]\!]$, and so are $p \wedge \top(q_0, \dots, q_i)$ and $p \vee \bot(r_0, \dots, r_j)$. Symbols have the same effect.

9.4.2 Symbol algebra's

In the *Tractatus* there are no classes, nor any other means to turn a totality of equivalent propositions into an abstract object. Instead, the effect of equivalence classes is obtained using symbols as canonical representatives of a proposition's meaning. Symbols provide for a perfect notation.

The tractarian analogue of a Lindenbaum algebra is called a 'Symbol algebra', and has the form:

$$\langle \{\Sigma[\psi] \mid \psi \text{ a proposition}\}, N^\star \rangle,$$

The domain of a symbol algebra is the totality of symbols Σ, and its sole truth-operation $N^\star$ is the composition of Σ with N, to ensure that the result is a symbol:

$$N^\star(\Sigma_1, \dots, \Sigma_n) = \Sigma[N'(\Sigma_1, \dots, \Sigma_n)].$$

That symbols can replace equivalence classes, follows from the fact that Lindenbaum algebra's are isomorphic to Symbol algebra's.

Proposition 72 *The Lindenbaum algebra of a frame Λ, π is isomorphic to its Symbol algebra.*

Proof. Let Λ, π be a frame. The propositions of Λ, π determine a Lindenbaum algebra

$$\langle \{ [\![\psi]\!] \mid \psi \text{ a proposition} \}, N \rangle,$$

and a Symbol algebra:

$$\langle \{ \Sigma[\varphi] \mid \psi \text{ a proposition} \}, N^* \rangle.$$

To establish isomorphism, we show that the map:

$$F : [\![\varphi]\!] \mapsto \Sigma[\varphi]$$

is a bijection which preserves the operation N.

The truth-operational completeness result of section 5.7 makes clear that each complex of truth-operations reduces to a truth-table sign, and that no truth-table signs are left out in this way. So $\Sigma[\varphi]$ is indeed a symbol, and all symbols are obtained in this way. Also, proposition 65 implies that the truth-table signs of the propositions in an equivalence class reduce to one and the same symbol:

$$\chi \in [\![\varphi]\!] \Leftrightarrow \Sigma[\chi] = \Sigma[\varphi].$$

Therefore $F : [\![\varphi]\!] \mapsto \Sigma[\varphi]$ is a bijection. It is even an isomorphism:

$$
\begin{aligned}
F([\![N'(\varphi_1, \ldots, \varphi_n)]\!]) &= \Sigma[N'(\varphi_1, \ldots, \varphi_n)] \\
&= N'^*(\Sigma[\varphi_1], \ldots, \Sigma[\varphi_n]) \\
&= N'^*(F([\![\varphi_1]\!]), \ldots, F([\![\varphi_n]\!])).
\end{aligned}
$$

The second step uses that φ_i and $\Sigma[\varphi_i]$ are equivalent, and that symbols of equivalents are identical. The third step is by induction hypothesis. $\square$

That symbols form an algebra with N^* its sole operation is an abstract way of saying that the sense of logically complex propositions is compositional (5.2341a).

This finishes our (non-tractarian) reflection to show that symbols are a perfect notation for the sense of proposition. In projection $\Sigma[\varphi]$ shows *the* sense of φ without redundancy. [*End of reflection*]

10

The impact of Russell's paradox

Russell's work is of extraordinary depth and diversity. In many ways it is the corpus of a modern researcher, who prefers publishing over polishing. Sometimes Russell presented his ideas at an early stage, before they reached full maturity, but always in an elegant style that evoked others to take their stand. Perhaps for this reason, the relationship between the work of Russell and that of Wittgenstein, which is the result of striving for perfection, is quite peculiar and hard to trace.

In the *Tractatus*, we mainly find statements of critique on Russell's logic and philosophy. This should not blind us for the fact that Russell's work was more than rich enough to help Wittgenstein devising the tractarian system; partly in opposition, and in a more coherent, single-minded manner. In section 4.4, I have suggested how Wittgenstein's may have transformed Russell's theory of judgments into the tractarian ontology. In a related manner Wittgenstein profited from Russell's considerations on his paradox. Although Wittgenstein was against Russell's theory of types, his own solution utilizes a notion of proposition whose core is derived from Russell.

In this chapter I will describe how Russell found his paradox, and how his approach to solve it had influence on Wittgenstein.[80] I hold that in reaction to the paradox Wittgenstein blocked vicious circles, just like Russell did, but instead of typing he used bounds on the extent structures can be embedded into each other. Combined with the system being universal – the essentials of any language are captured, – the system is seen to be non-reflexive: it is unable to describe its own sense conditions. Non-reflexivity turns out to be strict, for even the modern route to reflexivity via coding is unavailable. With a view to the system's inability, Wittgenstein took resort to an ostensive philosophy in which gestures surrounding the system show an understanding party what its sense conditions amount too, and so what its limits are. Such philosophy points toward a metaphysics of description without describing it; it is a swansong of metaphysics (Hacker (1996), 38). The chapter ends with a 'choreography for a swansong', which gives a detailed sequence of gestures that may have the envisioned ostensive effect.

---

[80] In this overview I have profited much from Russell (1987), Russell (1907), and Burgess (2005).

10.1 Paradox found

For years Russell felt the urge 'to find some reason for supposing mathematics true' (Russell, 1987, 65), and he set himself the aim to analyze the basic concepts of mathematics in a logical way. But only in 1900 he found a method to do so, at the International Congres in Paris where he started learning the notation of the Italian mathematician Giuseppe Peano. Since 1894, his fourth year as a student at Cambridge, Russell had copies of Cantor's *Mannichfaltigkeitslehre* and Frege's *Begriffschrift*. But as to Frege he would later confess: '...I could not make out what it meant. Indeed, I did not understand it until I had myself independently discovered most of what it contained' (Russell, 1987, 65).

Having found a formal approach to logic, Russell continued his logicist program to show that mathematics is analytic and can be derived from logical principles alone. In the course of this work he returned to Cantor's work and in 1901 noticed that it was inconsistent. (Cantor had noticed this himself, but had kept it private.) Russell found the paradox by combining his belief in a universal class, a class of all things there are, with Cantor's theorem that there is no greatest number; that is, the number of subclasses of a class is always of greater cardinality than the number of elements in the class.[81] The observation is an immediate consequence of Cantor's proof. See Russell (1987), 150. The proof is also in Russell (1907), 32.

Theorem 73 (Cantor's Theorem) *For each class S: $|S| < |\wp(S)|$.*

Proof. It is clear from the injection $s \mapsto \{s\}$ that $|S| \le |\wp(S)|$. To proof inequality, assume for a contradiction that $|S| = |\wp(S)|$. By definition of equicardinality, there is a bijection $* : S \longrightarrow \wp(S)$. Consider the class $S' := \{s \in S \mid s \notin s^*\}$. Since $*$ is a bijection, there is a $s' \in S$ such that $s'^* = S'$. Is $s' \in S'$ or not? One has:

$$s' \in S' \Leftrightarrow s' \notin s'^* \Leftrightarrow s' \notin S',$$

a contradiction. Therefore, $|S| \ne |\wp(s)|$. Combined with $|S| \le |\wp(S)|$ this gives $|S| < |\wp(S)|$. $\square$

Russell observed that Cantor's proof gets a special twist if one considers the universal class V of 'all things there are'. For all classes S in V one has:

$$S \in V \Leftrightarrow S \subseteq V.$$

Indeed, if $S \in V$ then each element of S is in V, because V is universal. And since S as a subclass of V is itself an object, again by universality it is one of V's elements. Now notice that the identity function id : $s \mapsto s$ on V would establish $|V| = |\wp(V)|$. But Cantor's proof shows this cannot be. In case of the function *id* the class S' in this proof becomes the class of classes that do not contain themselves:

$$id(S') = \{S \in V \mid S \notin S\},$$

which implies a contradiction:

$$S' \in S' \Leftrightarrow S' \notin S'.$$

[81]I use the terminology of the time. The now current distinction between classes and sets is of a later date.

At first Russell thought the paradox was a triviality, and assumed to have overlooked a crucial step that would block it. He hoped that a study of Frege's *Grundgesetze der Arithmethik, I* would provide a way out of the predicament. Frege's major work made a deep impression on Russell: here was a philosopher and logician who had been working for decades on the same problems that Russell had just started working on, and who had developed notions similar to his. Yet Russell's hope to find a solution for his paradox in Frege's work turned out to be in vain. By letter Russell notified Frege of his admiration and of his paradox (Russell, 1902). It was Frege who first published a detailed reconstruction of how to derive the contradiction in his system, together with a proposal for a solution (Frege (1903a), *Nachwort*, 253–265). Frege's proposal looked promising at first, but later turned out to be inconsistent as well.[82]

10.2 Tracing the root of paradox

In order to finish his logicist program, Russell continued to analyze the root of his paradox. In a note read to the London Mathematical Society on December 14th, 1905, Russell observed that the paradoxes involved classes, but also the propositional functions that define them. A propositional function is an expression $\phi(x)$ that for all values of x gives a proposition (similarly in case of more variables).[83] It was clear that not all propositional functions define classes. But which criterion singles out the predicative ones that do from the non-predicative ones that do not? Russell suggested three approaches to search for an answer.

1. Perhaps propositional functions determine sets only 'when they are fairly simple, and fail to do so when they are complicated or recondite'. A first attempt to develop this suggestion led to 'exceedingly complicated' axioms that 'cannot be recommended by any intrinsic plausibility', and Russell left it at that (Russell, 1907, 38-39).

2. Another option would be to follow Jourdain's suggestion: only allow classes whose size can be measured against the ordinals (Russell, 1907, 43 ff.). This idea is basic to current versions of set-theory.

3. Finally, Russell proposes the no-classes theory in which classes are banished altogether. On this theory it is irrelevant whether a propositional function determines a class or not. Instead, the construction of mathematics should be based on the totalities of values for which a propositional function is true, without assuming 'that these values collectively form a single entity which is the class composed of them' (Russell, 1907, 46).

In printing, Russell added the following note to his article: '*February 5th, 1906.*—From further investigation I now feel hardly any doubt that the no-

---

[82] The episode is absent in Russell's autobiography. Perhaps Russell found it inappropriate to discuss his rôle in the fate of Frege's great works. Recent work has shown that large parts of Frege's work can be salvaged; cf. Burgess (2005) for an overview. Burgess (2005), chapter 1, also has details on how to derive the paradox in Frege's system. A more extensive correspondence between Russell and Frege can be found in Gabriel (1980), 47–100.

[83] Russell's propositional functions are more like Frege's senses than like Frege's reference. E.g. other than the True and the False, propositional functions have structure.

classes theory affords the complete solution of all the difficulties stated in the first section of this paper.' The first section of Russell (1907) was on paradoxes, but to solve them without classes there was a hurdle left.

10.3   Ramified Theory of Types

The classes associated with propositional functions, when taken too broadly, are not the only threat for inconsistency. As Russell noted himself, propositional functions themselves require care if they are rich enough to express semantical notions. Consider, for instance the predicate $\phi(x)$ (from Burgess (2005), 35–36):

'x is a formula with one free variable that does not satisfy itself'.

Then, $\phi(x)$ satisfies itself if and only if $\phi(x)$ does not satisfies itself, and it remains to ban viciously circular specifications like this one that do not involve any class. Russell aimed to do so in his theory of types, which is based on propositional functions coming in different levels.

As I have explained in section 2.1.2, Russell's approach consists in ramifying types. Objects have simple types – so there are individuals, propositional functions, properties, relations, propositions, etc., – but to preclude paradox the types are ramified. They are also subdivided in levels according to the complexity of their definition, and in such a way to avoid vicious circles.

> Vicious circle principle Whatever involves *all* of a collection must not be one of the collection.   (Whitehead and Russell, 1910, ch. 2, §1)

Roughly, the idea is that a propositional function ϕ^n of level n may contain bound variables of level at most n and free variables of level at most $n + 1$. The leveling must ensure that each propositional function does not quantify over entities of the same kind as itself. Clearly, this approach precludes a class and its elements to share a definition. According to Russell all classes have a defining property, so a class cannot have itself as a member. Unfortunately, ramification comes with infinite inflexibility. E.g., for each level n the validity $\forall x^n(x^n = x^n)$ most be asserted separately. A clarifying formalization can be found in Laan (1997), chapter 2 and 3.

In Russell's theory of types, classes are a manner of speech that can be eliminated in favour of propositional functions. In particular, there is no assumption that the class of things satisfying a certain propositional function exists as an object (Whitehead and Russell, 1910, p. 24). To compensate for the loss of 'real' classes and their extensionality and to loosen the inflexibility of ramification, Russell allowed for the axioms of reducibility, infinity and choice. Reducibility had the effect to introduce for each higher level concept an equivalent first level one; infinity ensured sufficiently many objects to get enough of mathematics; and choice enabled to prove some apparently different notions to be equivalent. A first version of the Theory of Types was published in Russell (1908). It formed the basis of the monumental Whitehead and Russell, *Principia Mathematica*, Volumes 1-3.

10.3.1 *A priory* typing is superflous

Russell had a strong influence on the architecture of the tractarian system; e.g., on its distinction between the superficial and the essential logical form of a proposition, and on the structured approach of elementary propositions. But quickly after the collaboration of Russell and Wittgenstein had started, it was clear that serious discrepancies would come to the fore. Perhaps the most serious difference between the philosophies of Russell and Wittgenstein concerned their ideas on typing. Russell had worked for years to develop the ramified theory of types, which became the cornerstone of the *Principia Mathematica*. But Russell's considerations on typing failed to persuade Wittgenstein for long.

More in general Wittgenstein held the cluster of techniques that Russell used to be untenable for quite a few reasons. He was against employing semantic notions in the rules for syntactic combination, as whether an expression denotes a specific type of object should be immaterial for sign-formation (3.331). He was against regarding some propositions, called 'axioms', as more basic than others. Russell used axioms to infer the possibly non-evident from the self-evident, but according to Wittgenstein self-evidence is irrelevant, and he held all propositions to be on a par (5.4731, 6.4). He was against Russell's (and Frege's) treatment of logical objects and constants (5.4), as complex logical structure is about nothing. He was against Russell's theory of identity, which states that two objects are the same if they share all properties of the appropriate type. Russell's had adapted this theory from Leibniz, but Wittgenstein held that identity cannot be stated: it must show itself in the signs of names and their meanings (5.53). Wittgenstein also took Russell's treatment of identity to be wrong, because in the tractarian ontology objects are not extensional but rather hyper-intensional: they can have the same internal properties (have the same form) and yet be different (2.02331). Wittgenstein was against Russell's axiom of infinity. Instead of being stated the crux of the axiom should show itself in the totality of different meaningful names (5.535). He was against Russell's axiom of reducibility, which was not logical and could hence be false (6.1232).[84]

Perhaps most radical was Wittgenstein's rejection of types. He rejected ramified types: there is no hierarchy of forms of elementary propositions (5.556), and all logically complex propositions remain on one and the same level (6, 6.4). Wittgenstein also held simple types to be of limited value. In chapter 2, I have argued that a theory of simple types is no part of the system, as simple types are inessential to showing how descriptive language is meaningful. Wittgenstein does distinguish between propositions and objects. But although objects may have different forms, they are not classified further. If simple types were to be used at all, they should result from analysis rather than being presupposed prior to analysis.

Given Wittgenstein's critique on Russell's theory of types, he must have seen a simpler way to block vicious circularity, or this is what I shall argue

---

[84]In this regard, Wittgenstein was among the first to hold a position that is now quite common. See for instance Quine's introduction to Russell (1908) in Heijenoort (1976), or Burgess overview of the ramified theory of types (Burgess, 2005, chapter 1).

now. Rather than trying to classify propositional functions into those that are or are not circular, Wittgenstein's idea is that insight into the general form of propositions is required.

10.4 Wittgenstein's solution to Russell's paradox

Early Wittgenstein agreed with Russell that vicious circularity was at the heart of paradox, but he solves Russell's paradox in a different way than Russell did. In the *Tractatus* an analyzed propositional sign is a truth-table based on elementary propositional signs. The signs are defined without any reference to semantics, as in the criticized Russellian approach.

3.33 In logical syntax the meaning of a sign should never play a rôle. It must be possible to establish logical syntax without mentioning the *meaning* of a sign; *only* the description of expressions may be presupposed.

It is rather the structure of the sign as it figures in meaningful use that should block reflexivity – i.e., the ability of self-description, – and with this vicious circularity. Thesis 3.332, which dates back to the Notes on Logic (1913), gives the main insight:

3.332 No proposition can convey something about itself, because the propositional sign cannot be contained in itself (that is the entire 'Theory of Types').

The question is: in what sense is it impossible for a sign to convey something about itself? The first kind of impossibility that may present itself is empirical. The propositional sign enables thoughts to be sensed (3.1), and following our perhaps naive intuitions its is natural to hold that no sensible matter can contain a copy of itself; in particular a propositional sign can not. This kind of impossibility, however, is not at stake. The impossibility named in 3.332 should be explained in configurational terms. It is not a contingent matter that a sign cannot contain itself, the impossibility is structural and necessary. Thesis 3.333 gives a bit more detail.

3.333 The reason why a function cannot be its own argument is that the functional sign already contains the prototype of its argument, and it cannot contain itself.

Let us assume that the function $F(fx)$ could be its own argument; then there would be a proposition '$F(F(fx))$' in which the outer function F and the inner function F must have different meanings, since the inner one has the form $\varphi(fx)$ and the outer one the form $\psi(\varphi(fx))$. These two functions have just the letter F in common, which in and of itself is meaningless.

This immediately becomes clear, if instead of '$F(F(fu))$' we write $\exists \varphi \, (F(\varphi u) \wedge \varphi u = Fu)$.

With this Russell's paradox is nullified.

The terminology 'prototype' that is used in 3.333 seems to derive from Russell (1907), who uses it in the same way. A 'prototype' replaces a constituent in a proposition the highlight a form. In the *Tractatus* the term 'variable' is reserved for: a sequence of signs that exhibit the same form. See section 11.3.1.

Like Ishiguro (1981, 52), I take the prototype in 3.333 to be the unit 'fx', in which f and x should not be regarded separately. The notation should suggest 'fx' to be a prototype of the function '$F(x)$'.

At first sight, thesis 3.333 seems to concern a more general setting than that of propositions, as it states that no function can be reflexive. This reading is mistaken. The functions named are propositional; 3.333 is about elementary propositions as functions of the expressions they contain (3.318, 4.24). Rather than being too general, the scope of 3.333 is too strict: all propositions should be non-reflexive, not just the elementary ones. In other words, 3.333 needs to be supplemented with an argument showing that irreflexivity of elementary propositions suffices.

Proposition 74 *If proposition φ in logical form makes a statement about itself, then at least one of its elementary propositions makes a statement about itself.*

Proof. Assume φ is logically complex, for else there is nothing to show. If φ makes a statement about itself—for example, using such phrases as *this statement, the statement at line number n,* etc.,—φ ascribes a property to itself. As each result of analysis must comply with the system, the ascription takes the form of a truth table based on elementary propositional signs. The table allows the property that φ ascribes to itself to be logically complex, but the ascription of the elementary parts making up the logically complex property would mainly occur at the elementary level. This means the analyzed version of φ would have the form: $\varphi(\alpha_1(\varphi), \ldots, \alpha_n(\varphi), \ldots)$. In fact, not all of the elementary signs α_i need to concern φ, but there must be at least one that does, say, $\alpha_r(\varphi)$, where φ in $\alpha_r(\varphi)$ can only be an expression of α_r describing φ. Since α_r occurs in φ, α_r is also of form: $\alpha_r(\varphi(\alpha_r))$: that is, of form: $\alpha'(\alpha_r)$. So, φ ascribing a property to itself, implies that each of the α_i that concern φ is reflexive. □

It follows from proposition 74 there are no reflexive propositions.

Proposition 75 *There are no reflexive propositions.*

Proof. According to proposition 74, it is sufficient to show there are no reflexive elementary propositions. So assume there is a reflexive α_r. In the tractarian system, meaning involves a projection π, which is an isomorphism between the sign α_r and the state of things it is about. So, if α_r ascribes a property to itself, the propositional sign must have form $\alpha'(\alpha_r)$ with α' the property ascribed. That is, α_r is the material function $\alpha'(u)$ with α_r as its argument. As 3.333 points out, this is impossible. It would mean that α_r occurs in two ways: (i) as the propositional sign $\alpha'(\alpha_r)$, and (ii) as an expression requiring the context $\alpha'(u)$ to form a propositional sign. These occurrences have different meanings. For instance, the first occurrence is propositional and can be true or false, while the second is a proper subexpressions of an elementary proposition that, by definition, cannot have a truth value. Reflexivity would only occur if both instances were identical in all respects. □

To my taste the last step in the argument is too doctrinal. The step is based on the feature of the tractarian system that by definition elementary propositions do not have propositional parts, else they would not be elementary. The argument would gain in strength if it were phrased in fully structural terms without appeal to this feature of the system. For finite elementary propositions such an argument exists. Then again the logical picture theory of meaning implies that the sign of a reflexive elementary proposition must have a strict part that is an isomorphic copy of itself. Since the sign is finite, this would mean there is a projection, and thus a bijection, from a finite number of objects onto a smaller number of objects. *Quod non.*

In the chapters 11 and 12, I shall present an infinite variant of the tractarian system. So, the question remains to what extent the structural argument can be salvaged if infinite signs are allowed? While writing the *Tractatus* Wittgenstein must have had his moments in which he abhorred infinite signs and meanings, especially elementary ones: 'Yet, the *infinitely* complex state of things appears to be an absurdity!' (NB 23.5.15(8)). But in the end he succumbed. Perhaps a proper balance between his fascination and his repugnance is found when truth-functions are allowed to be infinite, while in line with the quotation elementary propositions are kept finite. Then, a structural solution of Russell's paradox can be given without restricting the size of a sign too much. In this generalized setting the irreflexivity of finite propositions continues to hold as before, and a quick check of proposition 74 shows it remains true for infinitary truth-functions. In an infinite reflexive propositions a structural copy of itself would have to be part of a finite elementary proposition, which is impossible.

10.5   Reflexivity via coding?

I have shown the system does not allow for reflexive propositions. The system is unable to describe its own sense conditions. Still, one may wonder whether a more sophisticated form of reflexivity can be introduced based on coding? Coding requires:

1. A domain of meaningful elements,
2. A code for each of the elements, and
3. An interaction between the codes and the meaningful elements.

A code for an element is an object or structure that can be decoded: the coded element can be reconstructed from it. Codes and their interaction with elements make that elements can be about each other or even about themselves. See Barendregt (2005). In the tractarian system the elements are elementary propositions, and the interaction between element and code can only consists of the code occurring in the element. That is, a sign containing a code would have form $\alpha(c)$ with c the code and $\alpha(-)$ a material function. It remains to find a suitable coding, if there is one.

What are the options? Clearly, a code cannot be a name, for names only refer to unstructured objects. The code cannot be a tractarian numeral either, for such numerals are exponents of an operation (6.021) and do not interact with elementary propositions; they do not occur in the context where properties are ascribed. Within the system the only option remaining seems to be: an expression that is defined syntactically using the full sign to be coded.

Theses 3.261 and 4.241 state that what is so defined should be replaced by its definition to make the meaning of the expression more explicit. But as soon as such substitutions are effected one argues as for proposition 75 to see that this kind of code-based reflexivity is again impossible.

The tractarian possibilities for coding are quite different from the now common ones, which have a semantic aspect to it. Then it is the *interpretation* of a code that allows us to see it as a look-alike of a propositional sign. However, introducing such coding in the tractarian system would result in a curious mixture of different philosophies and in unwelcome ambiguity. To be more specific, let us engage in a thought-experiment and suppose that all signs have the form in (n) with the object e configured, say, in a linearly ordered expression of length n.

(n) $\alpha(\underbrace{e, \ldots, e}_{n \text{ times}})$

Furthermore let logical space be finite and an enumeration be fixed via a simply listing. Under these circumstances, could not the linear expression be taken as a code, so that its order of length n corresponds to the n-th sign in the finite list? If so, proposition (n) would mean that proposition n – itself! – has property α.

Much as in the study of self-description in section 3.4.1, it is seen that coding will not do. The meaning attributed to the linearly ordering is of a different kind than as presented in the *Tractatus*, and rooted in a different philosophy. More importantly, in this thought-experiment the elementary proposition, hallmark of univocality, is ambiguous. Besides the meaning obtained via coding, there is another meaning via projection of proposition (n) onto a state of things. I conclude that coding is unavailable in the tractarian system.

10.6   The ineffability of semantics

We have just seen how the tractarian system blocks reflexive propositions. More in general: since the tractarian system is universal, the semantics of language, of any language, cannot be described at all. This insight goes back to Frege (1892a), but it is sharpened in the light of Russell's paradox.

A system which *can* describe its semantics must be able to describe its signs, and to attribute semantic properties or relations to them. In the *Tractatus* this is impossible. In chapter 3, I have argued that fully analyzed elementary propositions cannot describe each other in a sensible way. But in case of logically complex propositions the situation is even worse. Since their logical structure lacks objects, it cannot be described at all. Also, the section on coding has indicated that within the tractarian system a mechanism like quotation is missing, and adding it along the lines of Gödel, would result in an undesirable mixture of philosophies. Finally, even if description of signs were available, it would still be impossible to describe its projection (interpretation). Projection is rooted in the identity of the forms of object configurations, but like logical structure the identity itself has no configuration that could be described.

The logico-semantic insight that there is no meaningful way to describe sense is at the heart of the distinction between saying and showing. What can be said must concern contingent object configuration without detour, and semantics lies outside this domain. Semantics largely concerns the purely formal, which can only be shown not described. In this way Russell's paradox helped Wittgenstein to sharpen his philosophy of semantic ineffability. The theses of the *Tractatus* just indicate the system in a non-descriptive, ostensive fashion, presuming that a few theses are enough for 'someone who has himself already had the thoughts that are expressed in it—or at least similar thoughts' (preface). To make this position more convincing it should shed light on the nature of tractarian nonsense, and present its choreography of gestures in more detail.

10.6.1 Nonsense

The tractarian system leaves no room for meta-linguistic description. All descriptive symbols are immediately about the non-symbolic. These observations, or so I think, give an indication of the kind of nonsense Wittgenstein is alluding to in what resolute readers call 'framing remarks', such as 6.54.

6.54 My propositions elucidate in this way: he who understands me will in the end recognize them as nonsense, after having used them to climb up on and passed them. (He must, so to speak, throw away the ladder after having climbed it.)

He must get over these propositions, then he will see the world aright.

Resolute readers take the framing remarks to support a deflationary philosophy: the tractarian system must be seen as therapeutic make-believe, which helps one realize that all philosophy is mere nonsense. By contrast, if the *Tractatus* is read as ostensive philosophy, the nonsense alluded to in the 'frame' is understood differently. Then, the system of representation is left intact, but due to its non-reflexive nature, its essentials can only be made manifest in a non-descriptive manner. In this way, the *Tractatus* is a swansong of metaphysics (Hacker (1996), 38): it gestures toward a metaphysics of description to reveal the essentials and the limits of representation.

Let us elaborate a bit more. The *Tractatus* comes with signs used for description and for logic, signs used for non-descriptive ostension, and merely nonsensical signs. (Yet other uses of signs, as e.g. in arithmetic, are ignored here.) The 'person who read and understood' the book – addressed in its first paragraph, – will see meaningful language in terms of the perfect notation of signs, in which possible situations in logical space are represented by means of truth-functional combinations of elementary propositions. Nonsense is an attempt to use a sign descriptively while it does not comply with the perfect notation: either it fails to be a truth-functional complex, or it appears to use names that cannot be configured to form an elementary sign. Since the system of perfect notation is non-reflexive and unable to describe the necessities of its forms, the language of the *Tractatus* itself relates to the system in an entirely different way than everyday language does when it tacitly uses the system in its everyday descriptions. In the *Tractatus* most sentences are used, not to describe, but to highlight the main characteristics of representation.

This intimate relationship to the system of representation makes tractarian nonsense of a special kind. Philosophy does not describe, truly or falsely. It is an activity in which signs are shown in a ostension, to give an understanding addressee sharper insight into the workings of language. In this way, it points to living the good life, in which philosophy is elucidating action.

Diamond (2000), p. 150, has the subtle observation that in 6.54 the phrase 'he who understands me' shifts our attention from the text to the reader and the author. Diamond's main focus is on 'me', the author, but on an ostensive reading the 'he' of the reader is just as important, as is the indication of the activity 'to climb up on and passed them'. I take it that 'he' links anaphorically to 'who has already had the thoughts expressed in it himself – or at least similar thoughts' with which the book begins. The pronoun paves the way for going full circle in the coda of philosophical ostension: What we cannot speak about we must pass over in silence (7).

10.6.2   A choreography for a swansong

It is one thing to hold that the *Tractatus* mainly consists of language used in a non-descriptive, ostensive way. But it is quite another thing to argue that there is a textual basis for such tractarian ostensions, or to clarify how they would sharpen a reader's understanding of the system as a whole. For one, doesn't ostension require an unjustified leap from showing a few sentences to showing the system *in toto*?

In the text, the basic gesture uses elementary propositional signs in quotation. Thesis 4.012 has a prime example.

4.012 It is obvious that we perceive a proposition of form 'aRb' as a picture. Here the sign is obviously a simile of what is signified.

In the end the tractarian system leaves no room for quotation. If indeed any suggestion of meta-description should be banned, even the simple:

'aRb' is a sentence

must be recognized as unacceptable *if* used as description. See section 3.4.6. But one would be hard-pressed to deny its non-descriptive, ostensive value for an understanding addressee. Also, showing but a few sentences might well suffice. In the context of descriptive essentialism, one is not after an inductive law based on observing a sufficiently rich sample. What is at stake is philosophical insight into representation, and here it is common for some instances of a totality to show the essence of all.

3.3421 A particular mode of signifying may be unimportant, but it is always important that it is a *possible* mode of signifying. This anyway is how things stand in philosophy: time and again the individual case turns out to be unimportant, but the possibility of each individual case gives us a clarification of the essence of the world.

This principle is at the heart of ostensive philosophy. It allows showing some sentences to give sufficient indication of what elementary propositions are.

Elementary propositions lack a general form generated by elementary operations. Still, examples such as that in 4.012 indicate sufficiently that each elementary sentence should be seen as a specific configuration of names.

With a notion of elementary sign at hand, quotation can be used once more to hint at their sense. Sense is based on a determinate uniformity between elementary signs and what they signify. An example, adapted to the case at hand, can be found in 5.542:

$$\text{'}aRb\text{' says } aRb.$$

This use of quotation stresses in a slightly different way than 4.012 that 'the sign is obviously a simile of what is signified'. The gesture can be taken in two related manners. Firstly, it indicates that quotation does not extend the descriptive capabilities of language: the quoted sentence says essentially the same as the unquoted one.[85] But secondly, the gesture indicates that each name in the quoted sentence is one-to-one related to the objects in the config- uration it describes. In description, it is essential that the sign-configuration is correlated to the state of things it signifies via the correlation of names with objects.

At this point the understanding reader has sharpened his insight into the totality of elementary signs, and into representation as isomorphism of sign and what is signified. Thus, he has reached the next step of the ladder: at least as far as elementary symbols are concerned, he will have insight into logical space itself. Logical space is the totality of all possible configurations of objects, the configuration that are used in elementary description. Given this much, it is also straightforward to indicate what truth and falsity of an elementary proposition is: it is true if what it signifies is indeed realized, else it is false.

The next step is to see how the truth conditions of a logical complex are made manifest in a truth-table *sign*. That elementary propositions lack logical structure indicates that they are logically independent of each other. After all, no combination of names logically precludes or implies any other combination of names. Since elementary propositions mirror the world, such dependencies should also be absent in what is described. Due to the logical independence of elementary propositions, all possible combinations can be listed in a table to indicate the truth-conditions of the full proposition. Since there are no logical objects that the logical structure refers to, the resulting truth-table signs do not alter the states of things complex signs are about. By contrast to the case of elementary signs, each specific instance of a truth-table sign is recognized as of a general form. Again a few signs suffices to exhibit the essentials of all logically complex signs. And what may appear non-truth-functional construc- tions, like quantifiers, can be understood in terms of 'simple' truth-table signs as well.

The sequence of ostensive gestures have now shown what the general form of propositions is. The form seems far removed from everyday language, but for an understanding recipient it is apparent they capture the essence of mean- ingful language. It should suffice to indicate – by means of definitions, sub- stitutions and sign-transformation, – how the general form arises from some typical analyses of everyday language. This will clarify how the descriptive

---

[85]This view is also typical of the middle period. E.g., in 'The Big Typescript' (TS213 I.9) one reads: On the question 'what do you mean' the answer should be: p; and not 'I mean what I mean with 'p' '.

strings are clothed for everyday use, and how they compress elaborate logical structures into manageable units. This final step completes showing the essential features of the system, and with it the limits of language and the world. Then, it is clear that anything of value in life is intrinsic to how we act; reflection, description, or prescription is nothing to do with it.

Insight into the representational system allows us to view the forms of language that are basic to ostensive philosophy as non-descriptive. Prime examples are sentences with formal concepts, like 'α is a sentence', 'n is a name' or 'C is a truth-function'. These sentences are non-descriptive manners of speech that should be understood in terms of tractarian variables; i.e., sequences of meaningful propositions that all exhibit a common form. For instance, 'n is an name' reduces to the sequence $\alpha_0(n), \ldots, \alpha_n(n), \ldots$ of all elementary propositions containing the name n. In this way, n is shown to be a name. Similarly C is shown to be a truth-function by indicating all complex signs $\varphi(C)$ in which C occurs.

Here, the choreography ends.

To summarize, in reaction to Russell's paradox, early Wittgenstein views the essentials of description to be given in a non-reflexive system of representation. The system can present the sense conditions of description, but it can not describe them because it blocks vicious circles. A philosophy that aims to describe the essentials of description is left empty handed. Instead, philosophy consists of ostensive action; it indicates 'what cannot be said by displaying what can be said clearly' (4.115).

The way in which the *Tractatus* presents its system is *un*like heating a hot air balloon until flames shock us out of our philosopher's dreams. Instead, the acts of showing the main features of the system are like climbing the steps of a ladder, which the reader must pass and get over until he sees the world aright (6.54). Once there, it will be apparent that philosophy is nonsense to the extent it purports to be descriptive. Still, only the route is dismissed, not where the route has taken us. Philosophical ostension leaves the system intact and helps the reader to view clearly what the essence and the ethical dimensions is of our limited capabilities to describe.

One may object, like Wittgenstein did in later years, that the system displayed is too truth-functional, too determinate, too limited in scope, too dogmatic, but this is not what is at stake here. The sign-based choreography was to clarify how the ostensive philosophy of the *Tractatus* presents its system of representation; it had no intention to evaluate it. In doing so, it has found a proper balance, or so I think, between Wittgenstein's continuous distrust for meta-description on the one hand, and his livelong fascination for logics, grammars and systems of representation on the other.

11

Toward infinite propositions

With the benefit of hindsight it is not too difficult to judge that many aspects of the tractarian system are problematic. In his *Remarks on Logical Form* (1929), Wittgenstein observes that the system cannot be universal. The now well-known example of color-exclusion indicates there may be elementary propositions that are logically dependent on each other, contrary to what is claimed in the *Tractatus*. Besides there are other examples concerning, e.g., propositional attitudes or generalized quantification, which do not have obvious analogues in the system, if at all. This means the system does not capture the essence of all meaningful language. Such observations initiated a shift toward an entirely new philosophy that had no place for the descriptive essentialism of a universal system.

Rather than universality, the current book investigates whether the truth-functional system presented in the *Tractatus* is coherent. Even in this much more modest field there is an obvious tension. Although the size of logical space should be inessential, the text implies it is infinite. Yet, the system presented appears to be finite and so possibly not rich enough to describe all that may be the case. One must find a midway between crudely dismissing the system, due to its apparent incoherence, and saying too much by suggesting that the near silent philosopher was logically more subtle than textual evidence allows.

The best approach is to develop an infinitary system that preserves as many traits of the finite system as possible, and to note it's logical and philosophical pros and cons from a tractarian point of view. In this way, taking seriously the infinitary aspects of the system will act as a 'stress-test'. Wittgenstein's neglect of the infinite makes the truth-functional system appear coherent, but what happens if such subtleties *are* taken into account? If justice is done to the uniqueness of the system, any infinitary variant will force a distinction between aspects that are compatible with the finite system and those that are not. The latter must be held inessential or, if no reasonable exegetic maneuver is at hand, must go.

Even if infinitary tractarian logic and philosophy cannot be brought in full harmony, it is still more interesting to look for a coherent infinite fragment than just to note the mismatch. Since Wittgenstein has paid little attention to infinitary aspects, most of what is offered on such an approach generalizes main features of the finite system.

My views on the infinitary system are spread over this chapter and the next. The current chapter first recalls the theses on the infinite together with some of their consequences. Next I indicate the main traits of the infinitary system based on a strict analogy between graphical signs and analytical tableaux. To judge whether the resulting system is sufficiently 'tractarian' in spirit should be a matter of debate. The reasons why I think it is – e.g., why in my opinion decidability is inessential, – can be found at the end of this chapter. The technical details of the infinitary system are left for the next chapter.

11.1 Theses on infinity

To get a feel for the 'size' of the system, if it has any, let us begin with revisiting the theses that mention infinity in a systematic way.[86]

Theses 4.2211, 5.43, and 5.535c mention infinity in a hypothetical context, but allow no conclusion as to the system's size.

4.2211 Even if the world is infinitely complex, so that each fact consists of infinitely many states of things and each state of thing is composed of infinitely many objects, even then there would have to be objects and states of things.

Notice that 4.2211 mentions two possible sources for logical space to be infinite: the substantial complexity of states of things and the logical complexity of facts. Indeed, 4.2211 mainly stresses the importance of objects and states of things, which are fundamental regardless.

Thesis 5.43 names infinity to highlight an apparent absurdity concerning logical consequence.

5.43a That from a fact p infinitely many *others* would follow, namely $\neg\neg p$, $\neg\neg\neg\neg p$, etc., is even at first sight hard to believe. And it is no less remarkable that the infinite number of logical (mathematical) propositions would follow from half a dozen basic laws.

We have seen in chapters 5 and 9 that due to the use of truth-operations the absurdity is absent in the system; negation as such is no source for infinity.

Next there is thesis 5.535c, which is crucial in the current setting: it points out that infinity cannot be assumed or implicated, as Russell does when he holds the axiom of infinity to be true. According to Wittgenstein, size should rather show itself in the diversity of names and reference.

5.535c That what the axiom of infinity intends to say, would express itself in language through the existence of infinitely many names with different reference.

Thesis 5.535 makes plain that we should not expect any feature of the system that would force it to have a certain size. In particular, no such conclusion can be drawn from 6.4311, in which infinity is used as a metaphor indicating what the good life is like.

---

[86] Marion summarizes the main findings of Wrigley's unpublished manuscript 'Infinity in the *Tractatus*' (Marion, 1998, 34), who holds that the *Tractatus* allows for infinite totalities. Unfortunately, I have no access to Wrigley's manuscript, but see also footnote 87, p. 177.

6.4311b,c If eternity is not understood as infinite duration but as timelessness, then he who lives in the present lives eternally.

Our life is just as endless as our range of vision is without bound.

I tend to read 6.4311b,c as Wittgenstein's personal, more powerful rendering of some leading thoughts in the *The Gospel in Brief*, Tolstoy (1894), e.g. from its chapters 8 and 12: 'Therefore the true life is independent of time: it is in the present.' 'Therefore, he who lives by love in the present, through the common life of all men, unites with the Father, the source and foundation of life.'.

All in all, there are just three theses that explicitly state logical space to be infinite:

2.0131a Spatial objects must be situated in infinite space. (A spatial point is an argument-place.)

4.463c The tautology leaves reality the entire – infinite – logical space.

5.511 Why does the all-embracing logic that mirrors the world, need such particular difficulties and manipulations? Only because they are all connected in an infinitely refined network, the great mirror.

Even though the theses present logical space as infinite, they give no idea of the kind of infinity at stake, as they should not in the light of 5.535. Yet, given 2.0131, 4.463 and 5.511, it is natural to assume logical space to be infinite.[87]

The above overview has made plain that due to 5.535, on the axiom of infinity , one should not expect the system to have features that force it to have a certain size. It may even be a bit curious to try 'measuring' it. Logical space is the totality of states of things, i.e. configurations of objects, but the space is not an object itself. In particular, it is not a set or a class in the modern sense of the word, and so, in a way, it lacks a specific cardinality. Indeed, prior to analysis we only see the most abstract aspects of states of things, those which enable descriptive language to have sense. But we do not know which specific states of things there are, or which forms they have, let alone their number.

4.128 Logical forms are *without* number.

Hence there are no distinguished numbers in logic, and hence there is no philosophical monism or dualism, etc.

---

[87]In his talk 'Tractarian Heresies' (*Tractatus in Holland* meeting, December 20th, 2010, Leyden), Göran Sundholm presented an argument, learned from Wrigley, why logical space must be infinite. The axiom of reducibility is a contingent proposition (6.1232-33). In a finite logical space, however, all predicates are predicative. So, the axiom of reducibility can only be contingent in an infinite space.

I think it is quite surprising that Wittgenstein presents the axiom of reducibility as a proposition. It seems more in line with his philosophy to take the 'reduction' from impredicative to predicative properties as a feature of representations that must show itself, cannot be stated. In this regard observe that by contrast to the axiom of reducibility, the axiom of infinity *is* a pseudo-proposition (5.535). Also, Wittgenstein did not comment on the third problematic axiom of the *Principia Mathematica*: the axiom of multiplicity (now called: the axiom of choice). It is hard to imagine Wittgenstein considering the axiom of choice a proposition.

Be this as it may, even if one tries to keep the notion of logical space as open as possible, it clearly makes a difference for a detailed interpretation whether the system is viewed as finite or as infinite. Introducing the infinite will have impact on the complexity of about all aspects of the system: propositions, perfect notation, consequence,.... For this reason I have refrained from incorporating infinity from the start. To do so would have introduced a conceptual chasm between the text of the *Tractatus* and my reading of it, which was likely to distract from the more important issues at stake.

11.2   A size of logical space?

For philosophical reasons, Wittgenstein did not introduce logical space as of a specific size. Let us now see how this idea is captured in the four features of the system that are size related; namely, the totality of states of things, logical complexity, the ratios of probability, and Wittgenstein's treatment of arithmetic. It will turn out that only arithmetic suggests something more specific.

States of things In line with Wittgenstein's thoughts on the axiom of infinity, there is clearly no *a priori* foothold as to how many states of things there would be. The size of logical space can only show itself in the number of different names with different meanings, and nothing forces one to assume them to have a particular number. An indication of the size of logical space can only be obtained from analysis but not prior to it.

Logical complexity Modern logic would allow an *a priori* argument for infinity from syntax alone. Then, just one elementary proposition p is enough to generate an infinite sequence:

$$p, \neg'p, \neg'\neg'p, \ldots, \underbrace{\neg'\ldots\neg'}_{\text{n times}}p, \ldots$$

However, in devising his perfect notation Wittgenstein ensured the inference was blocked (5.43). In chapter 5 and 9 we have seen that the above sequence swaps between the two propositions $(\text{TF})(p)$ and $(\text{FT})(p)$. More in general, the logical structure of signs cannot induce infinity. Chapter 6 recalls that logical constants lack substance – they are purely a matter of form, – and so have no effect on the size of logical space.

Probability In 5.15 Wittgenstein introduces probability in a logical way. The degree of probability that φ gives to ψ is identified to be the ratio of truth-grounds for $\varphi \wedge \psi$ to those of φ:

$$\frac{W_{\varphi \wedge \psi}}{W_{\varphi}}$$

For instance, the independent elementary propositions p and q give each other probability $\frac{1}{2}$, because among the four truth-possibilities for p and q there is one truth-ground for $p \wedge q$ and two for p (or for q). Similarly, the tautology gives each proposition degree 1, and the contradiction degree 0. For the current context, it is not so much Wittgenstein's ideas on probability that matter, but rather that the degrees may be infinite in number. However, it is clear in case logical space turns out to be finite, the number of degrees will be finite as well, and again nothing is forcing us to assume infinity of any kind.

Arithmetic Given Wittgenstein's view on the axiom of infinity, he must have been eager to ensure that his treatment of numbers is *in*dependent of the size of logical space. Whatever the specifics of logical space are, the system should come with a suggestion of the countable. To see this, recall that tractarian numbers are not objects but indices of operations (6.021). An operation is a rule to transform signs into a sign, and takes any sign as its argument. In this regard an operation is different from a material function. A material function is abstracted from a sign in which it occurs, and so imposes quite strict requirements on the form (type) of its arguments.

5.251 A function cannot be its own argument, whereas an operation can take one of its results as its own base.

There is clearly no end to applying an operation to the result of its previous application. In the *Tractatus*, the successive application of an operation is core to how arithmetic is understood.

5.2523 The concept of successive application of an operation is equivalent to the concept 'and so on'.

For example, numbers as indices of the operation of negation are obtained as follows:

$$\neg^{0'}p \;=\; p$$
$$\neg^{n+1'}p \;=\; \neg'\neg^{n'}p$$

Observe that although the *result* of successive application remains finite, as before, the possibility of successive application is without bound. A number is the (non-material) abstraction common to all such successive applications using any operation. This abstraction has form $[0,\eta,\ \eta+1]$ (6.03).

What is important for the current discussion is that even if repeated application of an operation leaves no substantial traces, it does hint at infinity as the countably enumerable, and this is the only infinity hinted at.

The above 'transcendental suggestion' brings the countable infinite to the fore. Of course such meager basis gives insufficient ground to take the size of logical space to be of this kind. In keeping with Wittgenstein's view on the axiom of infinity (5.535), there is no requirement for logical space to be infinite at all. At best, the suggestion is to restrict the discussion of the infinite as much as possible to the countable, or this is what I will do.

11.3   An infinitary logic of everyday description?

If a coherent reading of the infinite system can be had one thing is for certain: in it the propositional sign takes center stage, just as it has in the finite system. The *Tractatus* offers a descriptive essentialism – to be is to be essential to description, – and signs with their ontology are core to such philosophy.

Infinitary signs tend to be complex. They may disturb the harmony in the *Tractatus* between the two rôles of logic. At the one extreme there is pure logic, with it's untainted necessity; on the other extreme there is logic as a frame of description, hidden in the rumpled clothing of everyday language use. It is far from obvious whether tractarian philosophy allows for an satisfactory balance between the two extremes. Whatever the balance may be, it requires a notion of infinitary sign that could still be regarded 'tractarian', while at the same time its operations and structure are perspicuous enough for human reflection.

One reason not to give up too early is the semi-formal nature of the system. An infinite system should stay close to the general form of a proposition in thesis 6.

> 6 The general form of a truth-function is $[\bar{p}, \bar{\xi}, N(\bar{\xi})]$.
> This is the general form of a proposition.

In an infinite setting the gist of 6 is that given a basic, countable sequence $\bar{p}$ of elementary propositions, each propositional sign results from a finite iteration of applying the truth-operation N to a propositional variable ξ; i.e., an at most countable sequence of propositions obtained before. The $N(\bar{\xi})$ in thesis 6 indicates an *instruction* to arrive at a propositional sign; strictly speaking it is not a propositional *sign* itself. In a finite setting there is ground to be a bit sloppy about this distinction, for truth-operational completeness ensures that the truth-operational process is equivalent to the truth-functional sign that results (6.1261). As explained in section 5.7.2, the finite system has an equilibrium between process and result that may be absent in a more general context. But perhaps such equilibrium is not necessary.

To determine the coherence of the infinitary tractarian logic, 6 makes clear two aspects have to be be considered in sufficient depth:

- propositional variables (the '$\bar{\xi}$'), and
- truth-functionality (the truth-operation N).

The notion of propositional variable is semi-formal, but we should at least indicate in more detail what it could amount to. Similarly, although 5 requires a proposition to be a truth-function of elementary propositions, the infinite setting may allow for alternative views on the truth-operation N, and we should make up our minds how to go about. As said, to get more clarity in this respect an analogy between graphical signs and analytical tableaux will help. In all this, concerns about size should be respected. It seems reasonable to restrict infinitary methods to those involving 'supertasks': at each stage of a process at most countably many steps can be taken, and the process itself should involve no more than countably many stages. As we shall see in chapter 12, within this perimeters a notion of infinitary sign and symbol is available that preserves crucial features of the finite system.

We start with elucidating the two main features of 6: propositional variables and signs for infinite truth-functionality.

11.3.1 Tractarian variables

The tractarian philosophy aims to indicate the limits of description. Beyond the limit lies what 'is passed over in silence' (7). Philosophers cannot talk about subjects that transcend the limits, but sometimes they may take resort to acts of philosophical ostension that show what the essentials of these subjects are. In these acts, formal concepts and relations are used that are made manifest in terms of propositional variables.

4.127 The propositional variable signifies a formal concept, and its terms signify the objects that fall under the concept.

To be sure, formal concepts and relations cannot be combined into descriptions, but given their roots in the system of representation they are the closest one may get to descriptive assertions.

In the *Tractatus* the word '*Variable*' is used in two different ways. Its most prominent use concerns so-called propositional variables. Also, there is its use to indicate a prototype of a constituent of a proposition (after which a propositional variable results). Both uses are different from the now current notion of variable as a placeholder that is assigned values in the context of interpretation.

Propositional variables are first described in the context of symbols – see 3.31 and its offspring, – but the fullest description is in 5.501.

5.501 A bracket-expression whose terms are propositions I indicate – if the order of the terms within the brackets is immaterial, – by a sign of form $(\overline{\xi})$. 'ξ' is a variable whose values are the terms of the bracket-expression; and the line over the variable indicates that it represents all its values.
(For instance, if ξ has values P, Q, R then $(\overline{\xi}) = (P,Q,R)$.)
The values of a variable are set.
The values are set by description of the propositions that the variable represent.
How the terms of the bracket-expression are described is inessential.
We *may* distinguish three types of description: 1. Direct enumeration. In this case, the variable may simply be replaced by its constant values. 2. Give a function fx, whose values for all values of x are the propositions to be described. 3. Give a formal law according to which the propositions are constructed. In this case the terms of the bracket-expression are all terms of a sequence of forms.

As said, the notion of propositional variable, like the notion of elementary proposition, is semi-formal. In 5.501 Wittgenstein indicates three ways in which a propositional variable may be described, but he does not claim they give all possible ways of description. I think this freedom of analysis is deliberate: to determine all possible ways in which a propositional variable may be given is inessential to the nature of propositions.

Let us have a brief look at the different kinds of variables. From 5.501 it is clear a propositional variable is a sequence of propositions, but not just any sequence is allowed. Broadly speaking, the number of terms in the sequence should either be finite, in which case they can be enumerated, or they should share a common form. This common form, in turn, can be given in two ways: via a function fx specified using a prototype, or via a rule governing a certain operation.

Prototypes Variables as prototypes – i.e. the second use of 'variable', – are clarified in 3.315.

3.315a If we change a constituent part of a proposition into a variable, there is a class of propositions all of which are values of the resulting variable proposition.

In 3.333 what replaces the constituent is called 'prototype'. Prototypes give the second way to describe propositional variables in 5.501; namely by means of function fx. For example, let $n_i R m_j$ indicate a sequence of elementary propositional signs. Then, prototyping the names may result in:

- $n_0 Rx$, which yields the sequence: $n_0 Rn_0, \ldots, n_0 Rn_i, \ldots$
- xRn_0, which yields the sequence: $n_0 Rn_0, \ldots, n_i Rn_0, \ldots$
- xRx, which yields the sequence: $n_0 Rn_0, \ldots, n_i Rn_i, \ldots$.

The specifics of these sequences depend on the available states of things in logical space. (Recall from section 3.2 that I regard elementary propositions as states of things used in projection to picture what they are about.) The general form of propositions – discussed in chapters 5 and 7 – makes clear there are different kinds of constituent that can be 'prototyped': names, sub-propositional expressions, material functions, logically complex functions, propositions, truth-operational complexes,... In this book I consider only prototypes for names.

Prototypes *versus* modern variables Prototypes differ in at least three respects from the now common variables in, say, the syntax of predicate logic. Firstly, the prototypes are no part of fully analyzed propositions. In the end, propositions consist of elementary propositions, which are configurations of names, joined in a truth-functional context. By contrast, the prototypes are part of the pre-final stages of analysis, in which they are used to highlight commonalities in propositional form. Secondly, a prototype x is not assigned a single name n_i. Rather, it is assigned a single name *per* term in the sequence, and the sequence itself has all such assignments possible. For example, the form xRy could yield the sequence: aRb, bRa. But then, in the first term x is assigned to a and y to b, while in the second term x is assigned to b and y to a. Indeed, each term in the sequence corresponds to a possible instance of the prototypes, and the assignment of instances varies per term. Thirdly, prototypes are used injectively. In line with the tractarian idea that identity is best captured as identity of signs – so *not* via a quasi-relation of identity among terms, – different prototypes must be assigned different constituents. For names we have, e.g., 5.531.

5.531 Thus I do not write '$f(a,b).a = b$' but '$f(a,a)$' (or '$f(b,b)$'). And not '$f(a,b).a \neq b$', but '$f(a,b)$'.

Consequently, different variables x and y indicate different names a and b; the form xRy will not have aRa as term. Similarly remarks can be made for other types of constituent.

The repercussions of using prototypes in logic is best considered in the context of quantification. See chapter 13.

Variables from rules The notion of variable as determined by prototypes is quite open, but it is bounded by the possibilities inherent to the general form of propositions. In this respect the third kind of variable in 5.501, which is obtained from a formal law according to which it is generated, seems to be restricted even less. Although each term in such a variable must reduce to a truth-function of elementary propositions (5), the laws that generate such variables could be diverse. From a modern point of view it is natural to interpret 'law' in terms of recursion, but there is little foothold in the *Tractatus* for this formal interpretation. Due to its semi-formal nature, any attempt of a full characterization, if there is one, will be arbitrary.

A simple example of a law-based variable is the one giving the general form of finite propositions. To paraphrase:

6.*fin* The general form of a finite truth-function is:

$$[\bar{p}, (\overline{\varphi_0, \ldots, \varphi_n}), N(\overline{\varphi_0, \ldots, \varphi_n})].$$

This is the general form of a finite proposition.

In words, given a countable sequence $\bar{p}$ of elementary propositions, the following rule generates the finite truth-functions of elementary propositions:

If $\varphi_0, \ldots, \varphi_n$ are propositions, then so is $N(\varphi_0, \ldots, \varphi_n)$.

In this example, the truth-operation $N(\varphi_0, \ldots, \varphi_n)$ will reduce to a finite truth-table sign, as discussed in chapter 5. Since the propositions generated are added to the pool from which finite sequences are formed, all finite propositions will eventually be generated.

In chapter 13 some further examples of propositional variables are discussed. Here I continue with the diagonal argument, which indicates that not all sequences of propositions are variables.

11.3.2 On diagonals and descriptive essentialism

In the tractarian system, sense is prior to truth. Since the sense of a proposition is made manifest in its sign, signs take center stage. What remains to be done is to find a perspicuous kind of sign generated from truth-operations. As said, the variable in 6 is our main foothold:

$$[\bar{p}, \bar{\xi}, N(\bar{\xi})].$$

Since we have sufficient understanding of the elementary propositions $\bar{p}$ and of the variables ξ, it is left to determine what the truth-operation N is as soon as its input is infinite. Here a Cantorian hurdle needs to be taken.

Finite truth-operations are understood as rules that transform truth-table signs into truth-table signs. Cantor's diagonal argument shows that on a naive approach variables for infinitary truth-table signs tend to be too complex. The infinite truth-operation N must be handled with care.

Let us recall why unrestricted infinite truth-table signs are unsuitable to make the sense of a proposition manifest. The first step toward infinite tables consists in introducing a countable sequence of elementary propositions:

$$p_0, \ldots, p_k, \ldots.$$

Similar to the finite case, infinite canonical signs come with truth-possibilities as sequences of T's and F's. But Cantor has proved there are uncountably many such truth-possibilities, which seems to make infinite truth-table signs infeasible.

Proposition 76 *Countably many elementary propositions allow uncountably many truth-possibilities, and so uncountably many truth-table signs.*

Proof. Let a countable sequence $p_0, \ldots, p_k, \ldots$ of elementary propositions be given. Suppose that the corresponding truth-possibilities are countably enumerable. Then this totality can be arranged in a countable matrix, as below, where each τ_j^i is either 'T' or 'F':

$$
\begin{matrix}
\tau_1^1 & \cdots & \tau_m^1 & \cdots \\
\vdots & \ddots & \vdots & \ddots \\
\tau_1^m & \cdots & \tau_m^m & \cdots \\
\vdots & \ddots & \vdots & \ddots
\end{matrix}
$$

Now consider the truth-possibility δ defined by:

$$
\delta_i := \left\{ \begin{matrix} \text{T} & , \ if \ \tau_i^i = \text{F} \\ \text{F} & , \ if \ \tau_i^i = \text{T} \end{matrix} \right.
$$

Clearly, the resulting truth-possibility: $\delta_0, \ldots, \delta_m, \ldots$ does not occur in the matrix, for it differs with each truth-possibility: $\tau_1^i, \ldots, \tau_m^i, \ldots$ at the i-th place. This contradicts the assumption that *all* possible truth-possibilities for $p_0, \ldots, p_k, \ldots$ could be enumerated: there are uncountably many truth-possibilities. Since the number of truth-table signs is even larger, they are uncountable as well. □

The problem the diagonal argument brings forward is not so much one of size. The difficulty is rather that 6 is a variable, but of what kind? Truth-table signs do have a common form, but it is not of the prototypical kind that is shown by abstracting over a given form. And Cantor's argument shows truth-table signs cannot be enumerated by a law either. To give the general form of propositions via a variable, as 6 purports to do, seems infeasible in the infinite case. Cf. also Anscombe (1959), p. 137.

In his later work Wittgenstein objected to the diagonal argument, see for instance Wittgenstein (1967), 54-63. The question is: could he have done so within the framework of the *Tractatus*? Proposition 76 has the form of a reflection on signs. According to the *Tractatus* such reflection on the system of representation, which the *Tractatus* itself to a large extent consists of, is barely admissible. Whether Cantor's argument exceeds what is permitted is hard to judge. But perhaps we do not have to pursue this point further, for the effect of the argument will be limited anyhow. It shows a particular kind of representation not to be in line with 6. From this it does not follow there are no other kinds of representations that are, and we are well advised to look for those.

11.3.3 Sign transformations and tableaux

According to Frege 'W. lays too great value upon signs'.[88] But the rule-based nature of signs and their symbols can be of great value to come to grips with the infinite in a tractarian way. Although worlds do exist independently of language, there are no worlds that do not figure in description. Language and world should be in perfect harmony.

As said, the variable in 6 is our guide in finding appropriate signs and

---

[88] See the motto of chapter 5.

symbols. I will argue in particular that 6 leads rather naturally to the use of so-called analytical tableaux, which strike a 'tractarian' optimum between perspicuity, rich logical structure and descriptive capabilities. Analytical tableaux are much like unraveled graphical signs (6.1203), and Wittgenstein reasoned with graphical signs as we do with tableaux.

As it happens, the finite truth-table signs used earlier are already a kind of tableau. As a first step toward the introduction of infinitary tableaux, I shall begin with making the tableau-like rules explicit that finite truth-operations use implicitly.

In chapter 5, the rule for finite truth-operations at once collects all elementary propositions that are involved in generating the resulting truth-table sign. This approach ensures that the elementary propositions cannot be in logical conflict. But if all steps in their truth-functional combination are made explicit, different occurrences of the same elementary proposition may allow for inconsistent combinations of their truth- and falsity 'poles', and rules must be introduced to handle the inconsistencies. A simple example should suffice to make the point clear.

Consider figure 12. On the left-hand side it has proposition $p \leftrightarrow q$ as a truth-function of itself. On the right-hand side there is the familiar table that results from applying the operation for equivalence.

$p \leftrightarrow q$			p	q
T	T		T	T	T
F	F		F	T	F
			F	F	T
			T	F	F

FIGURE 12

What happens if instead of $p \leftrightarrow q$, $p \leftrightarrow p$ is considered? As figure 13 shows, due to the doubling of p, a straightforward application of the previous transformation brings us outside truth-table signs proper. The reason is of course

p	p			p	
T	T	T		T	F
F	T	F	$\checkmark$	T	T
F	F	T	$\checkmark$		
T	F	F			

FIGURE 13

that p has double occurrence with at the marked rows inconsistent truth-assignments. The remedy is simple. Inconsistent rows cannot be realized, so must be eliminated, and with all inconsistency removed the remaining occurrences of p can be contracted into one. This results in the tautology at the right-hand side of figure 13, which cannot be simplified further. Observe that the rules applied are basically the same as those of tableaux: branches with inconsistent formulas are closed and multiple occurrences of the same proposition at a branch are unified.

The similarity with tableaux is even more prominent when looking at the graphical sign for $p \leftrightarrow p$, which Wittgenstein drew in an early letter to Russell (using 'a' instead of 'T' and 'b' instead of 'F'). See figure 14, from 6.1203.[89]

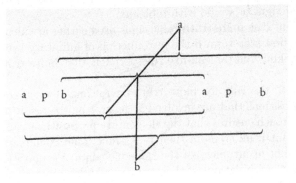

FIGURE 14

The graphical sign can be unraveled into the two trees. To do so, start from an outer pole, and let branching from a pole indicate branching in the tree, and let being connected with a bracket indicate remaining in the same branch. To keep track of which pole of which proposition is considered, it is best to make the poles part of the proposition as one of the signs 'T' or 'F'. For the case at hand, we have for the outer poles either $T(p \leftrightarrow p)$ or $F(p \leftrightarrow p)$. Indeed, rather than considering the poles simultaneously we consider them apart, leaving it understood that it is essential to propositions to have two poles. In this way, the pair of trees in figure 15 results.

FIGURE 15

In the tableau at the left-hand side the identical occurrences could of course have been unified, but I did not do so to highlight the similarity with truth-table signs. Cf. figure 16.

FIGURE 16

---

[89] Or see Wittgenstein (1995), letter 32.

Wittgenstein rightly notices that the method used to reason with the (un-raveled) graphical signs is based on a rule to distinguish logical from contingent propositions.

> [...] ONE *symbolic rule* is sufficient to recognize each of them [propositions, JDO] as true or false. And this is the *one* symbolic rule: write the prop[osition] down in the ab-notation, trace all connections (of poles) from the outside to the inside poles: Then if the b-pole is connected to such *groups of inside poles* ONLY *as contain opposite poles of* ONE *prop[osition]*, then the whole prop[osition] is a true, logical prop[osition]. If on the other hand this is the case with the a-pole the prop[osition] is false and logical. If finally neither is the case the prop[osition] may be true or false but is in no case logical. (Wittgenstein, 1995, letter 30, November 1913)

The rule is that of tableaux. In 6.1203 the same rule surfaces, be it in highly condensed form. In the now current notation and terminology the rule becomes:

- To show that a proposition φ is valid, show that all branches of the systematic tree for $F\varphi$ close;
- To show that a proposition φ is inconsistent, show that all branches of the systematic tree for $T\varphi$ close;
- If neither the systematic tree for $F\varphi$ nor the systematic tree for $T\varphi$ closes, φ is contingent.

Given the strict analogy between reasoning with graphical signs and with tableaux, it is natural to exploit the rules for the truth-operation N in terms of infinite tableaux. Wittgenstein's rule can then be seen to hold universally. The restriction in 6.1203 to non-quantificational propositions is unnecessary, and seems due to the use of graphical signs that, though an interesting step in the right direction, are too unwieldy to represent infinite propositions (or complex finite ones, for that matter).

11.3.4 Rules for truth-operation N

A tableau is a tree of propositional signs that is generated according to rules based on the forms of the signs. To formulate the transformation rule for the infinitary truth-operation N, we must assume an appropriate notion of variable to be given:

$$\xi \equiv \varphi_0, \ldots, \varphi_n, \ldots$$

Since variables are semi-formal, in general we cannot be more precise than this. The transformation rule for N is now in figure 17.

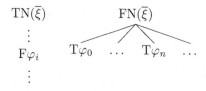

FIGURE 17

In words, each open branch in a tableau at which a node $TN(\bar{\xi})$ occurs can be extended with any node $F\varphi_i$, φ_i in ξ. And each open branch in a tableau at which a node $FN(\bar{\xi})$ occurs may branch into the paths with nodes: $T\varphi_0, \ldots, T\varphi_n, \ldots$; all φ_i in ξ.

It is immediate that the infinite notion of transformation is different from the finite one presented in chapter 5. The finite truth-operations transform so-called systematic tableaux – i.e., tableaux that contain all relevant truth-possibilities, – into a new systematic tableau.[90] By contrast, the truth-operations presented here, which would work for finite and infinite signs, transform tableaux in stages rather than at one fell swoop. They do not take full tableaux as input, but construct them based on the form of the propositional signs at its nodes, and they are finished only if no further transformation is possible.

Let us check what may be gained by using infinite tableaux instead of truth-table signs. Compared to full truth-table signs, infinite tableaux are much like truth-tables signs reduced to a minimum. In particular, the truth- and the falsity-tree for finite uses of N correspond to a simplified truth-table sign with $n + 1$ instead of 2^n rows.

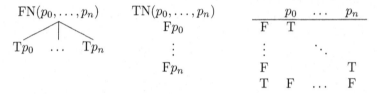

FIGURE 18

The economy of notation is particularly helpful for infinite variants. Depending on which 'pole' is considered, a systematic application of the truth-operation N gives one of two countable trees at the left-hand side of figure 19, so that the truth-table analogue in figure 20 remains countable.

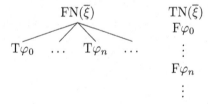

FIGURE 19

That simplified truth-table signs are sufficient is due to the fact that a row *constrains* truth-possibilities to comply with it, it does not have to define a full world. E.g., in the finite case it suffices to indicate that $N(p_0, \ldots, p_n)$ is false if some p_i is true, and that $N(p_0, \ldots, p_n)$ is true if all p_i are true $(0 \leq i \leq n)$. One does not have to enlist all truth-possibilities for which this is so. Similarly for the infinite case.

These simple examples already make clear that Cantor's diagonal argument should be applied with caution. Although there will be uncountably many

---

[90]See Smullyan (1968) §V.3 for a discussion of systematic tableaux.

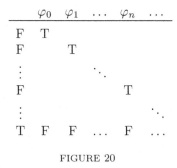

FIGURE 20

truth-possibilities, this is not to say that the sign has to represent them all explicitly. The sign only needs to make those aspects of truth-conditions manifest that *constrain* the worlds a proposition describes. Uncountable aspects of a proposition's truth-conditions should only be represented when forced too. Section 13.7 has an example that requires uncountability. But even then the uncountably many paths are represented in a countable sign generated in countably many steps.

11.3.5 Main features of an infinitary system

In chapter 12, I shall continue to develop the technical details of an infinitary system. This system preserves the following features of the finite system:

1. The system is sign-based.
2. Each propositional sign has two opposing poles.
3. Only signs show logical structure overtly.
4. Only names refer. Logical constants do not refer. Names refer in the context of elementary propositions.
5. Logical operations are rule-based and their results are obtained systematically.
6. The sign of a proposition is derived logically from its composing parts, and shows its sense, if it has any.
7. Logical propositions are characterized from among all propositions.
8. The characterization is based on a non-axiomatic technique in which the rules only concern signs, not their sense.
9. The system offers a perfect notation of sense (symbols). The perfection is of first degree: all logical parts are eliminated (cf. section 9.1).
10. Signs are rich enough to represent each possible world, but not every possible world can be described (defined uniquely).

As has been noted before, each reconstruction of the infinitary system is speculative. The characteristic uniqueness of the system can be preserved to the extent that it is based on a single totality of elementary propositions in logical space, but the notions of propositional variable and truth-operation allow for speculative variation. Also, it is impossible to retain all features of the finite system. Candidate features that may have to to go are: the decidability of logic and symbols; the extreme perspicuity of symbols; the ability to uniquely describe any world; the balance between logical and human abilities. Yet, an infinitary system may deepen our understanding of early Wittgenstein's philosophy, since it forces one to choose among features that should be essential, and those that are more like side-effects of the finite.

At this point, readers who are interested in the technical details of the infinitary system may wish to proceed to the next chapter. Prior to a more technical discussion, we now reflect on its metaphysical and human aspects.

11.4 Metaphysical and human aspects of logic

An infinite logical space suggests such undesirable consequences that an infinitary tractarian system has been deemed incoherent or even impossible. Anscombe (1959), 134-7, combines a discussion of infinitary truth-table signs with Church's result on the undecidability of predicate logic to conclude that the theory of the *Tractatus* has been 'clearly and cogently refuted'. More recently, Potter (2009), section 24.4, and Sundholm (2009), section 6, draw similar conclusions. These judgments, however, are based on broad observations that may loose force if more detail is considered.

In this section, I'll reflect on the tension between the independence of logic and its human applicability. To this end I will consider the (in)decidability of the system, the characterization of logical propositions, the independence of logic, and the perspicuity of logical structure. I hold that the metaphysical subject rather than the empirical subject is central to the meaningful use of language. In keeping with the idea that the metaphysical subject and the essentials of logic are closely related, it should suffice to have a notion of signs and operations that is perspicuous enough to show the logical nature of propositions in a way that is quite independent from human affairs. Features of logic that concern human feasibility, such as decidability, are desirable at best, but not essential.

11.4.1 Metaphysical and empirical subject

In the *Tractatus*, about all problematic aspects of the infinitary system are related to the troublesome relationship between the human empirical subject, which is a totality of thoughts rather than an object that can be described or referred to (5.631), and the metaphysical subject, which is not in the world but a limit thereof (5.632). Logic, sense, arithmetic, ethics, the mystical, and perhaps even some forms of scientific laws, are all shown without detour in the metaphysical subject. It keeps their essentials necessary and untainted by the empirical. But how does this strong philosophy of logic relate to common everyday activities? The metaphysical subject is constitutive of meaning, logic, the good life. Still logic often seems too complex for humans to be grasped immediately; it is prior anyway to human empirical and epistemological abilities.

Wittgenstein's phrasing in theses on the metaphysical subject and human activity is sometimes misleading. That *we* picture facts to ourselves (2.1) suggests empirical activity. But prior to analysis it is often opaque how we succeed (4.002), and once made manifest sense is not to be found in experience but rather in its limit. Again, the logical forms of propositions show the metaphysical subject whether one is a logical consequence of the others (5.13). If so, *I* can infer (*schließen, folgern*) one from the others (5.132). Yet in the end logical consequence is an internal relationship between propositions that holds independent of whether we are able to recognize it.

11.4.2 Decidability

Theses 6.126 and 6.1262 – on our ability to recognize logical propositions among the contingent ones, – are often quoted to argue that the tractarian system is incoherent and untenable. Cf. Anscombe (1959), 134-7, or Potter (2009), section 24.4. They seem to require decidability, which is without doubt problematic in an infinite setting.

6.126 Whether a proposition belongs to logic can be calculated by calculating the logical properties of the *symbol*.

And this is what we do when we 'prove' a logical proposition. For without troubling ourselves about sense and meaning, we form the logical proposition by mere *rules that deal with signs*

The proof of a logical proposition consists in letting it emerge from other logical propositions through the application of certain operations, which from given tautologies produce new ones over and over again. (And surely from tautologies only tautologies *follow*.)

Of course this way of showing that the propositions of logic are tautologies, is not at all essential to logic, if only because the propositions from which the proof starts must show without proof that they are tautologies.

6.1262 Proof in logic is just a mechanical expedient to simplify the recognition of a tautology, where it is complex.

The question is: to what extent do the theses *force* the system to be decidable? An unconstrained infinitary system is impossible if decidability is *essential* to it. I think, however, that to require decidability gives the theses too much weight.

It is indeed natural to read 'to calculate' (*berechnen*, 6.126a,b) and the 'mechanical expedient' (*mechanisches Hilfsmittel*, 6.1262) in the constructive tradition that had Wittgenstein's sympathy in his later work.[91] But natural language is often a subtle arbiter. In the famous BBC documentary, the mathematician Wiles calls his proof of Fermat's last theorem a 'calculation'. Also, the first sentence of Gödel (1931) uses '*mechanisch*' in a sense that does not imply decidability. We should be cautious anyway to read the theses as about the current notion of decidability. Decidability in this sense requires a recursive algorithm to check in finitely many steps whether a proposition is either a tautology, a contradiction, or a contingency. By contrast, Wittgenstein's use of '*mechanisch*' is rather to focus on the formal aspects of the signs and symbols involved.

Löwenheim had proved monadic predicate logic to be decidable in 1917, but it was only in 1928 that Hilbert put the modern notion of decidability on the logical agenda, and in doing so inspired a sequence of (un)decidability

---

[91] Göran Sundholm stressed this point in his talk 'Tractarian Heresies' (*Tractatus in Holland* meeting, December 20th, Leyden).

results concerning fragments of predicate logic. Wittgenstein had noted the decidability of finite propositional logic as early as 1913, but decidability does not seem to be part of his problem space. In the *Tractatus* his observation surfaces in 6.1203, when introducing graphical signs.

6.1203a To recognize a tautology as such one may use, in case the tautology has no indications of generality, the following graphical method [*here an explanation of the method follows*, JDO]

For the current discussion it is important to observe that Wittgenstein indicates decidability not to hold in general. He restricts the method to the fragment of propositions that lack quantifiers and other indications of generality (*Allgemeinheitsbezeichnungen*). Thus it remains open whether the most general method to check logicality, which is based on truth-operations, should be included in 6.126a,b and 6.1262 and so be decidable as well.

If decidability were essential to the *Tractatus*, it would be grossly incoherent, for clearly the most natural generalization into the infinite is undecidable. There is reason to believe, however, that early Wittgenstein was not too keen on decidability. Both states of things and logical space are allowed to be infinite (4.2211). This means it may be impossible to decide whether or not a configuration of names is an elementary proposition, or whether a complex of operations and elementary propositions is a proposition. If such basic notions are not required to be decidable, why should logicality be? In line with Wittgenstein's restriction on the applicability of his decision-method, it is as natural to give the phrases in 6.126a,b and 6.1262 a much weaker reading. The method to check logicality is mechanical in that it concerns the form of a proposition disregarding its semantic content. Calculations that are mechanical in this formal sense do not have to decide recursively whether a proposition is logical or not.

Even if Wittgenstein did intend recursive decidability – but who is to decide?, – it is still more interesting to gauge what insights of the *Tractatus* can be retained rather than to disregard the system offhand. The weaker reading of 6.126a,b and 6.1262, in particular, gives the valuable insight that logical operations should disregard sense. Then, 6.126 and its offspring is erroneous at best, due to Wittgenstein's gross underestimation of infinite complexity, but they are not lethal to the system and its philosophy, for which decidability is inessential.

11.4.3 Characterizing logical propositions

Rather than decidability, I think the characterization of logical propositions is essential to early Wittgenstein's system. About November 1913, he wrote to Russell: 'All propositions of logic are generalizations of tautologies, and all generalizations of tautologies are proposition of logic. There are no other logical propositions.' (Wittgenstein (1995), letter 32). Without inclination to proof it, Wittgenstein seems to have sensed the soundness and the completeness of finite propositional logic. In the *Tractatus* the observation is condensed into 6.1.

6.1 The propositions of logic are tautologies.

Wittgenstein adds that the characterization of logical propositions should be in the context of all propositions.

6.112 The proper explanation of logical propositions must give them a unique position among all propositions.

For this it is enough to have forms and formal rules that show logical propositions to be empty structures with all descriptive content dissolved.

6.12 That the proposition of logic are tautologies *shows* the formal – logical – properties of language, of the world.

That *in this combination* their constituents yield a tautology characterizes the logic of their constituents.

If propositions are to give a tautology when connected in a certain way, they must have specific structural properties. That *in this connection* they yield a tautology shows therefore that they have these structural properties.

Such a logic does not have to be decidable; soundness and completeness would suffice. Thesis 6.12 states that logical propositions show the formal properties of the world, which may well remain immanent, prior to human error or proof. 'In a certain sense, we cannot make mistakes in logic.' (5.473) This indicates, I think, that the question of how logical complexity is made manifest to humans did not have early Wittgenstein's prime interest.

11.4.4 The independence of logic

Decidability has the ring of human feasibility about it. But in spite of some turns of phrase that suggest otherwise, in the tractarian philosophy logical necessity and contingency take center stage. Logic is non-empirical throughout.

5.473 Logic must take care of itself

6.1222b Not only must a proposition of logic not be capable of being refuted by any possible experience, but it must also not be capable of being confirmed by any possible experience.

Logic makes itself manifest in the metaphysical subject, the limit of experience, where showing is akin to a transcendental '*Wesenschau*' concerning sign-forms.

5.552 The 'experience' that we need to understand logic is not that something is so and so, but that something *is*; that, however, is rather *no* experience.

Logic is *prior* to any experience – that something is *so*.

It is prior to 'How?', not prior to 'What?'.

In the end, all aspects of logic that suggest empirical survey, like decidability, are inessential.

11.4.5 Perspicuity

Taken to its extreme, the independence of logic leads to a position that is even more radical than requiring decidability. The general form of logical propositions must be so perspicuous and uniform it is recognized immediately, leaving no room for surprise. Thesis 6.126c,d – quoted on page 191, – points in this direction. Good notation *shows* logicality, and thus it must be traced where the limits of showing lie.

From a non-transcendental, empirical point of view even finite truth-table signs may not be perspicuous enough. Small tables can still be regarded perspicuous, but the complexity of truth-table signs increases rapidly. How to check if a table with 2^k truth-possibilities is a tautology if $k = 10^{10^{10}}$? Once on this path, it is clear that the notion of logical inference that 6.126c sketches is hardly of help for humans. There is no principled distinction between 'small' tautologies that show their true nature without proof and 'large' tautologies' that must be inferred from smaller ones by means of truth-preserving operations. (Presumably, not all 'large' tautologies can be obtained in this way.)

Thesis 6.126d suggests that all showing is immediate. In the tractarian philosophy, this ideal can only concern the metaphysical subject; the limit of experience that encompasses logic as a frame of all that is possible, without proof. The hint that tautologies show their true nature 'without proof' eliminates all proof from logic. This radical philosophy has little to do with traditional logic and formal reasoning. The untainted necessity that is safeguarded in this way comes with the high price of leaving in the dark how logic relates to us, humans, for whom its mechanics are crucial. I think the *Tractatus* does not force one to adopt such a radical position. As in philosophy, we may engage in non-descriptive reflection on the system of representation. This reflection may help us, say, to the non-verbal insight there are patterns in the transformations of propositional signs that characterize logical propositions among all other propositions. In the final analysis such logical reflection, like philosophy, lacks the sense of contingent description. But it does leave room for infinite propositions, as long as their signs and transformations are perspicuous enough.

11.4.6   The finite as philosophical strategy

The above discussion makes one wonder whether Wittgenstein had not noticed at all that when generalized into the infinite some basic properties of the finite system can no longer be assumed? The serious problems that result from not noting, are often taken to support the view that other than Hertz, Frege or Russell, Wittgenstein was primarily a philosopher and writer, not a formal logician.

It is well-known that despite his keen logical insights, Wittgenstein had little taste for technicalities. In a letter of 1913 to Russell in which he explained his graphical notation, he wrote: 'I beg you to think about these matters for yourself: it is INTOLERABLE for me, to repeat a written explanation which even the first time I gave only with the *utmost repugnance*.' (Wittgenstein (1995), letter 32.) And in one of the first letters to his supervisor a brief sketch of a new technical idea is followed by: 'The rest I leave to your imagination.' (August 16th, 1912) This book resulted from finding myself in the same situation.

Early sources indicate that Wittgenstein did try to find a perspicuous infinite extension of finite signs, but apparently he did not succeed. A trace of his attempts can be found in the exclusion of generalities from the graphical signs introduced in 6.1203. It is natural to assume that after Wittgenstein had tried a few infinite variants without achieving the elegance of the finite case, he left it at that and continued concentrating on philosophy. After all, with the notions of truth-operation and propositional variable at hand, to arrive at a satisfactory infinitary representation of sense only seemed to require patience and an extended focus on detail; traits that Wittgenstein lacked when it came to formalization.

In later years Wittgenstein suggested his restriction to the finite was a matter of philosophical strategy, not of failure. He seems to have remarked to Georg Kreisel 'that he had put down his system in a finite setting without bothering about the infinite case, assuming that if a problem was to be found in the infinite case, then there would have already been a problem in the finite case' (Marion, 1998, 34). The acuteness with which Wittgenstein criticizes his early work in *Philosopische Untersuchungen* testifies that the paradigm shift that had taken place in the meantime had hardly effected his insight into where he had come from. So, Kreisel's testimony may well be correct. This would mean that the lack of detail on the infinite is due to logical temperament, not to philosophical insight, and that the parts of the *Tractatus* which are invariant under generalization should be valued most.

12

Infinitary logic and symbols

Although the *Tractatus* allows a speculative reconstruction of its infinitary
system at best, it is important to attempt one, if only to highlight the characteristics of the philosophy that are invariant under generalization and that
may therefore be considered essential. This chapter is about the technical
details of such an infinitary logic.

In the infinitary system signs and symbols retain center stage, but tableaux
take over the rôle that truth-table signs had in the finite system. Likewise, the
truth-operation N is viewed as transforming tableaux rather than truth-table
signs, but now in a less total, more piecemeal manner. Given this basic set-up,
it is inquired whether the resulting tableau can still be used to determine a
perfect notation to represent sense. Since it must fix a world to either 'yes' or
'no' (4.023), sense is shown to be determinate prior to the truth or falsity of a
proposition. Next, in terms of sense, truth and falsity are defined. The insights
obtained are used to shed light on the characterization of logical propositions,
on logical consequence, on descriptive completeness, on compactness, and on
interpolation.

Although there are no hard and vast principles in this regard, the infinitary methods are restricted to those involving 'supertasks': at each stage of
a process at most countably many steps can be taken, and the process itself
should involve no more than countably many stages. Within this perimeters,
a notion of infinitary sign is available that allows for a perfect notation of
first degree (cf. section 9.1). It eliminates logical parts to bring the sense of
contingent propositions to the fore, and characterizes the logical propositions
from among the totality of all propositions.

All in all, the system proposed preserves the features of the finite system
listed in section 11.4, page 189. Since signs and symbols have center stage, we
begin with investigating how the notion of a systematic tableau can be used
to extend signs and symbols into the infinite.

12.1 Truth-operation N

This section continues the discussion of tableaux that was started in section 11.3.4. To formulate the transformation rule for the infinitary truth-
operation N, we must assume an appropriate notion of variable to be given:

$$\xi \equiv \varphi_0, \ldots, \varphi_n, \ldots$$

Since variables are semi-formal, we cannot be much more precise than this, but it is assumed that the sequences are invariant under ordering and under repetition. The invariance under ordering is from 5.501. The invariance under repetition seems to be implicitly assumed in the *Tractatus*; repetitions are inessential to sense and thus hamper the perfection of the notation. The rules for the truth-operation N are now in figure 21.

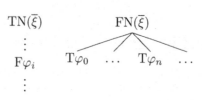

FIGURE 21

In words the rule TN amounts to: each open branch in a tableau at which a node $TN(\bar{\xi})$ occurs can be extended with any node $F\varphi_i$, φ_i in ξ. And rule FN reads: each open branch in a tableau at which a node $FN(\bar{\xi})$ occurs may branch into the paths with nodes: $T\varphi_0, \ldots, T\varphi_n, \ldots$; all φ_i in ξ. When formulated in this way, the rules TN and FN are indeterministic. They leave choice (i) which proposition to handle and (ii) in case of TN which formulas to use in the expansion of a branch.

Truth-operations generate tableaux according to the rules for signed formulas $T\varphi$ or $F\psi$, but by contrast to graphical signs they consider one pole at a time. Basically, each proposition φ will generate two trees: a tree with $T\varphi$ at its root, and a tree with $F\varphi$ at its root. In line with this, the rules for N have one case for each pole.

Applying the rules for infinitary N makes part of the truth-conditions of a proposition manifest. At top-level the process knows two phases:

- A phase in which the full tableau is unfolded using truth-operation N. This phase also sieves the consistent branches in a tableau from the inconsistent ones.

- A phase in which all information that was needed to arrive at a tableau is cleaned up to leave a proposition's symbol.

The truth-operation N generates a tableau by extending its branches. By definition, a branch is a linear order of children and their immediate parents until the proposition at the root of the tableau is reached. Consider, for instance, the simple tableau in figure 22.
The tableau has root $TNN(p, N(q, N(q)))$ – i.e. $p \vee (q \wedge \neg q)$ – and it has two branches; the first open, the second closed:

1. $TNN(p, N(q, N(q))) > FN(p, N(q, N(q))) > Tp$;

2. $TNN(p, N(q, N(q))) > FN(p, N(q, N(q))) > TN(q, N(q)) > Fq$
   $FN(q) > Tq > \times$.

In general, a branch is either open or closed. A branch is called *open,* iff for no proposition χ both $S\chi$ and its conjugate $S°\chi$ occur at it.

$$\text{TNN}(p, \text{N}(q, \text{N}(q)))$$
$$\text{FN}(p, \text{N}(q, \text{N}(q)))$$

$$\text{T}p \quad \text{TN}(q, \text{N}(q))$$
$$\text{F}q$$
$$\text{FN}(q)$$
$$\text{T}q$$
$$\times$$

FIGURE 22

Definition 77 *The conjugate $S°\chi$ of a signed proposition $S\chi$ is defined to be:*

$$S°\chi = \begin{cases} \text{T}\chi & \text{if S is 'F';} \\ \text{F}\chi & \text{if S is 'T'.} \end{cases}$$

A branch is *closed,* iff it is not open. In the process of generating a tableau, the propositions on an open branch may turn out to be consistent. The propositions on a closed branch are shown to be inconsistent, and receive no further attention.

12.2 Systematic signs

To ensure that the truth-conditions of a proposition are presented in full, its tableau must be generated in a systematic way. All (signs of) all subpropositions of a proposition must be handled. In particular, on each open branch at which a propositional sign $\text{TN}(\overline{\xi})$ occurs, all subpropositional signs $\text{F}\varphi_i$ with φ_i in ξ, must occur as well.

To generate a tableau systematically, we use an enumeration of the subpropositional signs of a propositional sign.[92] From now on I will often use '(sub)proposition' rather than '(sub)propositional sign', leaving it understood that tableaux concern signs only, not their sense-content. We first define an ordered tree of signed subpropositions, in which conjugates play an important rôle. We think of this tree as ordered, using the order of the sequences ξ. (A bracket-expression $(\overline{\xi})$ indicates that the order of ξ is immaterial, but for now I assume an order fixed.)

Definition 78 *The ordered tree $Tr[S\chi]$ of subpropositions of a signed proposition $S\chi$ is defined recursively. If χ is elementary proposition p:*

$$Tr[S\chi_s] = Sp_s.$$

If χ is the complex $N(\overline{\xi})$, $Tr[S\chi_s]$ is:

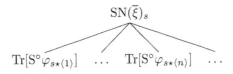

$$SN(\overline{\xi})_s$$
$$Tr[S°\varphi_{s\star\langle 1\rangle}] \quad \cdots \quad Tr[S°\varphi_{s\star\langle n\rangle}] \quad \cdots$$

Finite sequences s of positive natural numbers are used to indicate the order among the children of a parent. In particular, node $s \star \langle n\rangle$, with $\star$ for

---

[92]The systematic procedure is a special case of that in Linden (1968), which is based on that in Smullyan (1968).

concatenation, is the n-th child of parent s. The root has the empty $\langle\rangle$. □

A systematic tableau is defined in terms of an enumeration of the ordered tree of subpropositions. In the abstract the tree of sub-propositions looks as follows:

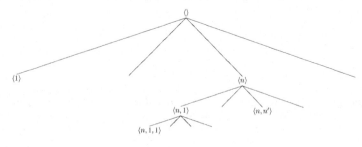

FIGURE 23

The idea is to obtain the enumeration from a linear ordering of countably many finite linear orders; namely for each n, a finite linear order of the sequences that sum up to n.

Definition 79 *The sum $\Sigma(s)$ of a sequence of numbers is defined by:*

$$\Sigma(s) = \begin{cases} 0 & \text{if } s \equiv \langle\rangle; \\ \Sigma(s') + n & \text{if } s \equiv s' \star \langle n\rangle. \end{cases}$$ □

Notice that if we had allowed sequences of natural numbers, the sequences s with $\Sigma(s) = 0$ would have been the only ones infinite in number.

Sequences s of the same sum are first ordered in terms of their length $lh(s)$, and next are ordered recursively by $<^*$:

$$\begin{cases} \langle\rangle \not<^* \langle\rangle & \text{if } s = \langle\rangle = s' \\ n < m \text{ or } (n = m \text{ and } s'' <^* s''') & \text{if } s = \langle n\rangle \star s'', s' = \langle m\rangle \star s''' \end{cases}$$

The enumeration of indices we are after is now defined by $s <^\circ s'$ iff:

$$\begin{aligned} & \Sigma(s) < \Sigma(s') \\ or \quad & \Sigma(s) = \Sigma(s') \text{ and } lh(s) < lh(s') \\ or \quad & \Sigma(s) = \Sigma(s') \text{ and } lh(s) = lh(s') \text{ and } s <^* s'. \end{aligned}$$

The subpropositions of a proposition are ordered via their indices. To generate a systematic tableau for $SN(\bar{\xi})$ we may therefore assume an enumeration of its signed subpropositions given:

$$SN(\bar{\xi})_{\langle\rangle}, S_1\varphi_{\langle 1\rangle}, \ldots, S_n\varphi_{\langle n\rangle}, \ldots$$

Since the sum of the index of a parent is always smaller than that of a child, parents occur before their children.

In terms of the enumeration the following systematic procedure can be defined.

Stage 0 Place the labeled proposition $S\varphi_{\langle\rangle}$ at the root of the tableau.

Stage $n \Rightarrow n+1$ Begin with checking whether the ordered tableau is closed. If so, stop. Else observe there are at most countable many branches, each finite (but possibly without finite upperbound). For each branch in the given order do the following:

- If the branch is marked closed, leave it as is.
- If the branch is not marked closed, check if it contains a proposition $S\varphi$ and its conjugate $S^\circ\varphi$. If so, mark it closed with '$\times$'.
- Consider the n-th subproposition $S\varphi_{s\star\langle m\rangle}$ in the enumeration. If its parent is of form $TN(\overline{\xi})_s$ and occurs at the branch, extend the branch with the n-th subproposition. Next, if the n-th subproposition is of form $FN(\overline{\xi'})_{s'}$, extend the branch with branches:

$$T\varphi_{s'\star\langle 1\rangle}, \ldots, T\varphi_{s'\star\langle k\rangle}, \ldots \ .$$

- Proceed to the next branch, if there is any, else proceed to the next stage.

The process may stop at a finite stage l to leave a well-founded tree (all paths finite but perhaps countably branching). In case the process continues at each stage n, all subpropositions in the enumeration will have been processed in its countable limit. Then, the systematic tree is finished. [*End of procedure*]

Systematic tableaux suffice to characterize the logical propositions among all propositions. But in a tractarian setting systematic tableaux should serve the more complex purpose of getting at a perfect notation of sense, capturing the contingent nucleus of a non-logical proposition. Contingent nuclei concern the truth and the falsity conditions of a proposition, and so require an extended notion of tableau that represents both. I will show how systematic tableaux can be adapted to serve this purpose.

12.3 Infinitary perfect notation

The *Tractatus* sets out to show that what can be said at all can be pictured with perfect logical clarity; in this way the limits of our descriptive capabilities are made apparent.

In chapter 9, a finite perfect notation was developed that satisfied the three degrees of perfection discerned in section 9.1. Here I start investigating to what degree an infinitary perfect notation can be had. A perfect notation of first degree is presented, which partially realizes the second degree but not the third. As far as I can see, the second and third degree of perfection require heavier means, which should not be interesting when interpreting the *Tractatus*. Instead, I will argue that in the context of a proper criterion of symbol identity, a perfect notation of first degree suffices.

Even less so than before, the suggestion that symbols simplify or even trivialize showing sense cannot be upheld. In section 11.4, I've argued that this idea did not have Wittgenstein's prime interest: sense is rooted in a logic that should take care of itself (5.473). Also, in spite of an increase in complexity, it should still be more interesting to see that the infinitary system allows for a perspicuous representation of sense than to dismiss it 'off the cuff'.

12.3.1 Extended tableaux

It is more complex to determine the symbol of a proposition than to determine its logicality. In case of logicality it suffices to consider one pole, but to determine its contingent nucleus both poles are needed. A proposition φ is logical iff the tableau for $F\varphi$ or the tableau for $T\varphi$ closes. But φ is contingent iff neither the tableau for $F\varphi$ nor for $T\varphi$ closes. Logical propositions result from combining contingent propositions in a certain way (6.124). Therefore, to determine the symbol of a proposition a stage is needed to eliminate the logical parts in either the T- or in the F-pole of a proposition.

Finite symbols are calculated from truth-table signs. Even though complexity increases, there is no essential difference when moving to possibly infinite signs. To determine the symbol of a proposition its *extended* tableau is used; i.e., the systematic tableau that involves both its poles. (This reminds one of Beth tableaux but the T- and F-parts are kept strictly separate.) A symbol is determined in a single process of alternating steps in the computation of either pole. The proposition is logical if one of its poles closes. Then the symbol has no elementary content; it is about nothing. A symbol has content only if the proposition is contingent, that is: if neither pole closes.

12.3.2 Tracing occurrences

To arrive at the symbol the idea is to determine whether a *sub*proposition occurs logical or not. To this end two things are needed:

1. A notation for occurrences of subpropositions,

2. A method to keep track of (non)logical occurrences.

We already have a notation for the occurrence of a subproposition, namely the sequences of numbers used to enumerate them. E.g., in the propositions $(p \wedge q) \vee (p \wedge \neg q)$ the first occurrence of p has code $\langle 11 \rangle$ and its second occurrence has code $\langle 21 \rangle$. To keep track of whether an occurrence of a subproposition is logical or not, we must define what it means for an occurrence to be logical. Since closed branches determine logicality, this is simple.

Definition 80 *An occurrence of a proposition φ_s in $\chi_{\langle\rangle}$ is logical, iff φ_s appears on a branch in the extended tableau of $\chi_{\langle\rangle}$ with its conjugate.* □

An occurrence of a (sub)proposition should be eliminated from a symbol as soon as its logicality is shown. To trace whether an occurrence of a proposition is logical, the function Δ is used:

$$\Delta[\varphi_s] = \begin{cases} () & \text{if } \varphi_s \text{ appears logical} \\ \varphi_s & \text{else} \end{cases}$$

The values of Δ are determined in the course of generating an extended systematic tableau. Each logical occurrence of a subproposition is found at one of the stages of the systematic tableau. This means the check for logicality stays within the bounds of supertasks.

The extended tableau $ST(N(\bar{\xi}))$ is obtained from the systematic tableau for $T\Delta[N(\bar{\xi})]$ and $F\Delta[N(\bar{\xi})]$ using the obvious analogues of the rules for N in figure 24. The notions of conjugate and closed branch are adapted accordingly. In particular, a branch with both T() and F() is closed.

Values of Δ are set as soon as possible, but at the latest when the extended tableau is finished. They can be set whenever a branch closes, for if a proposition occurs logical so do all its subpropositions. See figure 25.

The case where $T\Delta[N(\bar{\xi})]$ occurs in the branches below $F\Delta[N(\bar{\xi})]$ is similar.

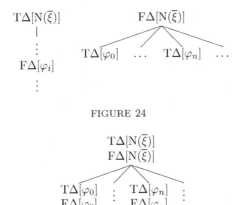

FIGURE 24

$$T\Delta[N(\overline{\xi})]$$
$$F\Delta[N(\overline{\xi})]$$

$$T\overline{\Delta[\varphi_0]} \quad : \quad T\overline{\Delta[\varphi_n]} \quad :$$
$$F\Delta[\varphi_0] \quad \quad F\Delta[\varphi_n]$$

FIGURE 25

12.3.3 Symbols defined

The method introduced to eliminate the logical parts of a proposition acts as a refined sieve for contingency and logicality. Logical propositions are those with either the T- or the F-part of their extended tableaux closed. By contrast, the extended tableau of a contingent proposition has open branches in both its T- and in its F-part. But the method ensures that all its logical sub-parts are identified and eliminated.

In effect, the symbol of a contingent proposition reduces to a disjunctive normal form of the elementary content at the open branches in the T-part of φ's extended tableau. (Proposition 84 below makes plain why the F-part may be ignored.) To be more precise, an open branch β comes with the truth-possibility θ_β of all signed elementary propositions Sp occurring at it. Call an elementary proposition or the negation thereof a literal. Each truth-possibility θ_β can be assigned a possibly countable conjunction $\bigwedge \xi_\beta$ of literals, with ξ_β the variable:

$$\xi_\beta(i) = \begin{cases} p & \text{if } Tp \text{ is in } \theta_\beta, \\ Np & \text{if } Fp \text{ is in } \theta_\beta. \end{cases}$$

Now the symbol (contingent nucleus) $\Sigma[\varphi]$ for contingent φ is:

$$\Sigma[\varphi] = \begin{cases} \bigvee_{\beta \text{ open in T-ET}(\varphi)} \bigwedge \xi_\beta & \text{if } \varphi \text{ is logically complex} \\ \varphi & \text{if } \varphi \text{ is elementary} \end{cases}$$

How the approach includes logical propositions is discussed shortly.

It should be stressed that thinking in terms of disjunctive normal forms directly is a bit misleading. If taken in a tractarian way, the arguments of the operations $\bigwedge$ and $\bigvee$ should be variables. Due to the systematic generation of tableaux, the ξ_β can be seen as generated by a rule, and so $\bigwedge \xi_\beta$ is still in line with thesis 6. But there is no obvious rule that generates all $\bigwedge \xi_\beta$ to yield the disjunction:

$$\Sigma[\varphi] = \bigvee_{\beta \text{ open in T-ET}(\varphi)} \bigwedge \xi_\beta.$$

Indeed, there may be uncountably many open branches. This is not to say that infinitary symbols are not tractarian. Although the cleansed T-part of an extended tableau may have uncountably many branches, the tableau itself is a countable structure that evolved from supertasks using simple truth-operational rules.

Contingent symbols as finished tableaux are analogous to the finite truth-table signs used earlier, but there are some important differences. As before, the notation is of first degree, since all logical parts are removed, but it lacks the second and third degree. Identical conjuncts in the conjunction $\bigwedge \xi_\beta$ are removed – since sequences are invariant under repetition, – and so are iterated negations. However, there may be multiple occurrences of the conjuncts $\bigwedge \xi_\beta$ themselves, and these conjunctions may be included in each other; there could be variables ξ_β and $\xi_{\beta'}$ such that each literal in ξ_β also occurs in $\xi_{\beta'}$. To attain the higher degrees of perfection seems to require heavier means than supertasks, and I will not investigate this topic further.

12.3.4 Symbol identity

A drawback of symbols of lower perfection than the second degree appears to be that they allow equisignificant propositions to have different symbols. One may wonder whether this is fatal to the notion of symbol? The crux of symbols is: (i) they give a representation of sense without logical redundancy; (ii) they allow establishing two apparently different representations as essentially the same. Thus the question becomes: what does 'essentially the same' mean in this context?

Finite symbols may already be infeasibly complex. Yet, to determine whether symbols are identical, finite resources suffice. One is inclined to take this as a major advantage, but at the same time this aspect of finite representations is very misleading. It may tempt one to highlight the visual or empirical usefulness of finite symbols, which strictly speaking is inessential. Independent of the finite resources that establish the identity of symbols, identity must imply logical equivalence. In the end, inhumane logic with its untainted necessity is to judge whether the identity is as required. Proposition 65 shows the finite symbols specified in definition 46 to be all right in this regard.

For infinitary symbols the situation is no different. They also call for a perfect notation of first degree, such as defined in the previous section, together with a logical criterion of identity. As explained, identity of infinitary disjunctive normal form does not suffice (the perfection is not of second or third degree). So another criterion of identity must be found, preferably one within the range of supertasks. As non-logical means must fall short to show logical identity, the best we can do, or so I think, is to require equivalence:

$$\Sigma[\varphi] = \Sigma[\psi] \text{ iff } \Sigma[\varphi] \leftrightarrow \Sigma[\psi] \text{ is valid.}$$

Using tableaux, supertasks suffice to establish the validity. Now, as for the finite system one has:

$$\vdash \varphi \leftrightarrow \Sigma[\varphi] \,,$$

$$\Sigma[\varphi] = \Sigma[\psi] \text{ iff } \vdash \varphi \leftrightarrow \psi.$$

The approach presumes logicality to be characterized completely and correctly. This is established in section 12.6.

12.3.5　Sense

With an appropriate notion of symbol in place, it remains to specify the form and content of sense. An immediate generalization of finite sense suffices (definitions 45, 47).

Definition 81 *The sense of a contingent proposition φ has form and content. The* form *of φ's sense is the form of its symbol $\Sigma[\varphi]$. In logical space Λ, π with π a projection in use, the* content *of its sense is the totality $\pi[\Sigma[\varphi]]$ of states of things $\pi[p]$, with p occurring in $\Sigma[\varphi]$.*

A $\bigwedge \xi_\beta$ in $\Sigma[\varphi]$ pictures a realization-pattern of φ's content. It shows $\pi[p]$ as realized, if p occurs in ξ_β, and as not realized, if Np occurs in ξ_β. The situation pictured by $\pi[\Sigma[\varphi]]$, is the disjunction of realization-patterns of all $\bigwedge \xi_\beta$ in $\Sigma[\varphi]$. □

Distinguishing the form and content of a contingent proposition allows showing the content of a proposition in logically incompatible ways. To see how sense is made manifest, it should suffices to give an example showing the main problem at hand.

12.3.6　Logical propositions

The symbol of a logical proposition, whether obtained from a finite or infinite sign, is either $(T)()$ or $(F)()$. For if a proposition φ is tautological, the F-part of its extended tableau closes. This means all its elementary propositions occur logically, and therefore its tableau reduces to the empty $(T)()$. *Mutatis mutandis* the same holds for a contradictory φ and its closed T-part, which results in the empty $(F)()$. In this way, the route via (infinitary) sense makes clear that logical propositions are limits without content and thus without sense (4.46 and its offspring).

Logical propositions have a unique position among all propositions (6.112). Contingent propositions show the form and content of a specific situation in logical space. By contrast, logical propositions are borderline cases of 'empty situations'. Lacking content, logical propositions concern the entire logical space (4.463). The empty $(T)()$ comes from the closed F-part of an extended tableau, and can therefore be seen as a disjunction of empty conjunctions. An empty conjunction is true iff all terms of its empty variable are true. In other words, empty conjunctions are true vacuously, and so the disjunction of these vacuous truths is true as well. This is how the tautology $(T)()$ shows all states of things as either realized or not realized without constraint. The empty $(F)()$ arises from the closed T-part of an extended tableau. It can be seen as an empty disjunction. An empty disjunction is true if some term of its empty variable is true; that is, it is always false. The contradiction $(F)()$ fills out logical space entirely to leave no states of things room to be realized.

12.3.7　A simple example

The main problem in determining a symbol is that logical occurrences may be distributed over a possibly infinite proposition. Consider for example:

$$(p \wedge \bigwedge_I q_i) \vee (p \wedge \bigvee_I \neg q_i)$$

In this proposition all occurrences of q_i, with i in I, are logical, but in its current form this will only be manifest after the (sub)propositions are considered. Instead of this infinite example, it suffice to consider its finite analogue in figure 26 that show the same point in a simpler way.

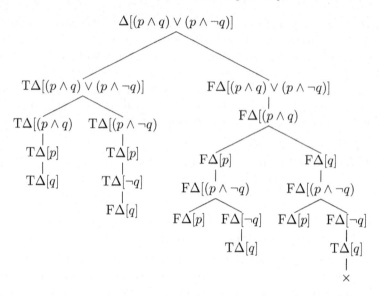

FIGURE 26

Observe that the logicality of the occurrences of q is shown no earlier than in the last step. Then all values of Δ are set and the tableau in figure 27 results (with the relevant occurrences made bold). The use of Δ ensures that the occurrences of q in the T-part of the tableau are recognized as logical and do not contribute to sense. (Similarly in the F-part.) To get at the symbol of the resulting tableau all complex and tautologous items in the T-part are ignored, and all identical branches identified (which is unproblematic in the finite case). In this way the single truth-condition Tp results:

$$\Sigma[(p \wedge q) \vee (p \wedge \neg q)] = Tp.$$

Along the same line one sees that:

$$\Sigma[(p \wedge p) \vee (p \wedge \neg p)] = Tp,$$

Although the occurrence $p_{(21)}$ would now become logical, too, the occurrence $p_{(11)}$ still is not.

For simple examples the method appears to be overkill, but the point is to indicate its general workings. Notice in particular that logical subpropositions in the T-part of the extended tableau are removed on the basis of information from the F-part. It works exactly the same if logical parts are distributed over infinite propositions, although it may take a supertask to detect them.

Our discussion of whether symbols can be generalized into the infinite is finished. The perfect notation of first degree ensures that all superfluous content in a proposition's infinitary sign is removed. This important aspect of

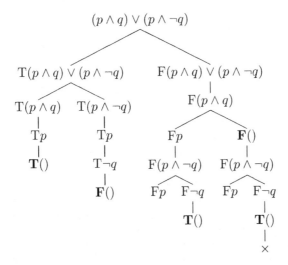

FIGURE 27

symbols sieves logical propositions from the contingent ones. The symbol of a contingent proposition constrains the possible ways worlds may be for the proposition to be true. Indeed, contingent propositions are equisignificant, iff their symbols are logically identical (equivalent). Finite and infinite symbols may be too complex to be humanly feasible. Whether such complexity bars symbols from being tractarian is open to debate. Cf. section 11.4.

12.4 Are symbols determinate?

With the sense of a proposition in place, it remains to verify what its representation shows about the way worlds may be. In particular, the sense of contingent propositions should be determinate; it must fix a world to either 'yes' or 'no', but never to both (4.023). Finite truth-table signs have this property, since they list all truth-possibilities and each truth-possibility is marked with either 'T' or 'F'. Reasoning semantically, it is clear the same must hold for extended tableaux. In a world, a contingent proposition is either true or false. But this must hold for one of its subpropositions, too, and so the world determines an open branch in either the T- or the F-part of its extended tableaux.

Semantic reasoning shows properties of sense in terms of ontology and truth. It is however a main theme of the *Tractatus* that sense is prior to truth. Although tableaux are closely related to semantics, in general they provide less information. The shift in focus from truth to sense therefore makes it worth our while to show sense to be determinate reflecting on the representation of sense alone. To show sense determinate I will study to what extent an open branch in a systematic tableau *enforces* a proposition. The notion of enforcing requires some aspects of tableaux to be discussed first.

In the previous section we distinguished between branches β and truth-possibilities θ. A branch is a branch in a tableau viewed as a tree at which both complex and elementary signed propositions may occur. A truth-possibility is an unordered sequence of signed elementary propositions with identical elements contracted into one. Truth-possibilities θ are open: there is no elementary proposition p such that θ contains both Sp and its conjugate $S^\circ p$. We have seen that an open branch in a tableau determines a truth-possibility. Just erase all logically complex propositions occurring at the branch to leave a truth-possibility $S_0 p_0, \ldots, S_n p_n, \ldots$ We say that θ is a truth-possibility for the proposition φ iff θ encompasses all elementary propositions occurring in φ, possibly among others.

Truth-possibilities do not describe a world completely; they constrain the part of a world a symbol is about. Be this as it may, the representation of sense should be so rich as to be determinate. Analogous to truth-table signs, each truth-possibility should either come from the T-part or from its F-part of a proposition's extended tableau but never from both. The notion of enforcing captures the corresponding idea that if a truth-possibility θ describes part of a world, $S\varphi$ is enforced to hold in that world as well. Recall that $\theta \sqsubseteq \theta'$ means: if Sp is in θ, Sp is also in θ'.

Definition 82 *For each proposition φ and each truth-possibility θ for φ, θ enforces $S\varphi$ – notation: $\theta \Vdash S\varphi$ – iff $\exists \theta' \sqsubseteq \theta[\theta'$ is determined by an open branch in a systematic tableau of $S\varphi]$.* □

The idea of enforcing derives from Hintikka's lemma which proves that each downward closed branch has a model. Each open branch β in a systematic tableau is downward closed:

- There is no elementary proposition p such that both Tp and Fp occur on β;
- If $TN(\overline{\xi})$ occurs on β, then so do $F\psi$ for all ψ in ξ;
- If $FN(\overline{\xi'})$ occurs on β, then so does $F\psi'$ for a ψ' in ξ'.

To prove sense determinate, we start with showing how 5.2341a can be understood in the infinite system.

5.2341a The sense of a truth-function of p is a function of p's sense.

Proposition 83 shows the sense of a proposition indeed to be compositional.

Proposition 83 *The sense of a symbol is compositional. For all relevant θ:*

$$\theta \Vdash Sp \quad \Leftrightarrow \quad Sp \text{ in } \theta$$
$$\theta \Vdash TN(\overline{\xi}) \quad \Leftrightarrow \quad \theta \Vdash F\varphi, \text{ for all } \varphi \text{ in } \xi$$
$$\theta \Vdash FN(\overline{\xi}) \quad \Leftrightarrow \quad \theta \Vdash T\varphi, \text{ for a } \varphi \text{ in } \xi$$

Proof. The basic case is elementary. Consider the complex $TN(\overline{\xi})$. We must prove for all θ for φ:

$$\theta \Vdash TN(\overline{\xi}) \Longleftrightarrow \text{ For all } \varphi \text{ in } \xi: \theta \Vdash F\varphi.$$

Let θ for φ be arbitrary. Assume $\theta \Vdash T\varphi$. By definition θ has a part θ' that is determined by an open branch β in the systematic tableau of $TN(\overline{\xi})$. Checking the stages that generate systematic tableaux, it is clear that each step in the branch β concerns a subproposition of one of $N(\overline{\xi})$'s subpropositions $F\varphi$. In this way, β intertwines branches β_φ in the systematic tableau of the $F\varphi$. Moreover, since β is in the systematic tableau for $TN(\overline{\xi})$, for each φ in ξ a

β_φ is involved. That β is open, implies that all β_φ are open, too. The β_φ determine truth-possibilities θ_φ with $\theta_\varphi \sqsubseteq \theta' \sqsubseteq \theta$. Therefore: $\theta \Vdash F\varphi$, for all φ in ξ. Conversely, assume $\theta \Vdash F\varphi$, for all φ in ξ. By definition, for each φ in ξ, θ has a part θ_φ determined by an open branch β_φ in the systematic tableau of $F\varphi$. The $F\varphi$ are subpropositions of $TN(\bar{\xi})$. This means in generating the systematic tableau for $TN(\bar{\xi})$ there will be a branch β in which each step is taken in the same way as the corresponding step in the given β_φ. Since the θ_φ determined by the β_φ can be joined into θ, it follows that the resulting branch β must be open. Therefore: $\theta \Vdash \varphi$, as required. The proof for the complex $FN(\bar{\xi})$ is much simpler because for all ψ in ξ the systematic tableau for $F\varphi$ branches immediately to $T\psi$. □

The proof shows how compositionality of infinitary sense is to be understood. In section 9.4.2, the symbol of a finite proposition is given as a function of the symbols of its parts. In the infinitary case the straightforward analogue of this procedure brings us outside the realm of supertasks, therefore the composition is determined in a different way. The symbols of the parts are not input *in toto* into a truth-operation to yield the symbol of the main proposition at once. Rather, the systematic tableau of the main proposition intertwines the symbols of the parts. Except for the first, each step in the systematic tableau is also a step in a 'sub-tableau'. Since intertwining may result in closed branches, not all steps of a part may be needed, but all consistent merges are composed in full. The end result can be viewed as the composition of the subpropositions in that it enforces the recursive definition of truth.

Proposition 83 is a strong indication that the symbol of a proposition as defined in section 12.3 represents the appropriate truth-conditions. Proposition 84 gives the basic property of enforcing.

Proposition 84 *For all propositions φ and all truth-possibilities θ over the elementary $\bar{q}$:*

$$\theta \Vdash S\varphi \iff not: [\theta \Vdash S°\varphi].$$

Proof. [*Elementary*]: If $\theta \Vdash Sq$, Sq is in θ. But then $S°q$ is not in θ, because θ is open. Conversely, if not $\theta \Vdash S°q$, $S°q$ is not in θ. Since q is in $\bar{q}$ and θ is over $\bar{q}$, Sq is in θ and so $\theta \Vdash Sq$.
[*Complex*]: We may reason as follows:

$$\theta \Vdash TN(\bar{\xi}) \quad \Leftrightarrow \quad \text{For all } \varphi \text{ in } \xi: \theta \Vdash F\varphi$$
$$\Leftrightarrow \quad \text{For all } \varphi \text{ in } \xi, \text{ not: } \theta \Vdash T\varphi$$
$$\Leftrightarrow \quad \text{Not: } \theta \Vdash FN(\bar{\xi})$$

The first equivalence comes from proposition 83; the second from the induction hypothesis; the third again from proposition 83. Since:

$$\theta \Vdash TN(\bar{\xi}) \Leftrightarrow \text{ not: } \theta \Vdash FN(\bar{\xi}),$$

of course: $\theta \Vdash FN(\bar{\xi}) \Leftrightarrow \text{not: } \theta \Vdash TN(\bar{\xi})$ as well. □

Proposition 84 shows the representation of sense to be determinate. If sense excludes inconsistent description – here: closed branches are non-descriptive, – the core of classical logic is already available. The bi-conditional:

$$\theta \Vdash S\varphi \iff \text{not: } [\theta \Vdash S°\varphi]$$

is equivalent to the conjunction of:

$$\theta \Vdash S\varphi \text{ or } \theta \Vdash S^\circ \varphi$$

$$\text{not: } [\theta \Vdash S\varphi \text{ and } \theta \Vdash S^\circ \varphi],$$

so each θ for φ enforces $S\varphi$ or its conjugate $S^\circ\varphi$ but never both. At the level of sense it remains to see why inconsistent description should be excluded. In a tractarian setting I think this can only be due to realism. Determinate sense is in harmony with worlds, and worlds are taken to fix propositions to truth xor to falsity.

As a corollary to proposition 84 it can be seen that logical propositions have extended tableaux with either the T- or the F-part closed and the conjugate part open. A contingent proposition has an extended tableau with both its T- and its F-part open. Its symbol will not be empty.

12.5   Worlds and truth

With sense sufficiently understood it is time to consider truth. To define the truth and the falsity of a proposition in terms of its sense, a symbol and a world must be linked via the truth-possibility obtained from the world. Here, a slight variation of definition 50 suffices.

Definition 85 *Let Λ, R, π be a world w $(= \Lambda, R)$ together with a projection π in use. The truth-possibility $\tau_w[\overline{p}]$ for the sequence of elementary signs $\overline{p}$ is defined by:*

- Tp_i *occurs in $\tau_w[\overline{p}]$, if w realizes $\pi[p_i]$;*
- Fp_i *occurs in $\tau_w[\overline{p}]$, if w does not realize $\pi[p_i]$.*   □

The truth of a proposition in a world is now defined as: the truth-possibility corresponding to the world enforces the proposition.

Definition 86 *Let φ be a proposition over $\overline{p}$, w a world in logical space Λ, and π a projection in use. Define φ is true in w (or: $w \models^{tlp} \varphi$) by:*

$$w \models^{tlp} \varphi \text{ iff } \tau_w[\overline{p}] \Vdash T\varphi.$$ □

As in definition 51 concerning the truth and falsity for finite propositions in terms of truth-table signs, definition 86 defines truth non-recursively. From propositions 83 and 84 it is immediate that the tractarian notion of truth is equivalent to the familiar recursive one.

Proposition 87 *Let logical space Λ, π be given together with a projection π in use. For all φ and all worlds w $(= \Lambda, R)$ in Λ, π:*

$$w \models^{tlp} \varphi \quad \Leftrightarrow \quad w \models \varphi$$ □

That there is a close relationship between the sense of a proposition and its truth is no surprise: possible truth is core to sense itself. In a truth-functional context the route to truth via sense may therefore appear cumbersome. But in this way any suggestion is precluded that the senseless logical parts of a proposition contribute to its being about part of logical space. In particular, logical propositions are about nothing. This distinction is absent on a purely extensional approach.

12.6   Characterizing logical propositions

Generalizing perfect notation into the infinite retains the trichotomy of tautology, contingent proposition, and contradiction. It remains to be seen, however, whether the characterization of logical propositions is correct and complete. Are logical propositions those and only those propositions that reduce to an empty symbol? This section shows the answer is 'yes': logical proposition can indeed be characterized from among all propositions in this way. Since φ is a tautology if and only if its negation is a contradiction, the characterization may concentrate on either. As is usual, we concentrate on tautologies (or: validities).

In line with Wittgenstein's insight concerning graphical signs we define:

Definition 88 *A proposition φ is provable – notation: $\vdash \varphi$, – iff it has a closed tableau with $F\varphi$ at its root.*   □

The idea is that if no world can be found in which φ is false, φ must be a tautology (or: valid). Here the notion is understood with regard to the unique logical space Λ.

Definition 89 *Let $\mathcal{F}$ $(= \Lambda, \pi)$ be a frame. A proposition φ in $\mathcal{F}$ is valid – notation: $\models \varphi$, – iff for all worlds w $(= \Lambda, R, \pi)$ in $\mathcal{F}$: $w \models \varphi$.*

It is an immediate consequence of proposition 83, the current analogue of Hintikka's lemma, that the tableau procedure to check logicality is correct and complete.

Proposition 90 *For $\mathcal{F} := \Lambda, \pi$ a frame, and for all φ in $\mathcal{F}$:*

$$\vdash \varphi \Leftrightarrow \models \varphi.$$

Proof. [*Soundness* ⇒:] We reason by contraposition. Assume that φ is not valid: there is a world w – i.e., a Λ, π, R, – such that $w \not\models \varphi$. By proposition 87: $\tau_w[\bar{p}] \Vdash F\varphi$. This means the F-part of the extended tableau has an open branch: φ is not provable.

[*Completeness* ⇐:] Again we reason via contraposition. Assume $\not\vdash \varphi$. Each tableaux with $F\varphi$ at its root has an open branch. This holds in particular for a systematic tableau. Choose an open branch β in this tableau. The truth-possibility θ determined by β has: $\theta \Vdash F\varphi$, and so for any world w with: $\tau_w[\bar{p}] \sqsupseteq \theta$: $\tau_w[\bar{p}] \Vdash F\varphi$. Such a world exists because θ is consistent. Therefore, $w \not\models \varphi$ and $\not\models \varphi$. The proposition φ is not a tautology.   □

12.7   Logical consequence

In chapter 8 it is argued that the structural consequence relation of 5.11 is most basic to the *Tractatus*.

5.11 If the truth-grounds that are common to a number of propositions are jointly the truth-grounds of a certain proposition, then we say that the truth of this proposition follows from the truth of those propositions.

This insight does not change in an infinite setting. Let Γ, φ consist of propositions over the elementary $\bar{p}$. The infinitary analogue of 5.11 is:

$$\Gamma \vdash \varphi, \text{ iff for all truth-possibilities } \theta \text{ over } \bar{p} :$$

if for all ψ in $\Gamma[\,\theta \Vdash \mathrm{T}\psi\,]$, then $\theta \Vdash \mathrm{T}\varphi$.

In this way it still holds that logical consequence is shown in representation. But even if logical consequence is restricted to countably many premises, as is done here, the complexity of inference may change considerably, and with that the tenability of 5.13, if taken literally.

5.13 That the truth of one proposition follows from the truth of other propositions, we can see from the structure of the propositions.

Since there could be uncountably many θ, a piecemeal check of all truth-possibilities may be far too complex for *us*. We do better to resort to 6.1221, which in the current setting becomes:

$$\Gamma \vdash \varphi \text{ iff } \vdash \bigwedge \Gamma \to \varphi.$$

Proposition 90 shows that whether or not $\bigwedge \Gamma \to \varphi$ is a tautology can be checked in a systematic, rule-based manner using tableaux; i.e., the unrestricted analogue of the graphical method in 6.1203. The reduction does presume that the premises are given as a tractarian variable, else $\bigwedge \Gamma$ cannot be formed. From a tractarian point of view this is natural anyway.

12.8   Descriptive and truth-operational completeness

Thesis 4.26 requires the system to be descriptively complete.

4.26 The specification of all true elementary propositions describes the world completely. The world is described completely by specifying all elementary propositions together with an indication which of them are true and which false.

Descriptively completeness holds for the infinitary system as well. With infinite proposition at hand it is tempting to strengthen 4.26 to: there is a proposition that describes the world completely. Just take the conjunction of literals that are true in a world. (Recall that a literal is an elementary proposition or the negation thereof.) This, however, will not do. Since elementary propositions lack a common form, the *Tractatus* has no variable to generate them, and the same goes for the list of literals required to define a world. Perhaps for this reason Wittgenstein used 'specification' (*Angabe*) rather than proposition (*Satz*) or propositional variable (*Variable*) of all true elementary propositions. Presumably 'the all-embracing logic [...] mirrors the world' in a different way (5.511). The best we can do is to appeal to determinateness of sense (proposition 84). The determinate sense of a proposition gives for each truth-possibility that the sequence of literals corresponding to it is either in the T- or in the F-part of the proposition's symbol. This means that somehow a symbol involves all truth-possibilities over $\overline{p}$. Since truth-possibilities constrain possible worlds, it follows that in a way all possible worlds are involved in description. Yet, a possible world may be too complex to be defined by a single proposition.

In the finite system there seems to be a perfect balance between truth-functional signs on the one hand, and the worlds and situations that can be defined on the other. The finite system is truth-operationally complete: truth-table signs are precisely those that can be generated by repeated application of the truth-operation N given a finite number of elementary propositions. See proposition 40. One could say: all finite worlds and all finite situations are definable. This way of putting it, however, is not in line with how situations (*Sachlage*) are understood here. Cf. page 117*ff*.

For an infinite logical space truth-operational completeness no longer holds.[93] But on the current descriptive approach, where all logical complexity resides in the sign, this hardly has any negative side-effect. In the infinitary system there is no longer an independent totality of truth-table signs that can be considered for truth-operational completeness; only those signs are available that the truth-operations generate. Worlds can still be realized independent of description, as they should be. And due to its determinate sense, each contingent proposition will somehow represent or constrain them. Situations, however, are a different matter. From a platonistic point of view, one could hold that most alternations of logically incompatible possible worlds will be too complex for description. But this way of putting it, is misleading in the light of how the *Tractatus* is understood here. Situations are no part of ontology, they are the logical shadows of signs projected into logical space. By its very nature, there are as many situations as there are symbols: truth-operational process and truth-functional result are equivalent (6.1261).

12.9 Compactness

Is the infinite system of the *Tractatus* finitely compact? That is, does it hold for each countably totality of propositions Γ that each of its finite parts Γ_e is true in a world, iff Γ itself is true in that world? The question seems far removed from tractarian considerations, but let us try to answer it anyway.

The first thing to note is that we are restricted to a unique logical space, and perhaps this space does not allow for counterexamples to compactness. This, of course, is insufficient reason to call the logic compact. From a tractarian point of view, logical space may as well turn out to allow such counterexamples. Consider, e.g., a logical space that is not well-ordered, and have the states of things:

$$\ldots, o_{n+1}Ro_n, \ldots, o_1Ro_0.$$

Then each finite initial segment of the variable

$$m_0Rm_1, \exists x[m_0Rx, xRm_2], \exists x, y[m_0Rx, xRy, yRm_3], \ldots.$$

will be true in a world; just let the last name n_l in an initial part of the sequence refer to o_0, its immediate predecessor to o_1, etc. But the variable itself cannot be true in world, since the space has no objects with infinitely many R-successors.

Given the close connection between compactness and properties of tableaux as trees, and thus with representations of sense, one would expect Wittgenstein to have concluded that the statement of compactness is a pseudo-proposition that tries to describe a regularity in the system of representation that can only be shown. Cf. Wittgenstein's position on the axiom of infinity

---

[93]The semi-formal notion of variable does not allow to proof this. But if we assume that each 'sensible' notion of rule would at each stage in the generation of signs restrict variables to countably many, at most countably many worlds and situations can be defined.

5.535.

12.10 Interpolation

In the treatment of logical consequence – 5.13 and its offspring, – Wittgenstein makes a sequence of interesting observations. We focus on 5.135, which generalizes the observation that elementary propositions are logically independent.

5.135 In no way the existence of a situation can be inferred from the existence of a situation entirely different from it.

The phrasing of 5.135 may appear a bit strange, since situations are naturally taken ontologically. Recall, however, that on the present approach situations consists of possible realization-patterns that a propositional sign projects onto the part of logical space it is about. Read in this way 5.135 states that if one situation can be inferred from another one, the signs describing them must have elementary propositions in common. The modern term for this is 'interpolation', first proved for first-order logic by William Craig in 1957.[94] In the *Tractatus* 5.135 remains a thesis, but from the details developed up till now it can be proved why 5.135 should be so.

Definition 91 *Let χ, σ be propositions. The proposition ϕ is an interpolant for χ, σ iff:*

1. *$\vdash \chi \to \phi$ and $\vdash \phi \to \sigma$;*
2. *The elementary propositions of ϕ are among those that χ and σ have in common;*
3. *An elementary proposition occurs positively (negatively) in χ (σ) if it occurs positively (negatively) in ϕ.*   □

Proposition 92 *If χ, σ are propositions over the elementary $\bar{q}$ such that $\vdash \chi \to \sigma$, there is an interpolant ϕ so that: $\vdash \chi \to \phi$ and $\vdash \phi \to \sigma$.*   □

Not all infinitary logics have interpolation, but for a proof of the countable case, refer to Keisler (1971, chapter 5). An abstract version of the result can be found in Linden (1968, Chapter V, §3).

---

[94] I owe the observation that 5.135 is related to interpolation to Göran Sundholm (private communication).

13

Quantifiers

A book on the *Tractatus* and its system is incomplete if it does not address the somewhat thorny issue of quantification. To arrive at a proper treatment of tractarian quantification the following questions must be answered:

- How are quantifiers represented? More in particular how is the truth-functional structure in their representation separated from the indication of generality? Cf. 5.52, 5.521, 5.522, 5.501.
- How is binding effected in line with the observations in 4.0411?
- What is the precise nature of the injective treatment to quantification argued for in 5.53 and its offspring?

Ever since Fogelin (1976, chapter VI) suggested that tractarian quantifiers lack a binding mechanism and cannot be nested, they have an aura of failure about them. Perhaps for this reason the first two questions did not receive the attention they deserve. The third question has been addressed, but mainly because it was of independent interest. Initiated by Hintikka (1956) injective quantification has been studied quite extensively as a variant of the now common first-order logic; see Landini (2007, Appendix A), and Wehmeier (2004, 2008, 2009).[95]

Fogelin's observation on the apparently problematic status of quantifiers led to a short but fierce debate. See Fogelin (1976, chapter VI), Geach (1981), Fogelin (1982), Geach (1982), Soames (1983), and Fogelin (1987). Fogelin (1987) gives an accurate summary, where he states that it remains unclear whether the system could deal with quantifier iteration without the extensions suggested by Geach (1981) and Soames (1983) to represent binding.

The early debate did not pay much attention to the details of how the *Tractatus* presents quantified propositions. E.g., the bracket-expressions introduced in 5.501, which are essential to the tractarian approach, only surface in an unfortunate comparison of Fogelin (1982, 126) with Geach' notation.[96]

---

[95] Landini offers a system Qe of injective quantification that implies the axiom of infinity. This is interesting but clearly not in line with early Wittgenstein's intentions. Whether or not logical space allows for a form of infinity should only dependent on the number of names; logical complexity has nothing to do with it (5.535).

[96] Fogelin suggests incorrectly that by contrast to Geach' notation bracket-expressions cannot be based on complex propositions, because the *Tractatus* only allows the successive application of a finite number of truth-operations (5.32). He fails to observe that if this were true, the *Tractatus* would not only have a problem with iteration but also with the

The first article that did stay close to the *Tractatus* is Von Kibéd (2001), of which the current approach is a refinement and extension. But Von Kibéd, too, did not give bracket-expressions their due weight.

Once it is clear that the notions and techniques in the *Tractatus* are quite different from that of Frege and Russell, it is not too difficult to see its treatment of quantification is all right. Other than Geach (1981), Soames (1983), and in the end even Von Kibéd (2001, App.2), I think tractarian quantification does not suffer from the problem Fogelin (1976) brought to the fore. Surely, it has its limitations, like e.g. first-order quantification has, but the inability to iterate is not among them.

With the questions concerning representation, generalization and binding answered, we may transfer the modern insights on injective quantification to the tractarian system. To simplify the presentation this aspect is disregard at first, but the injective usage has our full attention from section 13.6 onward.

13.1 Representation and generalization

Wittgenstein indicates his treatment of quantification with the usual terseness. The core thesis is 5.52 which concerns quantification over names; i.e., the only form of quantification considered here.[97]

5.52 If the values of ξ are all the values of a function fx for all values of x, then $N(\bar{\xi}) = \neg(\exists x).fx$.

Simple as it is, the example in 5.52 does show the two main ingredients of quantification:

1. The truth-operation N;
2. The variable ξ obtained from a function f with prototype x.

Also 5.52 highlights the special care that 'domains of quantification' require, and with that the substitutional nature of tractarian quantification. Let us discuss these aspects first.

Since logical space consists of states of things, not of objects, the tractarian system has no single domain of quantification. In particular, it does not make sense to require quantification over all names. Names go proxy for objects of different 'types'; 'types' that are manifest only if language is analyzed completely. Instead, a prototype x determines a domain of quantification ('a range of significance'); i.e., the totality of all instances for which fx has sense. Since

---

square of opposition. See section 13.1. Geach (1982, 128) retorts: 'Fogelin still appears, as in his book, to confuse the performance of *one* operation on a (possibly) infinite class of operands with the performance of an infinite number of operations'.

[97]The early sources show that Wittgenstein struggled long before coming to an understanding of completely generalized propositions, such as $\exists x, \varphi.\varphi(x)$ in 5.5261. Completely generalized propositions indicate that 'higher-order' forms of quantification should be available. A quick look at the general form of propositions, as discussed in chapters 5 and 7, makes clear there are different kinds of constituent that can be 'prototyped'. One has, for instance, prototypes x for names n; prototypes E for sub-propositional expressions ε; prototypes $f(-)$ for material functions $\alpha(-)$; prototypes $F(-)$ for logically complex functions $\varphi(-)$; prototypes p for propositions; prototypes $C(-)$ for truth-operational complexes. Each kind of prototype could be used in quantification, but may come with additional requirements over and above those for quantification over names. In this book such extensions are not considered.

there is perfect harmony between elementary signs and what they signify, it suffices to concentrate on prototypes and the names that can be substituted for them. Then, domains consist of the referents of all constituents that a prototype varies over. This approach to domains of quantification is more complex than the now usual one, because the domains of quantification may depend on the constituents of a proposition. For example, aRx may give x another domain than bRx. Yet the method is in line with the principle that sense is prior to truth. In case of $\forall x.fx$, e.g., the totality is presupposed of all x for which fx has sense, and for each instance a in that totality – no more, no less – fa is claimed to be true. Section 13.7 discusses the option of empty domains.[98]

In 5.52 both the truth-operation N and the propositional variable ξ are used in a contextual definition of the quantifier that is based on the way the variable is specified. This holds in general. In case of quantification over objects via their names, the following laws generate the quantificational forms in the square of opposition.

- If $\varphi(x)$ is a variable, $N(\overline{\varphi(x)})$ is a proposition.
  $N(\overline{\varphi(x)}) = \neg\exists x.\varphi(x)$ Def.

- If $\varphi(x)$ is a variable, $NN(\overline{\varphi(x)})$ is a proposition.
  $NN(\overline{\varphi(x)}) = \exists x.\varphi(x)$ Def.

- If $N(\varphi(x))$ is a variable, $N(\overline{N(\varphi(x))})$ is a proposition.
  $N(\overline{N(\varphi(x))}) = \forall x.\varphi(x)$ Def.

- If $N(\varphi(x))$ is a variable, $NN(\overline{N(\varphi(x))})$ is a proposition.
  $NN(\overline{N(\varphi(x))}) = \neg\forall x.\varphi(x)$ Def.

Like in 5.52, the quantifiers are defined contextually, as abbreviations of prototyped structures that indicate a common propositional form. Observe that prototyping is for real: a genuine occurrence of a name is prototyped to arrive at a propositional variable. Consequently, all quantifiers bind 'real' prototypes within their scope; the tractarian system leaves no room for empty quantification, such as the modern '$\forall x.P(y)$'. It is also assumed that in the abbreviated form no free prototypes remain, for else the result would not be a proposition. Prototypes are introduced only at stages where they are needed.

The scope of a quantifier is given with its bracket-expression, introduced in 5.501.

5.501a A bracket-expression whose terms are propositions I indicate – if the order of the terms within the brackets is immaterial, – by a sign of form $(\overline{\xi})$. 'ξ' is a variable whose values are the terms of the bracket-expression; and the line over the variable indicates that it represents all its values.

Bracket-expressions represent generality. For instance, if one wants to describe neither Chouf nor Oughd howling, the variable (Hc,Ho) is fed into the truth-

---

[98]The use of ranges of significance is adapted from Russell. See for instance Whitehead and Russell (1910, Introduction, Chapter II, §III and §IV). According to Russell $\varphi(x)$ is a proposition that 'ambiguously denotes' $\varphi(a), \varphi(b), \ldots$ (Whitehead and Russell, 1910, p. 40). For Wittgenstein, $\varphi(x)$ is not a proposition but a propositional variable consisting of the (unordered) sequence of propositions $\varphi(a), \varphi(b), \ldots$.

operation N. This variable leaves in the dark whether howling is restricted to these two things. So if one wants to describe that nothing is howling, the bracket-expression $(\overline{\text{Hx}})$ is used to form $N(\overline{\text{Hx}})$. Here, the bar indicates that all possible instances of the prototype x in Hx are considered (which may turn out to be just Chouf and Oughd).

Bracket-expressions effect the separation of truth-functionality and generality Wittgenstein was after.

5.521a I separate the concept *all* from the truth-function.

For example in $N(\overline{\text{Hx}})$ $(= \neg \exists x.\text{Hx})$ generality comes from the bracket-expression $(\overline{\text{Hx}})$ and truth-functionality from the truth-operation N. Similarly for the more complex cases. The point of separation is that in specifying a general proposition a notion of totality is involved that appears closely akin to the concept of quantification. For instance, a proposition $\exists x.fx$ presumes *all fx* to be given, else the one x that makes $\exists x.fx$ true may be left out of consideration. Similarly, if *some fx* are disregarded, $\forall x.fx$ may be held true incorrectly. At face value the observation seems to indicate the notion of quantification is circular: in the logical order of things the notion of *all* must be presumed before quantification can be made explicit. Or to make the same point somewhat differently, if the concept *all* were only to occur in general propositions $\forall x.\varphi(x)$, it would be hard to understand how the concept of generality figures in other forms of quantification, like $\exists x.\varphi(x)$ (5.521). According to Wittgenstein this line of reasoning is based on a failure to distinguish the truth-functional content of quantification from the generality needed to specify its domain. The domain of quantification should rather be given as part of the argument of a truth-operation.

5.523 The generality sign occurs as an argument.

5.522 It is characteristic of generality signs, firstly that they indicate a logical prototype, and secondly that they highlight constants.

It is the bar in a bracket-expression which indicates that all terms of the variable are considered. The bracket-expression also gives the logical prototype, which at the same time highlights the constants it varies over (Black, 1964, chapter LXI). In this way, the prototype indirectly indicates the domain of quantification. Due to the perfect harmony between elementary signs and what they signify, the domain consists of the referents of all constituents – here: names, – that can be assigned to the prototype.[99]

---

[99] I agree with Stokhof (2002, 168-170) that names cannot be dispensed with in favor of fully general propositions, as Ishiguro (1969, 43-45) argues. Tractarian quantifiers are contextually defined, and derive their sense from elementary, prototyped structures. Quantified sentences do not have an independent sense, as Ishiguro assumes.

Ishiguro's argument is based on the logical insight that (finite) worlds can be defined in terms of fully general propositions. See 5.526. However, 5.526 may fail to hold in an infinite setting, simply because elementary propositions may lack the general form that is required to yield quantified propositions. Still, infinite worlds can be described by means of the totality of elementary propositions (4.26). This difference is sufficient reason to distinguish strictly between the use of elementary propositions and fully general ones. (A similar distinction can be made with regard to logical space instead of worlds.) The distinction, however, is also pertinent for a finite system. Wittgenstein's observation in 5.526 requires

It remains to be checked whether the above contextual definitions indeed capture the intended quantifiers. To this end, it suffices to consider monadic quantification (involving just properties, no genuine relations). Suppose logical space Λ, π has the following states of things concerning a property P:

$$Pa_0, \ldots, Pa_m \ldots, \pi[Pa_0], \ldots, \pi[Pa_m], \ldots$$

Here, the projection π partitions the space in signs Pa_i and signified $\pi[Pa_i]$. Since $\neg\exists x.Px$ is short for $\mathrm{N}(\overline{Px})$, writing the bracket-expression in full this becomes:

$$\mathrm{N}(Pa_0, \ldots, Pa_m \ldots).$$

The truth-functional content is that of joint negation, so in a world Λ, π, R the proposition should be read:

$\pi[Pa_0]$ is not realized and $\ldots$ and $\pi[Pa_m]$ is not realized and $\ldots$

Since *all* instances of propositions with form Px are listed and the projection π is bijective, the infinite conjunction is equivalent to: there is no x such that Px, as required. Given this much, it is immediate that $\exists x.Px$, which is short for the negation $\mathrm{NN}(\overline{Px})$, indeed means: there is an x such that Px. Similarly, $\forall x.Px$ is short for $\mathrm{N}(\overline{NPx})$, or:

$$\mathrm{N}(NPa_0, \ldots, NPa_m \ldots)$$

In world Λ, π, R the proposition reads:

$\pi[Pa_0]$ is realized and $\ldots$ and $\pi[Pa_m]$ is realized and $\ldots$,

which is equivalent to: for all x it holds that Px. Again, $\neg\forall x.Px$ is short for its negation $\mathrm{NN}(\overline{NPx})$ and so means: not for all x Px.

At this point some readers may already be convinced that tractarian quantification is quite in order. Contextual definition takes over the effect of variable binding and captures the quantificational force of the proposition abbreviated. Still, quantification should be considered in much more detail, if only to understand why it was felt to be so terribly mistaken.

13.2 Iterated quantification

To get nested forms of quantification the rules introduced in the previous section are intertwined. To be more specific, suppose logical space has the

---

all instances of all names to be prototyped simultaneously in order to form the fully general propositions. In this process of abstraction the content of a name, and hence its identity is lost. Forms remain that may be shared by different names. Tractarian names, by contrast, do have content and form that contribute to the content and form of sense.

In section 9.2.2 we have seen that names with the same form can be substituted for one another, but the substitution is simultaneous ('holistic'). Even if names take over each others rôle in the web of dependencies in logical space, no name occurrence can be disregarded. Names, like objects, are instantiations of a form. (Since names occur in facts, I even think they must be objects used as names.) Therefore, names with the same form are still different from each other (cf. 2.0233), and this also if the difference cannot be described (cf. 2.02331). In fully generalized propositions the occurrence aspect of names is lost. If logically identical, the different forms are truly interchangeable. They only get the identity of an occurrence, the moment they are assigned a name in the way 5.526 indicates. That the assignment does not have to be unique indicates forms lack the identity name occurrences have. Prior to assignment, they figure as pure forms waiting to be instantiated. See 3.315.

sequence of elementary propositions:

$$a_0 R b_0, \ldots, a_n R b_m \ldots$$

Among more complex variants, this sequence allows for the following quantifier iterations:

$$\exists x \exists y.x R y, \ \exists x \forall y.x R y, \ \forall x \exists y.x R y, \ \forall x \forall y.x R y.$$

Let us consider the details of these iterated forms. In doing so, we disregard for now the niceties of domains of quantification. They get the attention they deserve as soon as injective quantification is considered.

The structure $\exists x \forall y.x R y$ is the result of several abbreviations that must be eliminated one at a time. To begin with, $\exists x \forall y.x R y$ is short for $NN(\overline{\forall y.x R y})$. Writing the bracket-expression $(\overline{\forall y.x R y})$ in full this becomes:

$$NN(\forall y.a_0 R y, \ldots, \forall y.a_n R y, \ldots).$$

Next, $\forall y.a_n R y$ is short for $N(\overline{N(a_n R y)})$, or in full:

$$N(N(a_n R b_0), \ldots, N(a_n R b_m), \ldots)$$

Using simplified notation to save space, with all abbreviations eliminated $\exists x \forall y.x R y$ becomes:

$$NN(N(N(a_0 R b_0), \ldots, N(a_0 R b_m), \ldots), \ldots, N(N(a_n R b_0), \ldots, N(a_n R b_m), \ldots), \ldots).$$

From the explanation of monadic quantification in the previous section, it can be checked that this is as required.

Now consider the structure $\forall x \exists y.x R y$. As before the abbreviations are eliminated in order. Firstly, $\forall x \exists y.x R y$ is short for

$$N(\overline{N(\exists y.x R y)}).$$

With the bracket-expression $(\overline{N(\exists y.x R y)})$ written out this is:

$$N(N(\exists y.a_0 R y), \ldots, N(\exists y.a_n R y), \ldots).$$

Recall $\exists y.a_n R y$ is short for $NN(\overline{a_n R y})$, or unabbreviated:

$$NN(a_n R b_0, \ldots, a_n R b_m, \ldots)$$

All in all, the analyzed $\forall x \exists y.x R y$ becomes:

$$N(NNN(a_0 R b_0, \ldots, a_0 R b_m, \ldots), \ldots, NNN(a_n R b_m, \ldots, a_n R b_m, \ldots), \ldots),$$

which indeed means $\forall x \exists y.x R y$ (or: $\neg \exists x \neg \exists y.x R y$) as required. The other iterations are treated similarly, and are left to the reader.

Besides eliminating abbreviations in a top down manner, it is helpful to consider the iterations in reverse analysis from the bottom up. Given the sequence of elementary propositions:

$$a_0 R b_0, \ldots, a_n R b_m \ldots,$$

$\forall x \exists y.x R y$ is generated via the steps:

- $a_i R y \ (= a_i R b_0, \ldots, a_i R b_m \ldots)$ is a variable, so $NN(\overline{a_i R y})$ is a proposition;
- Set $NN(\overline{a_i R y}) = \exists y.a_i R y$ Def.
- $N(\overline{\exists y.x R y}) \ (= N(\exists y.a_0 R y), \ldots, N(\exists y.a_n R y), \ldots)$ is a variable, so $N(N(\exists y(x R y)))$ is a proposition.

- Set $N(\overline{N(\exists y.xRy)}) = \forall x \exists y.xRy$ Def.

The other iterations are generated likewise.

It seems clear that the iteration of tractarian quantification is unproblematic. So, what happened for it to get such a bad press?

13.3 A fundamental error in Fogelin's *Wittgenstein*?

Fogelin (1976, *Wittgenstein* 1st ed.) argues there is a fundamental error in the tractarian treatment of quantification: it lacks a proper binding mechanism and with that the possibility to iterate. His argument is based on an important observation. Consider, for example, the purported representation of $\forall x \exists y.xRy$:

$$N(N(NN(\overline{xRy}))).$$

Since no binding mechanism is in place, the representation lacks unique readability. In particular, the scope of the variables x and y is ambiguous. (It is instructive to check how many readings can be discerned.)

Now, how could Wittgenstein have overlooked such a basic fact, and with him his supervisor Russell and his first translator and critic Ramsey? This is unexpected, especially since Wittgenstein was so acutely aware of the requirements a binding mechanism must satisfy.

4.0411 If we tried to express e.g. what $(\forall x)fx$ expresses by putting an index before fx – say like: 'Gen.fx' – it would not do; we should not know what was generalized. If we tried to signal it with an index '$_g$' – say like: 'fx_g' – it would also not do; we should not know the scope of the generalization.

If we were to try it by introducing a mark in the argument places – say like: '$(G,G)f(G,G)$' – it would not do; we could not determine the identity of the variables. And so on.

All these notations fail, because they lack the necessary mathematical multiplicity.

In my opinion Wittgenstein, Russell, and Ramsey did not overlook anything. The mistake, if there is one, lies in our inclination to take the notation suggested in a modern way, outside the context of tractarian analysis. Consider 5.52 once more:

5.52 If the values of ξ are all the values of a function fx for all values of x, then $N(\overline{\xi}) = \neg(\exists x).fx$.

The thesis makes clear that the scope of a quantifier strictly depends on the way in which the variable is specified. In 5.52 it would be uninformative to focus just on the definiens $N(\overline{\xi})$ without taking into account the remainder of the definition:

$$N(\overline{\xi}) = \neg(\exists x).fx$$

Despite appearances to the contrary, it is as uninformative to focus just on a form like $N(N(NN(\overline{xRy})))$ without being told what its variables are. To see this consider the simple example:

$$NN(\overline{xRy}).$$

It exhibits all 'problematic' aspects of tractarian quantification, and is rich enough to show how to 'resolve' them.

Fogelin (1976, pp. 70 ff) rightly observes that a form such as 'NN($\overline{xRy}$)' is ambiguous. But the ambiguity is absent when propositional variables are properly indicated, as they should be according to 5.52:

- If xRb_j ($= a_0Rb_j, \ldots, a_nRb_j \ldots$) is a variable, then NN($\overline{xRb_j}$) $= \exists x.xRb_j$.
- If a_iRy ($= a_iRb_0, \ldots, a_iRb_m \ldots$) is a variable, then NN($\overline{a_iRy}$) $= \exists y.a_iRy$.
- If xRy ($= a_0Rb_0, \ldots, a_nRb_m \ldots$) is a variable, then NN($\overline{xRy}$) $= \exists xy.xRy$.

This bottom-up approach highlights that at a stage the bar is restricted to the prototypes used at the same stage. Also, the bond between prototypes and bar must be codified immediately in a contextual definition. None of the above forms are ambiguous. Also, the ambiguity is resolved using tractarian means. It is for good reasons that 5.52 specifies (i) the values of the variable ξ via fx and (ii) abbreviates the end-result N($\overline{\xi}$) as the uniquely readable $\neg(\exists x).fx$. If Wittgenstein had left it at the unabbreviated N($\overline{\xi}$), he would have encountered the problem noted by Fogelin. But he did not, and in the context of the relevant definitions there are no such problems.

Note in passing that akin to the care taken in modern logic to avoid clashes between bound and free variables, one should ensure that prototypes are not confused with prototypes introduced at earlier stages. Bounded (or: barred) prototypes may be used elsewhere, but the unbounded ones used in the specification of a variable should be chosen carefully. For example, above we used prototype y to get at the variables a_iRy. If the variable is used to generate:

$$\text{NN}(\overline{a_iRy}) = \exists y.a_iRy \text{ Def.}$$

the prototype y should not be used again to form $\exists y.yRy$, which is not as intended.

13.4   Unique readability

Fogelin, Geach, Soames and Von Kibéd seem to have ignored the rôle of contextual definition in Wittgenstein's approach to quantification. But even if contextual definition *is* taken into account, one could argue that from a modern perspective unique readability of the definiens, even if not strictly required, is still to be preferred; simply because it makes the resulting forms independent of contextual definition. Prior to considering ways in which unique readability of the definiens can be achieved, I suggest an interesting alternative to the abbreviation used in 5.52, which indicates that the bar-notation is closely related to a form of abstraction.

Highlighted prototypes If a propositional form contains several prototypes, the way in which a propositional variable is defined could shield some prototypes from focus. For instance, one could set:

$$xRy = \varphi(x) \text{ Def.}$$

to present x as a prototype that varies over all its admissible values, while at this stage y acts as a parameter whose value is left unspecified. Then the bar in the bracket-expression

$$(\overline{\varphi(x)}) = (\overline{xRy})$$

indicates that all values of x are represented, but that here those of y remain out of scope. This strongly suggests that the combination of bar and prototype functions like a kind of lambda abstraction. Instead of '$\overline{(\varphi(x))}$', one may write '$\lambda^w x.\varphi(x)$' or '$\widehat{x}^w.\varphi(x)$', and in line with this paraphrase 5.52 as:

If the values of ξ are all the values of a function fx for all values of x, then $N(\overline{\xi}) = N(\widehat{x}^w.fx) = \neg(\exists x).fx$.

The notation '$N(\widehat{x}^w.fx)$' retains the separation of truth-functional structure and generality. At the same time it leaves the end-result as unambiguous as the notation '$(\exists x).fx$' does. It still requires however that one tracks the various definitions that highlight different parameters at different stages to get at an unambiguous end-result.

Binding without binders To keep the structure of truth-operations and bracket-expressions uniquely readable requires little to no effort. Based on the idea that the bar is similar to abstraction, the simplest manner to achieve unique readability is to bind a prototype to the bar that generalizes it. This can be done in several ways. For instance, each occurrence of a prototype could come with a superscript i to indicate that its scope is that of the i-th bar above it. Then,

$$NN(\overline{x^1 R y^2}) \text{ means: } \exists x.xRy,$$
$$NN(\overline{x^2 R y^1}) \text{ means: } \exists y.xRy,$$
$$NN(\overline{x^1 R y^1}) \text{ means: } \exists xy.xRy,$$

and so on. The same effect can be had using a technique familiar from 'no variable' approaches to logic: fix a sequence of prototypes $x_1, \ldots, x_n, \ldots$ and stipulate that the prototype x_i is bound by the i-th bar above it.

Indexed bars It is simpler to have the bars themselves effect the binding by adding prototypes as indices. Then,

$$NN(\overline{xRy}^{xy}) \text{ means: } \exists xy.xRy$$
$$NN(\overline{xRy}^{y}) \text{ means: } \exists y.xRy$$
$$NN(\overline{xRy}^{x}) \text{ means: } \exists x.xRy.$$

The indexed bar is essentially an abstraction operator. It codifies the binding of bar and prototype explicitly in the definiens, and not just in the contextual definiendum as on Wittgenstein's approach. Equivalent notations would be the more familiar:

$$\overline{(\varphi(x)}^x) = \lambda^w x.\varphi(x) \text{ Def.}$$
$$\overline{(\varphi(x)}^x) = \widehat{x}^w.\varphi(x) \text{ Def.}$$

Some readers will recognize in these abstraction operators the operator '$\ddot{x}$' that Geach (1981, 169) proposed. It differs from the now standard notion of lambda abstraction in that it is substitutional. The notations 'λ^w', '$\widehat{x}^w$' and '$\ddot{x}$' concern the substitution of names for prototypes. By contrast, '$\widehat{x}$' or 'λx' concern the assignment of objects to variables as terms.

Geach (1981) states that Wittgenstein does not supply an abstraction operator such as '$\ddot{x}$'. This is correct. Still, the introduction of a quantifier via contextual definition has the same effect as the one Geach was after. The notations considered here all extend the tractarian system, but the extensions are minor. Readers who object to such alterations should stick to Wittgenstein's definitional approach.

With the basic forms of quantification in place, we may now start detailing the injective nature of tractarian quantification. Before doing so, we engage in a short interlude to discuss Wittgenstein's critical remarks in 5.521 concerning Frege and Russell.

13.5 *Interlude*: contra Frege and Russell

The insight that in quantified propositions generality must be kept separate from truth-functionality, comes with a critical observation concerning Frege and Russell.

5.521 I separate the concept *all* from the truth-function. Frege and Russell introduced generality in association with the logical product or the logical sum. This made it difficult to understand the propositions $(\exists x)fx$ and $(\forall x)fx$ in which both ideas are embedded.

It is not quite clear which position of Frege and Russell Wittgenstein had in mind. Black (1964, chapter LXI) is silent on the matter. Anscombe (1959, p. 141) holds 'there is no ground' in texts of Frege and Russell 'for a direct accusation'.

In *Begriffschrift*, Frege introduced the following notation for the content of universal generality:

$(\star)$ $\underline{}\mathfrak{a}\underline{}\Phi(\mathfrak{a})$,

which for any argument $\mathfrak{a}$ makes the function $\Phi(\mathfrak{a})$ a fact (Frege, 1879, §11). According to Anscombe (1959, p. 142) Russell's treatment of quantification is akin, and thus she observes that Wittgenstein could only have arrived at logical products and sums by reading his predecessors in the light of his own insights. By contrast, I think there is an important difference between Frege and Russell, which indicates that if Wittgenstein's critical remark is in order it concerns Russell, but not Frege.

Wittgenstein criticism of Frege is justified at best with regard to the 'notational' approach to quantification in Frege (1879), if read in the truth-functional way that Wittgenstein advocates. In case of Russell the situation is different. Round about the time of collaboration, he argued that besides general propositions there are general facts.

> It is perfectly clear, I think, that when you have enumerated all the atomic facts in the world, it is a further fact about the world that those are all the atomic facts there are about the world, that is just as much an objective fact about the world as any of them are. It is clear, I think, that you must admit general facts as distinct from and over and above particular facts. The same think applies to 'All men are mortal'. When you have taken all the particular men that there are, and found each of them severally to be mortal, it is definitely a new fact that all men are mortal.   *(Russell, 1918, p. 236)*

Thesis 5.524 explicitly objects to such general facts, which confuse what is at stake with general propositions.

5.524 If the objects are given, with that we are already given *all* objects.

If the elementary propositions are given, with that *all* elementary propositions are already given.

The totality shows itself in the way the world is, not as a general fact in the world. General facts would make 'it difficult to understand the propositions $(\exists x)fx$ and $(\forall x)fx$ in which both ideas are embedded.' (5.521). For each general proposition comes with its own general fact, but the different facts may well concern the same totalities quantified over. This double rôle is absent, if the totality is argument to the truth-function (5.523).[100]

Wittgenstein's criticism is beside the mark if it concerns Frege's later views on quantification. In *Funktion und Begriff*, Frege (1891, p. 26–27) introduced a notion of quantifier that complies with Wittgenstein's standards; it even has merits that Wittgenstein's approach lacks. Instead of the notation in Frege (1879), which mainly identifies the argument generalized over, Frege now takes quantifiers to be second-level concepts that require first-level functions Φ as argument. To be precise, the first level function is given via its *Wertverlauf* $\acute{a}.\Phi(\mathfrak{a})$, introduced earlier in Frege (1891), which is the class of *all* objects that have Φ. Frege's *Wertverlauf* separates the second-level quantifiers *all*, *some*, etc. from the generality of the *Wertverlauf* in their argument, just as Wittgenstein deems necessary in 5.521.

Other than Wittgenstein, Frege does not restrict second-level concepts to truth-functional ones. This means Frege's logic allows for generalized forms of extensional quantification that are not obviously tractarian. Using modern notation, a famous example is:

$$\text{MOST}(\widehat{x}.\varphi(x), \widehat{y}.\psi(y))$$

This proposition is true iff the φ's that have ψ outnumber the φ's that do not:

$$|\{d \mid \varphi(d) \wedge \neg\psi(d)\}| < |\{d \mid \varphi(d) \wedge \psi(d)\}|.$$

In the tractarian system one should take resort to higher-order quantification to get these forms of quantification:

$$\exists f^{\text{inj}} \, \forall x [\varphi(x) \wedge \neg\psi(x) \rightarrow \varphi(f(x)) \wedge \psi(f(x))].$$

But this approach has some disadvantages. The existential quantifier must be instantiated with a material function in logical space. Even if the φ's that have ψ outnumber the φ's that do not, logical space may lack a function that exhibit this truth explicitly. Also, how to express injectivity of the function f in a system that disallows identity?

To summarize, Wittgenstein's critical remarks in 5.521 strike Russell at best. By contrast, Frege's approach satisfies all tractarian requirements. It is even superior to tractarian quantification in certain respects.[*End of interlude*]

13.6 Injective quantification

In 4.26, Wittgenstein introduces the idea of descriptive completeness.

4.26 A specification of all true elementary propositions describes the world completely. The world is described completely by specifying all elemen-

---

[100]The same arguments may be used against other attempts to make truth-functional notions part of ontology; cf. the discussion on negative facts.

tary propositions together with an indication which of them are true and which false.

According to Hintikka (1986, p. 99), 4.24 ensures that basic presuppositions of determinate description are fulfilled: the projection of names onto objects is surjective: each name has a referent, and each object is named. In 5.53 and its offspring, the idea is refined, where it is argued that identity should be captured as identity of signs, not as a 'quasi-relation' of identity among terms.

5.531 Thus, I do not write '$f(a,b).a = b$' but '$f(a,a)$' (or '$f(b,b)$'). And not '$f(a,b).a \neq b$', but '$f(a,b)$'.

As a consequence, the projection of names onto objects must be injective: different names go proxy for different objects. Combining the import of 4.26 and 5.531, projection becomes a bijective (even: isomorphic) correspondence between names and objects. See also chapter 3, especially section 3.4.7.

Since quantified propositions abbreviate truth-functional structures of elementary configurations of names, it is natural to expect that the bijective nature of projection has an influence on quantification. Somehow the abbreviations with their prototypes for names should respect bijective projection, and indeed in the context of quantification the same idea surfaces again. Continuing the observations in 5.531 Wittgenstein writes:

5.532 And analogously, not '$(\exists x, y).f(x,y).x = y$' but '$(\exists x).f(x,x)$'; and not '$(\exists x, y).f(x,y).x \neq y$' but '$(\exists x, y).f(x,y)$'.
        (So, instead of the Russellian '$(\exists x, y).f(x,y)$':
            '$(\exists x, y).f(x,y) \vee (\exists x).f(x,x)$'.)

5.5321 Thus, instead of '$(x){:}fx \supset x = a$' we write e.g.
        '$(\exists x).fx. \supset .fa : \neg(\exists x, y).fx.fy$'.
    And the proposition '*only one* x satisfies $f()$' will read
        '$(\exists x).fx : \neg(\exists xy).fx.fy$'.

How should this be understood more formally? I shall present a treatment of quantification that corresponds to the weakly exclusive reading in Hintikka (1956), which is defined in terms of modern logic.

13.6.1 Principles of prototyping

In the tractarian system the main clue to quantification should come from the way prototypes are handled. Prototypes occur at pre-final stages of analysis, to highlight commonalities in propositional form. In generalized propositions, prototypes only occur bound: they are used to form bracket-expressions that are argument to a truth-operation. To get at an analyzed proposition, prototypes should be instantiated with genuine propositional constituents. A minimal requirement should be there is a strict analogy between projection and instantiation.

Projection different names go proxy for different objects.
Instantiation different prototypes are instantiated with different names.

This much is clear from 5.532. Using modern notation, the proposition $\exists xy[f(x,y) \wedge x = y]$ should be inconsistent if read injectively. By contrast,

$\exists^w xy.f(x,y)$ is short for the consistent: $NN(\overline{f(x,y)})$ with $f(x,y)$ the [possibly infinite] variable:

$$f(b_1, c_1), \ldots, f(b_n, c_n) [, \ldots]^{101}$$

with for each i the name b_i different from the name c_i. So $\exists^w xy.f(x,y)$ is a suitable analogue for $\overline{\exists xy}[f(x,y) \wedge x \neq y]$. Similarly, $\exists^w x[f(x,x)]$ is short for the consistent $NN(\overline{f(x,x)})$ with $f(x,x)$ the [possibly infinite] variable:

$$f(a_1, a_1), \ldots, f(a_n, a_n) [, \ldots].$$

In the above, the name c_k may be identical to either a_i or b_j. Therefore, $\exists^w xy.f(x,y) \vee \exists x.f(x,x)$ is indeed equivalent to the Russellian $\exists xy.f(x,y)$.

13.6.2 The scope of prototyping

Besides injective instantiation, there are further differences in prototyping that may have an influence on the resulting logic. When prototyping there is choice between:

Prototyping some but perhaps not all occurrences of a name.

Prototyping all occurrences of a name.

Prototyping some is inconsistent with the *Tractatus*, as long as the system lacks a bookkeeping-mechanism that records which prototype is instantiated. If some but not all occurrences of a name are prototyped, the instantiation of the prototype can be chosen from among the names in the prototyped proposition. Consider for instance '$\exists^w xy[xRy]$' and let x be instantiated, say, with a. Since no bookkeeping-mechanism is in place – such as 'a^x' presenting a as instantiation of x, – on *prototyping some* the instantiation of $\exists y.aRy$ could continue with aRa. After all, when prototyping with y the first occurrence of a may have been disregarded. But the continuation contradicts the principle that different prototypes should receive different instantiations. By contrast, if it is assumed that all occurrences of a name are prototyped, an injective context requires that each instantiation is different from the names occurring in the prototyped proposition. To continue the example, '$\exists y.aRy$' should then result, say, in aRb, as required. From now on I will assume *prototyping all* to hold.

13.6.3 The scope of injective quantification

The next question is: should the instantiation of a prototype be different from all expressions in a proposition, or should it only be different from some, and if so, from which ones? With regard to the *Tractatus*, the scope of prototyping is best restricted to all occurrences of names in the scope of a quantifier definition. This includes names introduced by instantiating prototypes that also occur within this scope. Take the proposition:

$$\exists x^w[f(x) \wedge f(b)] \wedge f(c)$$

On the reading advocated, '$\exists x^w[f(x) \wedge f(b)]$' describes not just that there is some x such that $[f(x) \wedge f(b)]$ but the stronger: there is some x *different from* b such that $[f(x) \wedge f(b)]$, with difference restricted to the scope of $\exists x^w$. A

---

[101] I distinguish between '$\exists x$' and the tractarian '$\exists^w x$' until the end of this section. Afterward, '$\exists x$' represents injective quantification.

stronger reading would make the quantifier also depend on elements outside its scope; e.g., there is some x *different from* b and c such that $[f(x) \wedge f(b)]$. The varying readings result in various logics. E.g., on the last, stronger reading of $\exists x^w$ the following are equivalent:

$$\exists x^w[f(x)] \wedge f(b) \wedge f(c),$$
$$\exists x^w[f(x) \wedge f(b)] \wedge f(c),$$
$$\exists x^w[f(x) \wedge f(b) \wedge f(c)].$$

The propositions are unequivalent on the weaker reading, where difference is restricted to quantifier scope.

The weaker reading assumed here is relative to a top-down elimination of abbreviations: instantiations of prototypes occurring within the scope of a quantifier should be considered as well. Take for example $\forall x \exists y.xRy$. Eliminating quantifiers in a top-down manner, the quantifier $\forall x$ is independent; it has free choice in selecting an a such that $\exists y.aRy$. But for any a chosen the y should be instantiated with a b different from a such that aRb. Accordingly, y's range of instantiation may vary with the instantiation of x.

13.6.4 Grounds for weak injectivity

Does the *Tractatus* give ground to choose the weakly injective reading? The main observation is due to Hintikka (1956, p. 230). Consider the proposition in 5.5321a, which states that the Russellian

$$\forall x[f(x) \to x = a]$$

is equivalent to the Wittgensteinian:

$$[\exists^w x.f(x) \to f(a)] \wedge \neg\exists^w xy[f(x) \wedge f(y)].$$

Whatever its details, on a strong reading the quantified $\exists^w x.f(x)$ must be instantiated with a name different from a, which may result, say, in $f(b) \to f(a)$. The second conjunct states, however, there are no two instantiations of $f(x)$ *different from* a. Therefore, on a strong reading the purported equivalence fails: there could be two names n such that $f(n)$, not just one. See also Wehmeier (2009, p. 342). On the weaker reading that is assumed here, there is equivalence. If $\exists^w x.f(x)$ is true, it may instantiate, say, to $f(a)$. Now, the negated second conjunct means: there are no two different x and y with $f(x) \wedge f(y)$. Therefore, if $\exists^w x.f(x)$ is true, no other instantiation than a is possible. On the other hand, if $\exists^w x.f(x)$ is false, for no a it is true that $f(a)$. Thus the entire proposition means: at most one a has $f(a)$. A similar reasoning shows that:

$$\exists^w x.f(x) \wedge \neg\exists^w xy[f(x) \wedge f(y)]$$

indeed means: just one x satisfies $f(x)$.

Another indication that Wittgenstein intended the weaker reading is due to an observation of Ramsey. In a letter of November 11th, 1923, he asked Wittgenstein: 'Have you noticed the difficulty in expressing without $=$ what Russell expresses by $(\exists x) : fx.x \neq a$?' (Wittgenstein, 2008, letter 101). Slightly later, Ramsey thanks Wittgenstein for giving him the proposition:

$$(\neg f(a) \to \exists x.f(x)) \wedge (f(a) \to \exists xy[f(x) \wedge f(y)]),$$

and attributes his earlier question to assuming that 'if an x and an a occurred in the same proposition the x could not take the value a' (Wittgenstein, 2008, letter 102; December 27th, 1923). Ramsey's assumption corresponds to a stronger reading of the quantifiers. See Hintikka (1956) and Wehmeier (2009). Wittgenstein's suggestion is correct only on the weak treatment of injective quantification, which leaves '$\exists^w xy[f(x) \wedge f(y)]$' in the second conjunct independent of '$f(a)$'. It should not mean: there are x and y other than a such that $f(x) \wedge f(y)$; either x or y should be able to instantiate to a.[102]

13.7 Quantifier logic

We now extend the tableau rules of the tractarian system in chapter 12 to accommodate weakly injective quantification. Quantifiers are abbreviations. Thus it suffices to extend the tableau rules so that abbreviations can be uncompressed; the truth-operational forms that result can then be treated using standard rules. The elimination of a quantifier-abbreviation should be in line with its injective meaning. Since the reduction is toward a proposition in which each name receives a unique referent (5.53, 5.531), it is enough to ensure that for each term of the propositional variable used in the quantifier-abbreviation an admissible instantiation is chosen which is different from all other names in that term.

In more detail the rule for quantifier-abbreviations amounts to the following. In a tableau a general proposition occurs in signed form: $SQx.\varphi(x)$. The new rule allows us to rewrite such quantifier occurrences according to definition. E.g., in case of existential quantification the definition is:

$$NN`(\overline{(\varphi(x))}) = \exists x.\varphi(x) \text{ Def.}$$

In a tableau, $S\exists x.\varphi(x)$ may therefore be replaced by $SNN`(\overline{(\varphi(x))})$. Next, in applying the uncompressed result each instantiation of the prototype(s) should (i) be admissible and (ii) be chosen injectively. As to admissibility, in applying the rules the 'ranges of significance' have to be determined based on the propositional variable that is used in the abbreviation. Since the scope of $\exists x.\varphi(x)$ is $\varphi(x)$, it suffices to ensure that only those names n may instantiate x for which $\varphi(n)$ has sense. As to injective choice, any name used to instantiate x in $\varphi(x)$ should be different from all other names occurring in $\varphi(x)$. Figure 28 shows that the tableau rules resulting in this way are as expected, provided the restrictions (i) and (ii) are reckoned with. We leave it to the reader check the rules for other quantifiers.

Let us continue with a finite example. Assume logical space allows the following range of elementary propositions:

$$aRa, aRb, bRb, bRa, cRc.$$

To apply the quantifier rules we should be explicit about the ranges of significance. The range of significance of the variable $\forall y.xRy$ includes the names

---

[102]Wehmeier (2009, p. 342): suggests $\exists x[f(a) \vee \neg f(a) \to f(x)]$ as an equivalent for both the weakly and the strongly exclusive reading. This is correct for modern logic, but in Wittgenstein's perfect notation the logical $f(a) \vee \neg f(a)$ is about nothing, and is thus filtered out to leave the unequivalent $\exists x.f(x)$. Wittgenstein's alternative does not have this effect. From this point onward, we no longer distinguish between '$\exists x$' and the tractarian '$\exists^w x$', and let '$\exists x$' represent injective quantification.

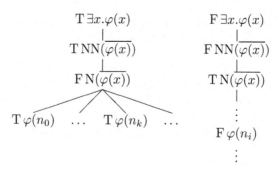

FIGURE 28 Rules for $\exists$

a and b. There are instances for aRy and bRy, and thus logical space allows $\forall y.aRy$ and $\forall y.bRy$. The question is: does the range also include c? In other words, does logical space allow $\forall y.cRy$?

In determining ranges of significance one must inquire after sense, not after truth, and the crux is that the range for y in cRy is empty: there is no n other than c with cRn. Now the following reasoning can be used to exclude c from the range of significance of the variable $\forall y.xRy$. Since logical space is unique, '$\forall y.cRy$' can be regarded an empty conjunction. From a modern point of view the empty '$\forall y.cRy$' is true, because all of its absent terms are. Since what is true vacuously is senseless, c is not in the range of significance for the variable $\forall y.xRy$. Although this conclusion is correct, the argument employs a kind of tautologous truth that is different from the structural notion advocated in the *Tractatus*. Tractarian logicality is rooted in signs that logically combine elementary propositions. Its logical propositions 'presuppose that names have reference and elementary propositions sense' (6.124); they cannot result from the absence of elementary propositions. Again, what matters is sense not (vacuous) truth. It is because in the given logical space there is no x for which cRx has sense that c should be excluded from the range of significance. Complex truth-functional structure, such as the purported '$\forall y.cRy$', simply cannot result from the absence of possible instantiations.

With c excluded, it remains to determine the ranges of significance for y in aRy and in bRy. The first range consists of the name b, and the second range of the name a. With all ranges determined, the tableau for '$T\exists x\forall y.xRy$' is as in figure 29. All in all, this tableau is equivalent to: $aRb \lor bRa$.

Mutatis mutandis the same method applies to infinite sequences of propositions. Consider available a countable sequence:

$$a_0Rb_0, \ldots, a_nRb_m, \ldots,$$

with for each i the sequence $a_iRb_j, \ldots, a_iRb_{k_i}, \ldots$ infinite. (Here, b_j is the the first b such that a_iRb.) As before, the variable $(\forall y.xRy)$ determines the range of all names a_i for which there is a b other than a_i with a_iRbs in logical space. And for each a_i in this range, the variable $(\overline{a_iRy})$ determines all names b_j other than a_i with a_iRb_j in logical space. Given this much, the extended tableau for the general proposition $\exists x\forall y.xRy$ amounts to a countable disjunction of countable conjunctions. As in chapter 12, each pole

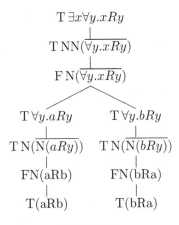

FIGURE 29

of the extended tableau receives its own sub-tableaux: figure 30 has the case for the T-pole, and figure 31 has an initial part of the more interesting case for the F-pole. The propositional variables involved ensure that the quantifiers are treated injectively, as required.

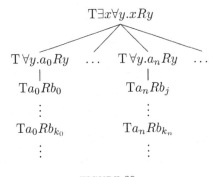

FIGURE 30

It is worth to observe how the tableau in figure 31 circumvents the 'cantorian' problem for truth-table signs discussed in section 11.3.2. Although the tableau allows for uncountably many paths, the tableau itself is a countable structure generated systematically from two simple rules. To see there are uncountable many paths, when cleansed from complex formulas each path in the tree corresponds to a sequence:

$$Fa_0Rb_{i(o)}, \dots, Fa_nRb_{i(n)}, \dots$$

with i some function from natural numbers to natural numbers. Since all these sequences are assumed countable, this means to yield the sequences of falsity-conditions we are dealing with all possible functions i from natural numbers to natural numbers. There are $\aleph_0^{\aleph_0}$ many such functions, which is uncountable. That the tree itself remains countable is due to the fact that truth-possibilities are no longer viewed independent from each other, as in truth-table signs, but are generated as overlapping branches in which elementary material is re-used across different paths.

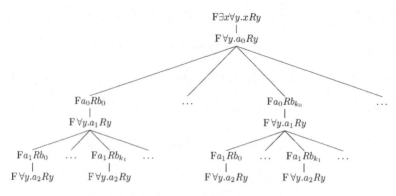

FIGURE 31 Partial tableau for $F\exists x\forall y.xRy$

With injective logic in place, the next step would be to check whether its logical propositions can still be characterized. This is immediate. The quantifier rules mainly determine the propositional variables involved in quantificational abbreviations. Therefore, all reasoning is within the context of the sound and complete logic of chapter 12. Even if we were to restrict logic to the part just concerning quantification over names, a quick check of the proofs show that the characterization is retained.

13.8   *Interlude:* The formal concept 'successor of'

In section 11.3.1, I have discussed different types of propositional variables as resulting from formal laws, prototyping, or enumeration. In the *Tractatus* the relation 'successor of' is rooted in a variable based on a formal law. Such variables give interesting combinations of the descriptive and the ostensive capabilities of the system of representation.

The formal relation 'is a successor of' surfaces in 4.1252.

4.1252 I call a sequence that is ordered by an *internal* relation a sequence of forms.

The sequence of numbers is not ordered by an external but by an internal relation.

Likewise the sequence of propositions 'aRb',
'$(\exists x):aRx.xRb$',
'$(\exists x,y):aRx.xRy.yRb$', and so forth.

(If b stands in one of these relations to a, I call b a successor of a.)

The sequence of forms:

$$aRb,\ (\exists x):aRx.xRb,\ (\exists x,y):aRx.xRy.yRb,\ \ldots$$

indicates the formal relation 'b is an R-successor of a' for a specific R. The sequence exhibits a general form that should be captured by a variable. But what kind of variable? The form cannot be captured by enumeration – since it is infinite, – nor by prototyping, – since the complexity of the form differs per term. Instead, the formal relation is obtained from a rule that the series exhibits. This rule should be given in two steps: (i) give the first term of the series, and (ii) specify how the next term in the sequence can be obtained from the previous one.

4.1273 If we want to express the general proposition 'b is a successor of a' in conceptual notation, we need an expression for the general term of the sequence of forms: aRb, $(\exists x){:}aRx.xRb$, $(\exists x, y){:}aRx.xRy.yRb,\ldots$ Only variables can express the general term of a sequence of forms, for the concept: term of this sequence of forms, is a *formal* concept. (This is what Russell and Frege overlooked; the way in which they want to express general propositions like the one above is therefore incorrect; it contains a vicious circle.)

We can determine the general term of a sequence of forms by giving its first term and the general form of the operation that produces the next term out of the proposition that precedes it.

To determine the general term in detail, suppose once more that logical space has the sequence of elementary propositions:

$$a_0 Rb_0, \ldots, a_n Rb_m \ldots.$$

The formal relation 'b is an R-successor of a' is now specified by:

1. The first term $\varphi_0(a, b)$ of the series is 'aRb';

2. The general form of the operation that produces the next term out of the preceding proposition is: If '$\varphi_n(a, x_n) \wedge x_n Rb$' is a variable – i.e., the n-th proposition $\varphi_n(a, b)$ with a new x_n replacing b conjugated with 'aRb' with the new x_n replacing a, – then the proposition $\varphi_{n+1}(a, b)$ is defined by:

$$
\begin{aligned}
\varphi_{n+1}(a, b) &= \mathrm{NN'}(\overline{\varphi_n(a, x_n) \wedge x_n Rb}) \\
&= \exists x_n[\varphi_n(a, x_n) \wedge x_n Rb] \text{ Def.}
\end{aligned}
$$

Due to injective quantification, $\varphi_{n+1}(a, b)$ is true in a world w within logical space Λ, π, iff there are n different objects $\pi[o]$ R-connecting $\pi[a]$ with $\pi[b]$. The reason is that in the definition of the operation, $\exists x_n$ has a, b and all previous existential quantifiers within its scope, and so enforces one to choose objects different from those chosen previously. (The R-sequence need not be unique.)

One would expect a close relationship between the notion 'is a successor of' and the concept of number. But which? Let us indicate the operation specified with Θ. Merging the definition of the existential quantifier with the succinct notation for operations in 6.01, its definition is:

$$\Theta'(aRb) = [\,\overline{\varphi(a, x^b) \wedge x^a Rb},\ \exists x(\varphi(a, x) \wedge xRb)\,](aRb) \text{ Def.}$$

In general, 'x^c' indicates that x has prototyped name c. According to 6.02 we have:

$$
\begin{aligned}
aRb &= aRb \\
\Theta'(aRb) &= \exists x_0[aRx_0 \wedge x_0 Rb] \\
\Theta'(\Theta'(aRb)) &= \exists x_1[\exists x_0[aRx_0 \wedge x_0 Rx_1] \wedge x_1 Rb] \\
\Theta'(\Theta'(\Theta'(aRb))) &= \ldots
\end{aligned}
$$

Since Wittgenstein takes numbers to be exponents of operations (6.02, 6.021), the relationship of numbers with 'is a successor of' now starts to emerge. More in particular Wittgenstein defines:

$$\Theta^{0'}(p) = p$$
$$\Theta^{n+1'}(p) = \Theta'(\Theta^{n'}(p))$$

The number n is what is common to iterating any operation n-times.[103] For the operation considered one has:

$$\Theta^{0'}(aRb) = aRb$$
$$\Theta^{1'}(aRb) = \exists x_0[aRx_0 \wedge x_0Rb]$$
$$\Theta^{2'}(aRb) = \exists x_1[\exists x_0[aRx_0 \wedge x_0Rx_1] \wedge x_1Rb]$$
$$\Theta^{3'}(aRb) = \ldots$$

This makes plain that numerals and propositions concerning successor-relations are related to each other: Θ is applied n-times iff n objects R-connect a and b. Cf. Potter (2000, chapter 6). Although a numeral n as an exponent is no constituent of a proposition, it has an intimate bond with propositions of form: there are n.... It should be stressed however that the relationship is *external*: any logical space would allow for all numerals but it is *not necessary* that a corresponding R-sequence of propositions is available.

In the tractarian system, the variable $\Theta'(aRb)$ helps to express the general proposition 'b is an R-successor of a', but a variable is not a proposition. In line with 4.1252: b is an R-successor of a, if b relates to a as described by one of the terms in the variable $\Theta'(aRb)$. To obtain the general proposition called for in 4.1273, the variable should be input to the operation $\bigvee$ (or: NN) to give:

$$\bigvee(\Theta'(aRb)),$$

or less formally:

$$aRb \vee \exists x_0[aRx_0 \wedge x_0Rb] \vee \exists x_1[\exists x_0[aRx_0 \wedge x_0Rx_1] \wedge x_1Rb] \vee \ldots$$

Whether this is a proposition or not, depends on the status of R. If R indicates a configuration (so: is not an object), one effectively has: b is an R-successor of a. On this view, the more general 'b is a successor of a' – so: independent of a specific R, – is seen as a formal relation abstracted from all propositions 'b is an S-successor of a' that for different configurations S are generated using the above operation Θ. If R denotes a tractarian object, the fully general: 'b is an successor of a' is obtained from existential quantification over R:

$$\exists R \bigvee(\Theta'(aRb)).$$

All in all, we have clarified most of 4.1273. The variable $\Theta'(aRb)$ was specified from its first term aRb using a rule-based operation. In terms of this variable,

---

[103]Since operations are non-material, the approach is independent of the size of logical space. As Marion (1998, 37) observes, Wittgenstein's approach is akin to that of Church. But of course a tractarian number n remains abstract: there is no object corresponding to it such as the Church numeral $\lambda f \lambda x.f^n(x)$, let alone systems with zero-terms, successor-terms, *etc.* For Church numerals, see Barendregt (1984, §6.4). For a tractarian approach, see Frascolla (1994).

'$\exists R \bigvee (\Theta'(aRb))$' was formed, capturing the general proposition: 'b is a successor of a'. In 4.1273 Wittgenstein claims moreover to have avoided a vicious circle. His claim is based on the approach of quantifiers as abbreviations. As we have seen in sections 13.1 and 13.2, the quantifier abbreviations reduce to truth-functional complexes of all their instances. These instances are simpler than the main proposition, and so no generalization will include itself within its scope. [*End of interlude*]

13.9 Variables and philosophical ostension

The use of variables to get at the formal relation 'b is a successor of a' makes plain that within the tractarian philosophy variables have two related purposes: they are used to generate logically complex descriptions, and they are used in philosophical acts of ostension to show formal concepts as rooted in the system of representation. Thus, variables are core both to meaningful description and to indicating its limit, beyond which lies 'what we must pass over in silence' (7). Especially variables that are based on a formal law intertwine genuine description with our means of showing, say, the fundamentals of logic and arithmetic in interesting, yet clearly distinguishable ways.

Philosophers cannot strictly talk about the fundamentals of logic, arithmetic, science, ethics,... Instead, they should take resort to philosophical acts of ostension to show what the essentials of these subjects are. In such acts the formal concepts and relations that are required – like: proposition, tautology, number, successor of, law, world – are made manifest in terms of propositional variables.

4.127 Propositional variables signify formal concepts, and their terms signify the objects that fall under the concept.

To be sure, formal concepts and relations cannot be combined into formal descriptions, but they do help to highlight the formal concepts and relations involved in genuine descriptions. For example, the proposition 'b is a successor of a' is true iff one of the terms in $\Theta'(aRb)$ is for a certain R. In this way, the proposition 'b is a successor of a' at once describes a possible state of things *and* instantiates a formal relation. This is the closest a formal ostension may get to everyday description. For more on Wittgenstein's early philosophy of mathematics the reader may consult, e.g., Frascolla (1994), Marion (1998), Potter (2000), and Landini (2007).

13.10 Wrapping up

By way of summary, it should be helpful to compare the main features of tractarian quantification with those of modern predicate logic.

1. Tractarian quantifiers are contextual abbreviations that only occur in the pre-final stages of a proposition's analysis. Since abbreviations are eliminated via the instantiation of prototypes, tractarian quantification is substitutional. Less abbreviated forms such as $N(\overline{\varphi(y)})$ for $\neg \exists y. \varphi(y)$ are but partly analyzed. The ultimate propositional structure has no occurrences of quantifiers, truth-operations or prototypes.

2. Tractarian quantifiers distinguish between truth-functional content and generality. In standard notations for predicate logic these notions are often combined (but see section 13.5).

3. The truth-functional aspect of quantification is given in terms of the truth-operation N. A truth-operation indicates an *instruction* to arrive at a truth-functional structure; it does not yet give the truth-functional sign (table, tableau) itself.

4. Tractarian quantifiers use prototypes, which are different from variables as placeholders. Prototypes indicate commonalities in propositional form and so give rise to propositional variables. Variables as placeholders are syntactic constituents of propositional signs that receive their semantic value via assignment.

5. Tractarian quantification has no single domain quantified over. Ontology consists of states of things, not of objects. Instead domains of quantification are given indirectly via the referents of the prototyped names in a propositional variable. The domains may be empty, or may depend on each other.

6. Different names go proxy for different objects. Likewise, tractarian quantifiers are interpreted injectively. The domain of a quantifier excludes the references of all names that occur within its scope.

The comparison indicates that tractarian quantification appears quite different from the now common approach. Yet, to a certain extent the differences are philosophical rather than logical. They highlight a preferred analysis of propositional and ontological structure. I must leave it for another occasion to investigate whether the two approaches ares inter-translatable, and so from a technical point of view essentially the same.

14

Endgame

The detailed interpretation of the *Tractatus* has been completed; it is time to summarize findings. This summary consists of two parts. The first offers the main themes in comparing the philosophies of Frege, Russell and Wittgenstein. The second gives an overview of Wittgenstein's system as it was specified pleat after pleat in the core of the book.

14.1   Frege, Russell, Wittgenstein

In the period from 1900 until 1916, the influence of Frege, Russell and Wittgenstein on each other can be pictured thus.

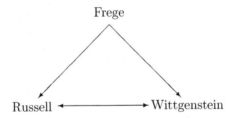

FIGURE 32 Influence

Frege had strong influence on both Russell and Wittgenstein, if only by being the first to give a perspicuous formal logic; to embark on the logicist program for arithmetic; and to expand some of his insights into a philosophy of language. Be this as it may, due to the paradox, Russell aimed for alternative logicist foundations, and he kept challenging some of Frege's basic ideas, like his notion of sense. Frege's influence on Wittgenstein, on the other hand, is often held to be 'pervasive and life-long' (Geach, 1976, 55). Despite fundamental differences, at the highest level Frege's ideas are basic to the architecture of logical picturing and meaning, notions that Wittgenstein partly inherited from Hertz.

Russell's effect on Frege is restricted to having noticed the paradox, with devastating effect on the logicist program Frege aimed for. There is no trace of any other logical or philosophical influence. Frege found the introduction to the *Principia* to unclear to finish it. Even in case of the paradox, Frege published his own proof because Russell had overlooked that functions only take objects as arguments. Convincing indications of Wittgenstein having an influence on Frege are absent, too. It is sometimes thought that Frege (1918) has such indications, but Potter (2009), pp. 260-261, makes clear that the manuscript was sent to the publisher before Frege received his copy of the *Tractatus*, and in the years before their contact had been shallow.

As we have seen, the relationship between Russell and Wittgenstein was intense at times. Wittgenstein quickly grew from Russell's student into his collaborator, who shortly afterward gained full responsibility for developing formal logic and its foundations. It is not always clear why Wittgenstein had this effect on Russell. (A similar effect on Whitehead seems to be lacking.) Wittgenstein had little taste for the technical aspect of logic, which would of course have been important if he were to improve on the foundations of all mathematics.

To compare the three philosophers in more detail, it is best to recapitulate some aspects of their theories of judgment or assertion first. Of course, judgment and assertion are not the same – a judgment of a proposition's truth-value need not be stated as in an assertion, – but they are sufficiently alike to allow for a comparison.

14.2 Frege on assertion

In Frege's philosophy of language, the rôle of sense (*Sinn*) is pivotal. In the context of a proposition, a sign has sense, which presents the sign's reference (*Bedeutung*) in a certain way. Sense is composite, and the way in which sense is composed should respect the basic typology of function and object discussed in section 2.1.1. If a sense presents a function, it must take senses that present objects as arguments. The sense of a propositional sign is a thought capturing its truth-conditions. The thought should not be taken purely subjectively. Although subjective imagery may obscure sense, its core consists of inter-subjective information that can be shared. A thought presents a truth-value, and so the propositional sign names the truth-value. See figure 33.

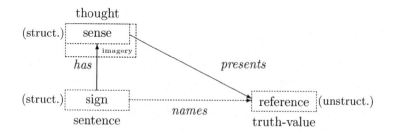

FIGURE 33 Frege on assertion

In Frege's philosophy, two principles are basic. Firstly, contextuality: one should only look for the sense of a phrase in the context of a proposition. Secondly, compositionality: the sense of a complex phrase is a function of the sense of its components and the way in which they are composed; *mutatis mutandis* the same holds for reference. Figure 34 has the example of the sentence 'John does not walk.'.

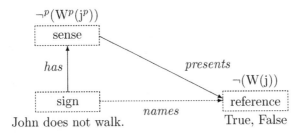

FIGURE 34   Fregean example

That sense presents a reference in a particular way is here indicated with adding the superscript '*p*'. In the example, the sense of 'John' presents the object *j*; the sense of 'walk' the concept *W*, and the sense of 'not' the truth-function of negation. The structure of sense indicates how the referents are to be applied to each other. So if $W(j)$ yields the True, the sentence 'John does not walk' presents: $\neg W(j)$, i.e., the reference False; and similarly conversely. Note that in spite of the structured notation '$\neg(W(j))$' this reference is an unstructured truth-value. A function and the objects of its arguments merge into the structureless object of the function's value for those arguments.

Frege offers a two-step theory of meaning. In the context of a proposition, the reference of a sign is obtained via its sense; an objective piece of content in a platonic third realm that contributes to inter-subjective information exchange. Sense is needed, Frege argues, to distinguish between the different ways in which the same reference can be presented; e.g. as Venus or as the Morning Star in the context of propositional belief. Sense also gives the content of an assertion. If a proposition is asserted, its sense, which is a thought, is presented as true; i.e., as referring to the True.

14.3   Russell on judgment

Some philosophers opposed to the intermediary notion of sense; in particular, Russell did. In the period from 1900 until 1920, Russell's work is of extraordinary depth and diversity. Although his thoughts on propositions were in constant flux, three lines of thinking kept returning. Firstly, logical analysis may show that expressions such as definite descriptions, which appear to denote in a certain context, should be analyzed away in favor of quite different logical structures. Secondly, reference is an immediate relation between a phrase and its referent; there is no intermediary sense. Thirdly, logical (and hence mathematical) notions should be taken as belonging to a platonic third realm. The specifics of Russell's position are not always clear, but all three aspects are used in the summary below.[104]

At the peak of Russell's collaboration with Wittgenstein, Russell argued that propositions should be understood in the context of a theory of judgment;

---

[104]See e.g. Landini (2007) for an alternative view.

in particular, true judgments involve propositions, false judgments do not. See figure 35.

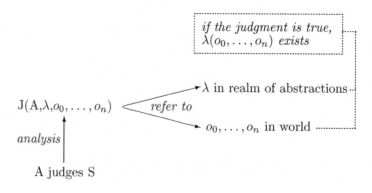

$$J(A,\lambda,o_0,\ldots,o_n)$$

if the judgment is true,
$\lambda(o_0,\ldots,o_n)$ *exists*

$\rightarrow \lambda$ in realm of abstractions

refer to

$o_0,\ldots,o_n$ in world

analysis

A judges S

FIGURE 35 Russell on judgment

In this picture λ is short for the logical form $R(x_0,\ldots,x_n)$, and $\lambda(o_0,\ldots,o_n)$ short for $R(o_0,\ldots,o_n)$. As in the theory of description, Russell's analysis presumes that the surface structure of a judgment does not have to be its logical structure: the judgment is seen as a multiple-relation between the judging subject A, the logical form λ and the real-world components $o_0,\ldots,o_n$ involved in the judgment. A reason for Russell to choose the analysis is to eliminate sense; cf. section 4.4 for more details.

In figure 35 the proposition $R(o_0,\ldots,o_n)$ that results from combining all components, requires the complex to be realized in the external world. Indeed, Russell's main worry concerns the unity of the proposition, if any. How to join the components in the judgment so as to indicate the complex that would exist in the world? In an earlier version of the idea, which only allowed real-world objects, no answer was forthcoming. But also platonic logical forms, introduced at a later stage to solve the problem, often do not specify a unique combination. Then extra information is needed to come to a full judgment, information which is absent in analysis. This problem is so serious, and Wittgenstein brought it to his attention so severely, that Russell decided to abandon the approach.

At the time, after about two decades of acute, hard work in both logic and philosophy, Russell felt it was Wittgenstein's responsibility to come up with an alternative. (By and large Wittgenstein's work was ignored again in the second edition of the *Principia Mathematica*.)

14.4 Wittgenstein on assertion

Early Wittgenstein's philosophy of language is best seen as a single-minded, creative re-working of ideas from both Frege and Russell to fill out the architecture of representation that Hertz had put forward in Hertz (1894). Wittgenstein re-establishes the pivotal rôle of sense (*Sinn*) to capture Hertzian pictures (models). But as I have argued in section 4.4, he understands sense in terms of an ontology that emerged in opposition to Russell's theories of types and of judgment. See figure 36.

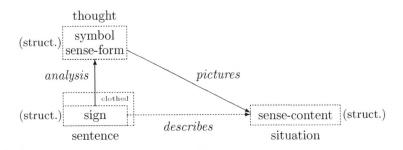

FIGURE 36 Wittgenstein on assertion

Like Frege, Wittgenstein took the sense of a proposition to consist of the conditions under which it was true or false. So in asserting a proposition, the thought the proposition expresses is claimed to be true. Other than in the philosophy of Frege or Russell there is no place for a platonic third-realm of sense or logical objects. Wittgenstein's 'symbolic turn' led him to consider ontology as a descriptive essentialism. Sense is understood in terms of the new ontology, but to be is to be essential to description. The *form* of sense is uncovered in the symbol of the sign – whose structure is often clothed in inessentials to enable human use. The *content* of sense is the situation its form is projected onto. It is what is pictured by the thought that has that form.

Wittgenstein's variant of the two-step philosophy of meaning results in sentences describing situations. It should be stressed, however, that the intermediate notion of sense is not to be located in an inter-subjective third realm. Rather, sense-form is *in* the sign itself; it is the symbol that logical analysis makes plain in perfect notation. The content of sense is given via the names in the sense-form that refer directly to their meanings. Finally, logical constants are understood as part of the symbol only; they lack referring constituents. Taken together, the three ideas show how semantic essentialism allowed Wittgenstein to interpret the Fregean notion of sense in terms of streamlined and unified notions derived from Russell: sense *is* the symbol used to describe a situation, where description is rooted in direct reference of the names involved.

The price to be paid for meaning to be a 'worldly' matter is that the essence of the world must be as abstract as the structures that other philosophers take to reside in a platonic realm. At the same time, this philosophy does allow Wittgenstein to view logic in an entirely new way, different from both Frege and Russell. Logic arises in the limit of contingency as the senseless necessities of form.

14.5 Comparison

Some interludes in this book have detailed comparisons of the work of Frege, Russell and Wittgenstein. Here I will highlight the main findings.[105]

---

[105]See Hacker (1984, Chapter 2) and Hacker (1996, §2.2) for similar overviews.

14.5.1 Universality

A major difference between the philosophies of language of Frege and Russell on the one hand, and of Wittgenstein on the other is their scope. Frege and Russell have a distrust concerning the logical nature of everyday language and so their ideas have been taken to apply to sufficiently rich fragments thereof. Then, the use of ideal, logical languages is best restricted to such areas as the analysis of quantification, arithmetic, geometry, science, propositional attitudes, metaphysics. By contrast, Wittgenstein's aim to demarcate of ethics and logic makes him consider all languages possible. He insists that from a logical point of view any meaningful use of any day-to-day language is perfectly in order. Wittgenstein's philosophy of language is universal, not fragmentary.

14.5.2 Intensionality

To the extent Frege and Russell have payed attention to the intensional nature of language – as it concerns ways the world may be, – both had a strong tendency to restrict interpretation to the actual world. This is so for Russell. But Frege, too, used his notion of sense to present referents not in a possible world, but in the actual world. Instead, Wittgenstein's philosophy is intensional throughout. Sense is prior to actual truth or falsity; it presents logically possible ways the world may be. Thus, Wittgenstein was the first to present a quite detailed intensional semantics of language.

14.5.3 Extensionality

Due to the mathematical origins of their work, Frege and Russell had a strong tendency to favor extensional interpretation. What matters in the end is the extension of certain linguistic expressions, the class of objects of which it holds, not the way in which this extension is given. (This is true as well for *Principia Mathematica*, which treats classes as a matter of speech.) Wittgenstein's interpretation is extensional, too, but his perfect notation blocks sense to be shown using synonyms. Instead, different expressions contribute differently to sense. In this regard the perfect notation is a rigid representation of sense that might as well be called hyper-intensional. At first sight his claim that each proposition is a truth-function of elementary propositions seems to give a principle of extensionality. But in the *Tractatus* truth-function signs get an intensional interpretation: they are used to show alternative realization patterns of the elementary propositions in different worlds. Elementary propositions, in particular, are hyper-intensional: any change leads to a different sense. And as long as a logical compound is contingent, only those substitutions are allowed that preserve this aspect.

14.5.4 Contextuality

Frege's work complies with the principle of contextuality: One should ask for the meaning of words within a propositional context, not in isolation; cf. Hacker (1979). Russell's position is unclear in this regard, but the radical atomism in, e.g., Russell (1918) seems to allow for expressions that have meaning independent of propositional context. Wittgenstein, by contrast, appears even more radical in applying the contextual principle than Frege: the sense of an expression, especially its form, is dependent on all propositional contexts in which it occurs (so possibly on a larger part of meaningful language than just a proposition). This radical holism is in line with Wittgenstein's position that the type of an expression is not manifest prior to analysis. Other than the notions of (elementary) proposition, name and object, the types of names and objects are inessential to see how descriptive language is meaningful.

14.5.5 Compositionality

Although the principle is not stated explicitly, Frege's work is in line with the principle that both sense and reference is obtained compositionally. Russell does not seem to have a stance on the topic, but Wittgenstein, too, takes the sense of logical complexes to be composed from the senses of its parts. Wittgenstein's position on the compositional nature of elementary propositions, however, is different. Wittgenstein must have noted that contextuality does not combine with compositionality, and so he does not require compositionality for elementary propositions. Due to the holistic nature of names and objects, elementary sense cannot be composed in a local manner. At best the *content* of elementary sense could be obtained compositionally in the context of meaningful use, but the *Tractatus* has no thesis to this effect.

14.5.6 Structured meaning

Frege held that propositions name truth-values, the structureless objects True or False. This position was available to him due to the intermediate level of structured sense that determines reference. Russell, by contrast, aimed to eliminate sense. As a consequence, the idea that propositions refer to truth-objects becomes highly unlikely.

In a world as the one we live in, structure abounds. Therefore, any theory of meaning must have a level that gives structure its proper place. Structure is one of the prime features of a proposition to indicate what the world must be like for it to be true; without structure its ability to describe the world is lost. Thus, Russell's elimination of sense forced him to develop the idea that propositions are structured, and that their components refer to things in the world. Once this step is taken, the truth of a proposition must be due to the fact that the structure of the components in a proposition corresponds to the realized structure of the things they refer to.

In my opinion Russell did not find a variant of this insight that was wholly satisfactory. Not only did he want to eliminate sense, he also had a strong tendency to restrict reference to the actual world. This combination leaves one with too little to get meaning of the ground. It took Wittgenstein's introduction of an intensional space of logically possible structures to arrive at a principled notion of sense and truth along these lines.

14.5.7 Logical differences

The logical framework Frege introduced is richly typed and expressive. It has truth-functions, quantifiers as second-order functions, functions taking senses, etc. In spite of its rather informal presentation the ramified theory of types is comparably rich. By contrast, Wittgenstein's treatment of logic and typing differs substantially from that of Frege and Russell.

As to typing, although Wittgenstein did allow objects to have different forms (roughly: types), he held that the range of types is not given prior to analysis. The only *a priori* typing that Wittgenstein allowed was between names and objects, and between elementary propositions and truth-functional combinations thereof. According to Wittgenstein the specifics of typing were largely inessential to philosophy of meaning: insight into the possibility of sense could be had without it.

Other than Frege and Russell, Wittgenstein held that logic is *about* nothing. Logical constants do not refer, they concern formal extensions of signs. Logical structure is used only to project a situation into logical space; i.e., a disjunction of different possible ways in which a proposition's elementary content can be realized. In this framework, logical propositions are understood as borderline cases of contingent propositions. Although they are formed of propositions with sense, their logical structure is such that all sense is dissolved. Logical propositions are senseless.

Wittgenstein's approach to logical connectives makes the ensuing system appear much weaker than Frege's. One has all notions that can be defined in terms of infinitary truth-functions of elementary propositions, such as different forms of quantification. But non-truth-functional operations that take propositions as arguments, are explicitly banned. Also, comparisons between variables – roughly: classes, – as in Frege's higher order treatment of quantification appear to be lacking as well.[106]

14.5.8 Thinking subjects

The logical analyses of propositional attitudes in the work of Frege and Russell have grammatical subjects – like 'Ravi' in 'Ravi thinks his father is the best.'. Such grammatical subjects appear to refer to a thinking subject. Russell's treatment is particularly clear in this regard. By contrast Wittgenstein holds that reference must be reserved to names that stand proxy for instances of grammatical categories. In logical space there are no such things as thinking subjects that can be named. The thinking, representing subject should rather be seen in a Humean way: its thoughts are a variation of states of things that are used to project sense onto what the subject represents. In a way, the thinking subject is the totality of all its thoughts, but this totality does not exists as a separate entity. Cf. Stokhof (2002), chapter 4.

In the *Tractatus* meaning can only be shown in the metaphysical subject. But Wittgenstein takes this subject to be infinitesimally thin; it is the world's limit – logical space viewed in total, – whose presence is noted as the eye is in a field of vision. On this view there is a threat of semantics to become entirely solipsistic. The notion of a metaphysical subject as the world's limit, however, allows Wittgenstein to identify solipsism with realism. And this, in a way, allows sense to remain public, not private.

---

[106]The observations in this paragraph should be proved by comparing the expressive power of the systems, which is not pursued here.

14.5.9 Ethics

In the *Tractatus* the symbolic turn delimits ethics from within the system of language. Here, philosophy is seen as a non-describing activity that reflects on the system of representation to indicate its limits. In the philosophies of Frege and Russell that are relevant here, such ethical motivation is absent.

This concludes the comparison of Wittgenstein with Frege and Russell. The next step is to summarize the system of the *Tractatus* as it emerged in the previous chapters.

14.6 The closing low pass

In this book I have compared the theses of the *Tractatus* with markings along an unorthodox route that force the reader to find his way from one to the other. For different aspects of the system – such as logical space, projection, elementary signs and symbols, contingent and logical propositions – I have often considered in detail which steps could be taken. My considerations may have helped to understand why I prefer one step over an other one, but at the same time they may have blurred a sight of the actual route taken. This is why the book closes with an overview of the resulting system. The overview highlights some features that Wittgenstein assumed without much argument.[107]

14.6.1 Ethics

The trail of reflection starts with realizing that the world I find myself in is ethically problematic, and that a way of living has to be found to come to grips with it. Wittgenstein's approach is to engage in a 'symbolic turn', based on the insight that the world is *my* world, and that the borders of my language are the borders of my world.

> F0 The ethical stance toward the world may come from insight into the symbols of language.

Unveiling the system of symbols results in the proper view on the world and the demarcation of ethics that comes with it. It dissolves the ethical problems rather than solving them. Instead of giving them their due in the light of an ethical theory, as has been customary in traditional philosophy, their philosophical, discursive aspect disappears entirely, thus paving the way to *living* the good live.

14.6.2 Descriptive language

Once the symbolic turn has been taken, the system that the *Tractatus* indicates can be seen as answering the question: 'How is meaningful language possible?'. This Kantian approach would be inherited from the work of Schopenhauer and Hertz.

So far we have spoken of any form of language, but for this broad category an answer to the above question may well be impossible. Instead, the *Tractatus* claims it is essential for meaning to represent the world. Insight into meaning

---

[107] A similar overview can be found in the rich biography McGuinness (1988), chapter 9, which according to the author (*pers. comm.*) gives his views on the *Tractatus* best.

can only be obtained for descriptive language, i.e., the language that describes the contingent, ever changing world. If at all, other forms of language should be understood in one way or the other in terms of the descriptive system of representation.

To demarcate ethics and logic, one must consider all possible symbolism; that is, all possible descriptive language. Despite vastly different surface appearances, the meaning conditions of descriptive languages give contingency primary stage. At the same time, its meaning involves the necessary aspects of the world's form that need to obtain for language to be possible. Both also help to recognize the transcendental nature of ethics and logic.

14.6.3 'Language internal' semantics?

Again, how is meaningful language possible? A radical answer would be take language itself as providing its own meaning conditions. Then, it is not in need of an extra-linguistic world to provide meaning. In the *Tractatus* such 'language internal' semantics is rejected.

F1 What makes language meaningful must lie outside it.

'Language internal' semantics, with language providing its own meaning conditions, seems to allow that *all* can be said. By contrast, the *Tractatus* has a strict distinction between saying contingencies about the world, and showing its necessities of form.

F2 Contingent configurations can only be described. The necessities of form can only be shown.

Meaning conditions reside in something non-linguistic; they are ineffable and can at best be shown in the form of representation.

14.6.4 Pictures

The crucial point in understanding how descriptive language has meaning, is to see that we picture facts to ourselves. The proposition is a model of reality as we think it is.

F3 Meaningful language and thought picture reality.

This idea is based on Hertz' notion of a model in science, but it is generalized to language as a whole using logical insights obtained from Frege and Russell.

In general a picture consists of elements that are structured in a pictorial form. The kind of form may vary per picture (spatial, temporal,...), but a picture can only picture those parts of reality that have the same kind of form. Despite the profusion in pictorial forms there is a most general kind of form that all pictures share.

F4 Pictures and depicted have a most general form called: 'logical form'.

Each picture is also a logical picture.

14.6.5 Analysis of descriptive language

The tractarian reflection on descriptive language starts with the perceptible signs that we utter, write, read, use to think,.... But perceptible signs cannot be taken at face value. They wrap up and compress the structure and content of their meaning, and in doing so conceal this structure and content. A meaningful sign still has to be analyzed to make its determinate meaning manifest.

F5 Propositions have a unique, determinate meaning that fixes reality to 'yes' or to 'no'.

F6 Meaning is prior to truth or falsity. Analysis, in particular, is independent of the truth or falsity of any proposition.

F7 Analysis determines the definitions that unveil the form and the contribution to meaning of the expressions in a proposition.

F8 Indeterminateness of meaning signals an incomplete analysis.

To hold that the analysis of a meaningful sign results in a unique structure and content is equivalent to holding that the analysis of a sign comes to an end. In a finite setting the analysis is well-founded, each of its paths ends after finitely many steps, but in general countable 'supertasks' may be needed. According to the *Tractatus* the end-result of analysis must show the meaningful sign to be a truth-functional complex of non-truth-functional structures, i.e., configurations of simple expressions that cannot be analyzed further. The simple expressions are called 'names', and the complexes of names are called 'elementary propositions'.

F9 In essence, the meaningful sign is a truth-functional complex of elementary propositions.

F10 Elementary propositions are configurations of simple names. They are the most elementary logical pictures.

Elementary propositions show what reality must be like for language to be meaningful.

F11 Ontology gives the essence of sense and picturing: to be is to be essential to description.

In the depicted part of logical space, however, there are only the states of things that elementary propositions depicts.

F12 Logical complexity only resides in the sign.

Logical structure consists of purely formal extensions of the form of elementary propositions. This logical structure shows the truth-conditions of a logical complex proposition, and enables to project them into logical space. In the depicted part of logical space there are just the states of things, either realized or not, that elementary propositions describe.

14.6.6 Ontology

What is the nature of the ontology that elementary propositions indicate? Just like elementary propositions are configurations of names, so the states of things they describe are configurations of objects. Objects and states of things in logical space are twin-notions. An object has content and form. Its form is the range of objects with which an object may combine into a state of things. Its content is an instantiation of its form as it occurs in a state of things. Conversely, a state of things is a configuration of object-contents that must comply with the form of the objects. The *Tractatus* presents the interdependent notion by first presenting logical space as the totality of all possible instances of object combination.[108]

Definition 1 *Logical space* Λ is the totality of states of things, which are configurations of objects. □

Conceptually, logical space is most basic: it is the semantical essence of the world I find myself in. States of things and elementary propositions *mutatis mutandis* satisfy the same radical form of contextuality, and the semantic holism that comes with it.

F13 A non-propositional expression only has sense in the context of all propositions in which it occurs.

F14 A configuration of objects can only be described in the context of all states of things in which it occurs.

On this route toward sense, the notions of object, object-form and identity of object-form are all presented in terms of logical space. The states of things in logical space are given abstractly in a semi-formal way.

Definition 2 A *state of things* (*Sachverhalt*) is an immediate combination of object-contents:

$$(o_0, \ldots, o_n)_c$$

The lower-case o's indicate object-contents, and the indexed parenthesis '$(-)_c$' indicate how they are configured. The sequence $o_0, \ldots, o_n$ is kept finite, and its object-contents may have multiple occurrences. □

This notion is semi-formal for principled reasons.

F15 Analysis makes the specifics of elementary propositions and the states of things they describe manifest.

F16 The specifics of elementary propositions and names, and of states of things and objects, such as their form ('type'), is not manifest prior to analysis.

The linear indication in definition 2 remains close to the notation used in the *Tractatus*. It indicates what the essentials of symbolizing are; namely (i) that a state of things has structure and (ii) that an object has certain rôles in state of things; that is, it contributes to structure in different ways. To be more detailed, requires one to assume that the specifics of states of things are known prior to analysis. *Quod non.*

Despite the abstract way in which the states of things are given, they have sufficient structure to abstract the form of an object from it.

Definition 3 The *form* $F(o)$ *of an object* o is defined by:

$$F(o) :\equiv \overline{(\alpha(o))}.$$

---

[108] In this summary, definitions have the same numbers as in the main text.

Here $(\overline{\alpha(o)})$ is the totality of all states of things α in Λ in which the object-content o occurs. □

An object is now given as content and form.

Definition 4 An *object* $\widehat{o}$ is a pair $o, F(o)$, with o the object's content and F the object's form. □

Although the notation indicates the essentials of an object's contribution to a symbol – namely: content and form, – it is a bit misleading: content and form should be understood as inseparable.

14.6.7 Pictures and projections

We picture facts to ourselves; this is how descriptive language symbolizes the world. It is the projection of signs onto what they depict that turns a sign into a symbol. Signs are used to project sense into logical space.

Projection has quite a few aspects to it. There are as many kinds of projection as there are kinds of pictorial form, but with regard to sense it is the most general kind of projection that matters: the projection of logical pictures. The essence of logical picturing consists in elementary, non-logical structures – elementary signs and states of things – that strictly mirror each other. Projection partitions logical space in two parts: one part with states of things used as elementary pictures, the other part the depicted states of things that give the sense of the elementary propositions its content.

F17 The difference between elementary propositions (pictures) and states of things is functional. An elementary proposition (picture) is a state of things used logically to project sense.

In all this projection and identity of form are closely related to each other.

F18 An elementary symbol is structurally identical to what it symbolizes.

Also, a projection must be possible; it must respect the form of all objects projected.

F19 A simple in a symbol goes proxy for a simple in what is symbolized.

F20 A simple is identical in form to the simple it goes proxy for.

The simple in a symbol is called 'name', and a simple in what is depicted is called object'. Since symbols are themselves facts, each name is also an object. It is an object occurring in a context that is *used* to symbolize in a logical way. The form of an object is given holistically in logical space, so projection is holistic, too, and encompasses logical space in full.

Definition 13 A *possible projection* π is an abstraction over logical space Λ so that:

1. π induces a bi-partitioning consisting of N, the occurrences of names in pictures, and O, the object-contents occurring in depicted states of things

2. If an n in N occurs in picture α, all other object-contents in α are in N. If an object-content o in O occurs in $\pi[\alpha]$, all other object-contents in $\pi[\alpha]$ are in O.

3. π maps the object-contents in N one-to-one onto the object-contents in O;

4. for each n in N: $F(n) \overset{\sim}{=} F(\pi(n))$, with identity of form as in definition 5 but via π. □

Definition 5 The forms $F(o)$ and $F(o')$ are identical, notation:

$$F(o) \overset{\sim}{=} F(o'),$$

iff there is a bijection ι from the different object-contents that occur in $F(o)$ to the different object-contents that occur in $F(o')$ so that:

i) $\iota(o) = o'$, and

ii) $(o_1, \ldots, o_n)_c$ in $F(o) \Longleftrightarrow (\iota(o_1), \ldots, \iota(o_n))_c$ in $F(o')$ □

In definition 5 the phrase 'there is a bijection' is a manner of speech. There is no bijection as an object, but the bijection is shown as soon as the sequences of its 'domain' and 'range' are brought in the appropriate one-to-one comparison. That a projection in use respects the form of all objects is equivalent to holding that the projection is an isomorphism from the elementary propositions in its 'domain' to the depicted states of things in its 'range'.

In use, a projection is unique, but in principle logical space may allow for different projections. That is why sense, which is determinate, involves more than possible projection.

F21 The sense of a proposition is shown via a certain projection in logical use.

Definition 15 Let w be the world: Λ, R with R the states of things that are realised. A *projection* is a projection π for Λ that is chosen so that all its pictures are realized. That is, all its pictures are among the realized states of things in R. □

The R is purely notational: it indicates whether a state of things is realized or not, but existence is not a property assigned to states of things.

14.6.8 Thoughts, propositions, symbols

In a perfect notation equivalent ways to express sense have the same representation, the same symbol. Thoughts are symbols; they are perfect signs that are used to show their sense without redundancy. Sense, in turn, is a logical projection of the sign's elementary parts into logical space, to show alternative ways in which they may be realized if the thought (proposition, symbol) is true.

F22 A thought is a sign in perfect notation that is used to show its sense.

Thoughts can be elementary or complex. A projected elementary sign is a symbol: all its aspects play a rôle in expressing its sense, and any change leads to a change in sense.

Definition 19 Let π be a projection in use in a world w. An *elementary proposition* is an elementary sign α together with the projection of α via π. The sense of α has form and content. The *form* of its sense is the form of its sign $(n_0, \ldots, n_k)_c$. The *content* of its sense is $(\pi(n_0), \ldots, \pi(n_k))_c$. The proposition shows its content as realized. $\square$

For logically complex symbols the situation is more subtle. In the finite system, truth-operations transform truth-table signs into a new one. But if the infinitary features of the system are taken seriously, the graphical notation in 6.1203 is more promising. Wittgenstein reasoned with graphical signs as we do with tableaux. See section 11.3.3. Thus, rather than from an infinitary truth-table, the sign of a proposition φ is obtained from its extended tableau; i.e., an unraveled graphical sign in which both the pole $T\varphi$ and the pole $F\varphi$ are considered. An analyzed, infinitary sign, in particular, has the form of a 'cleansed', extended tableau.

To get at the symbol of a sign, the logical occurrences of (sub)propositions are removed while generating the extended tableau. Wittgenstein just hints at this aspect of symbols by stating that the conjunction of a proposition and a tautology has the same sense as the proposition itself. See interlude 7.5 and section 12.3.

Once the tableau is finished, its truth-conditions are abstracted from the T-pole. Each open branch β comes with the truth-possibility θ_β of all signed elementary propositions Sp occurring at it. Call an elementary proposition or the negation thereof a literal. Each truth-possibility θ can be assigned a possibly countable conjunction $\bigwedge \xi_\beta$ of literals, with ξ_β the variable:

$$\xi_\beta(i) = \begin{cases} p & \text{if } Tp \text{ is in } \theta_\beta, \\ Np & \text{if } Fp \text{ is in } \theta_\beta. \end{cases}$$

Now, the symbol $\Sigma[\varphi]$ of a contingent proposition φ – its contingent nucleus, – reduces to a possibly infinite disjunctive normal form of the elementary content at the open branches in T-ET(φ); i.e., the T-part in φ's systematic extended tableau:

$$\Sigma[\varphi] = \begin{cases} \bigvee_{\beta \text{ open in T-ET}(\varphi)} \bigwedge \xi_\beta & \text{if } \varphi \text{ is complex contingent} \\ \varphi & \text{if } \varphi \text{ is elementary contingent} \end{cases}$$

The basis for considering only the pole $T\varphi$ is 4.022b: a proposition *shows* how things are *if* its true. That this approach is formally correct is due to the determinateness of sense: any truth-possibility for φ either enforces $T\varphi$ or enforces $F\varphi$.

Symbols obtained from extended tableaux are strictly analogous to symbols obtained from finite truth-table signs, but without further ado finite symbols are more perfect that infinite ones. In a finite symbol not only logical parts are eliminated but redundancies such as identical conjuncts are removed as well. Since symbols must be considered in the context of a notion of logical identity (equivalence), symbols that lack logical parts suffice.

Logical propositions are about nothing, and so in a way concern the entire logical space (4.463). The empty (T)() comes from a closed F-part of an extended tableau. It can be seen as a disjunction of empty conjuncts, which is true vacuously. It shows all states of things as either realized or not realized without constraint. The empty (F)() comes from a closed T-part of an extended tableau. It can be seen as an empty disjunction, which is always false. The proposition fills out logical space completely to leave no room for states of things to be realized. All this is in line with 6.112 that requires to give logical propositions a unique position among all propositions.

The emptiness of (T)() and (F)() indicate they are borderline cases without content and thus without sense (4.46 and its offspring). Only a contingent proposition has the form and content of sense.

Definition 81 The sense of a contingent proposition φ has form and content. The *form* of φ's sense is the form of its symbol $\Sigma[\varphi]$. In logical space Λ, π with π a projection in use, the *content* of its sense is the totality $\pi[\Sigma[\varphi]]$ of states of things $\pi[p]$, with p occurring in $\Sigma[\varphi]$.

A $\bigwedge \xi_\beta$ in $\Sigma[\varphi]$ pictures a realization-pattern of φ's content. It shows $\pi[p]$ as realized, if p occurs in ξ_β, and as not realized, if Np occurs in ξ_β. The *situation* pictured by $\pi[\Sigma[\varphi]]$, is the disjunction of realization-patterns of all $\bigwedge \xi_\beta$ in $\Sigma[\varphi]$. □

Distinguishing the form and content of a contingent proposition allows showing the content of a proposition in logically incompatible ways.

F23 A thought, a proposition, is the contingent nucleus of a sign, its symbol, that shows its truth-conditions in logical projection.

The alternative realization patterns, in particular, are logically incompatible with each other. This is why logical structure is in the sign but not in the part of logical space the proposition is about. Logical propositions lack specific content, and thus concern the entire logical space.

14.6.9 Compositionality

The end result of analyzing a proposition is a truth-function of elementary propositions. But at earlier stages of analysis, a logically complex proposition may be composed of non-elementary, logical complex propositions.

5.2341a The sense of a truth-function of p is a function of p's sense.

(Clearly, p is intended to indicate an arbitrary proposition.)Wittgenstein's approach to the compositionality of sense is to distinguish between truth-functions, which are finite truth-table signs or infinite tableaux, and truth-operations, which are rules with a 'truth-functional' character to transform truth-functions into a new one, e.g.:

$$\wedge' \left[(TF)(p), (TF)(q) \right] = (TFFF)(p, q).$$

Here the *rule* of conjunction transforms the truth-table signs $(TF)(p)$ and $(TF)(q)$ into the truth-table sign $(TFFF)(p, q)$. Similarly for more involved cases.

F24 Logical notation is a purely formal extension of elementary propositions that is about nothing.

The logical notation of truth-table and truth-operational signs has no reference. Yet, the logical notation of a sign does contribute to sense. The sense of a thought is its truth-conditions, and a truth-operation helps composing the sense of a complex in terms of the senses of the composing parts. For finite propositions this could be an operation that transforms completed truth-table signs into a new complete truth-table sign.

Definition 48 Let ψ be the complex: $\Omega'(\varphi_1, \ldots, \varphi_n)$. In a frame Λ, π, the sense of ψ has form and content. The form of $\Sigma[\psi]$ is the form of

$$\Sigma[\Omega'(\Sigma[\varphi_1], \ldots, \Sigma[\varphi_n])],$$

i.e., the symbol that results from the truth-table sign

$$\Omega'(\Sigma[\varphi_1], \ldots, \Sigma[\varphi_n]).$$

The content of $\Sigma[\psi]$ is the content ψ's contingent nucleus, if any. Cf. definition 81. □

For example, the tractarian analogue of $p \wedge (q \vee \neg q)$ is obtained thus:

$$\Sigma\left[\wedge'[(TF)(p), \Sigma[\vee'[(TF)(q), \Sigma[\neg'[(TF)(q)]]]]\right] = (TF)(p).$$

In general, the composition of sense has a somewhat different character. An infinitary truth-operation does not take completed infinite tableaux as input to yield their composition as a new completed infinite tableau. Instead, the truth-operation arrives at the symbol of a proposition in a staged process in which its rules are applied systematically to generate all logically possible combinations. In this process each step except the first is also a step in the tableau of one of its subpropositions. Since the steps of subpropositions are intertwined, inconsistencies may arise that are absent in the senses of the subpropositions themselves. Still, the end result can be viewed as the composition of the subpropositions in that it enforces the recursive definition of truth.

Definition 82 For each proposition φ and each truth-possibility θ for φ, θ *enforces* $S\varphi$ – notation: $\theta \Vdash S\varphi$ – iff $\exists \theta' \sqsubseteq \theta[\theta'$ is determined by an open branch in the systematic tableau of $S\varphi]$. □

Proposition 83 The sense of a symbol is compositional. For all relevant θ:

$$\theta \Vdash Sp \quad \Leftrightarrow \quad Sp \text{ in } \theta$$
$$\theta \Vdash TN(\overline{\xi}) \quad \Leftrightarrow \quad \theta \Vdash F\varphi, \text{ for all } \varphi \text{ in } \xi$$
$$\theta \Vdash FN(\overline{\xi}) \quad \Leftrightarrow \quad \theta \Vdash T\varphi, \text{ for a } \varphi \text{ in } \xi \qquad \square$$

Understood in this way, truth-operations realize the ideal notation Wittgenstein was after. Propositions have a truth-functional structure that connects to reality only via their elementary parts, which in the process of composition may (partly) dissolve due to the elimination of logical parts.

F25 Truth-operations yield symbols in a compositional way.

The process of applying truth-operations to signs results in the contingent nucleus of a proposition, if it has one, else it reduces to the logical $(T)()$ or

(F)(). Each feature of the notation captures an essential aspect of sense. In spite of using extensional means, tractarian symbols are hyper-intensional: any alteration in its non-logical structure leads to a change in sense.

It can be argued that the speculative generalization into the infinite retains the spirit of tractarian philosophy. See section 11.4. Without limitation, the infinite system allows different forms of injective quantification to be treated as abbreviations of infinitary truth-functions. See chapter 13.

14.6.10   Different kinds of language

The tractarian view on descriptive language is now before us in detail. In addition to descriptive language, the *Tractatus* discerns non-descriptive kinds of language. In use it is clear that a certain kind of language has sense or is without it. But analysis makes its philosophical status manifest in showing how it relates to the tractarian system. Logic, for instance, reflects on the fine-structure of the system of representation in terms of its signs and how they are transformed. Arithmetical identities concern regularities in the application of truth-operations. Probabilistic language indicates a certain ratio between its truth- and its falsity-conditions. Scientific language is essentially the same as descriptive language, except perhaps that within a certain field of research all its propositions share a common form... There is also pure nonsense: empty clothing that consists of signs made up of semantic inessentials only. In this way the different kinds of language, even if not strictly descriptive, can still be understood in different ways relative to the system of representation.

14.6.11   Ostensive philosophy

Besides the different kinds of language discerned above, there is the language of philosophy. The tractarian system solves Russell's paradox by pointing out that descriptive language cannot describe its own sense conditions; what meaningful language is about must lie outside it. See chapter 10. Still, large parts of the *Tractatus* appear to describe its own system of logic and language. To the extent it purports to be descriptive, philosophical language should indeed be recognized as a form of nonsense, but the descriptive use is just appearance. Philosophy is rather an activity that resorts to non-descriptive, ostensive uses of language, which are aimed at an understanding addressee to pinpoint insights that are already available. The insights concern the system of representation that captures the essentials of sense and projection.

The choreography of ostension starts with the observation that propositions 'aRb' are pictures, obvious similes of what they signify (4.012). Next, quasi-descriptions like: 'aRb' says aRb, indicate what the projection of elementary propositions amounts to. Finally, the independence of elementary propositions is shown in truth-table signs or in tableaux, which are all seen as instances of a general form (6). The general form is captured in a variable that shows where the limit of description lies.

F26 Ostensive philosophy makes the general form of propositions manifest, and with it the limit of description.

This is not a matter of theory, phrased in language. It is a philosophical act of ostension that lets the system and its limit be hinted at, not be described itself.

14.6.12   Ethics

The *Tractatus* views several areas as transcending descriptive language: logic, arithmetic, the nature of probability and science,... Still, these areas could be related in one way or another to the fine-structure of the system of representation. Ethics also transcends description. But reflecting on the system of representation *in toto*, the ethical stance is based on viewing the world as a living logic, as a variation of contingencies in a frame of necessity. Ethical value cannot be found in the world; it rather requires unifying with the world as a whole.

In the limit it turns out that all of value in life is inversely related to the system that delimits the space of description. Here it is seen that the world has a will of its own – i.e., the contingent variation of states of things, – and that this variation is independent of my will – i.e., a world whose states of things I deem desirable. In the end, this world does not even allow an 'I', a thinking subject, as a thing that can itself be represented or described. Instead the empirical subject is a totality of contingent states of things that are dressed up and compressed in more manageable structures for use as symbols. Although the empirical subject appears to bear sense, this is rather so for the metaphysical subject – i.e., my self as a projection of analyzed symbols into logical space.

With this insight the trail of reflection is about to end. To come to grips with the ethical problems of the world I find myself in, a symbolic turn must be made to delimit ethics from within.

F27 Ethics transcends meaningful language.

Thus ethics is safeguarded from theorizing and babbling. *Sub specie aeterni* the necessary aspect of the world is as transparent as its contingency, and here the necessity of ethical acts resides. To lead the good life the metaphysical subject, my self, must transcend the specifics of description to view this world, my world, as a whole, even unify with it. Only then my will can be reconciled with the will of the world, which is independent of it. Especially an insight into the world's contingency framed in its necessity should lead to renunciation, perhaps happiness, and acting just, in silence.

References

Anscombe, G.E.M. 1959. *An Introduction to Wittgenstein's Tractatus, 4th Edition*. London: Hutchinson. Second edition: 1971.

Barendregt, H. 1984. *The Lambda Calculus: Its Syntax and Semantics*. Amsterdam: North-Holland.

Barendregt, H. 2005. Reflection and its Use: from Science to Meditation. In C. Harper, Jr., ed., *Spiritual Information*, pages 415–423. West Conshohocken: Templeton Foundation Press. Perhaps still available at `ftp://ftp.cs.kun.nl/pub/CompMath.Found/reflection.pdf`.

Beth, E.W. 1955. Semantic Entailment and Formal Derivability. *Mededelingen Koninklijke Nederlandsche Akademie van Wetenschappen, Nieuwe Reeks* 18:309–342.

Black, M. 1964. *A Companion to Wittgenstein's Tractatus*. Cambridge: Cambridge University Press.

Burgess, J.P. 2005. *Fixing Frege*. New Jersey: Princeton University Press.

Chang, C. C. and H. J. Keisler. 1990. *Model Theory*, vol. 73 of *Studies in Logic and the Foundations of Mathematics*. Amsterdam: North-Holland. Third. First edition: 1973, second edition 1977.

Diamond, C. 2000. Ethics, Imagination, and the Method of Wittgenstein's *Tractatus*. In A. Crary and R. Read, eds., *The New Wittgenstein*, pages 149–173. London and New York: Routledge.

Dreben, B. and J. Floyd. 200x. Frege-Wittgenstein Correspondence. In E. D. Pellegrin and J. Hintikka, eds., *Festschrift for G.H. von Wright*. Publisher unknown.

Dummett, M. 1973. *Frege: Philosophy of Language*. London: Duckworth.

Edmonds, D. and J. Eidinow. 2001. *Wittgenstein's Poker*. New York: HarperCollins Publishers.

Engelmann, P. 1967. *Letters from Ludwig Wittgenstein with a Memoir*. Oxford: Basil Blackwell. Translated by L. Furtmüller. Edited by B.F. McGuinness (includes Editor's Appendix).

Floyd, J. 200x. Prefatory Note to the Frege-Wittgenstein Correspondence. In E. D. Pellegrin and J. Hintikka, eds., *Festschrift for G.H. von Wright*. Publisher unknown.

Fogelin, R.J. 1976. *Wittgenstein*. London: Routledge & Kegan Paul.

Fogelin, R.J. 1982. Wittgenstein's Operator N. *Analysis* 42:124–27.

Fogelin, R.J. 1987. *Wittgenstein, 2nd Edition*. London: Routledge & Kegan Paul. Chapter VI 'The Naive Constructivism of the *Tractatus*' rewritten.

Frascolla, P. 1994. *Wittgenstein's Philosophy of Mathematics*. London: Routledge.

Frege, G. 1879. *Begriffschrift*. Halle: Louis Nebert. Edition used: Georg Olms Verlag, Hildesheim 1998. Editor: Ignacio Angelelli.

Frege, G. 1884. *Die Grundlagen der Arithmetik*. Breslau: Koebner. Edition used: Reklam Universal-Bibliothek 8425, Stuttgart 1987. Editor: Joachim Schulte.

Frege, G. 1891. Funktion und Begriff. In G. Patzig, ed., *Funktion, Begriff, Bedeutung*, pages 18–39. Goettingen: Vandenhoeck & Ruprecht. Collection published in 1980.

Frege, G. 1892a. Über Begriff und Gegenstand. In G. Patzig, ed., *Funktion, Begriff, Bedeutung*, pages 66–80. Goettingen: Vandenhoeck & Ruprecht. Collection published in 1980.

Frege, G. 1892b. Über Sinn und Bedeutung. In G. Patzig, ed., *Funktion, Begriff, Bedeutung*, pages 40–65. Goettingen: Vandenhoeck & Ruprecht. Collection published in 1980.

Frege, G. 1893. *Grundgesetze der Arithmetik, I. Band*. Halle: Herman Pohle. Edition used: Georg Olms Verlag, Hildesheim 1998. Editor: Christian Thiel.

Frege, G. 1903a. *Grundgesetze der Arithmetik, II. Band*. Halle: Herman Pohle. Edition used: Georg Olms Verlag, Hildesheim 1998. Editor: Christian Thiel.

Frege, G. 1903b. Über die Grundlagen der Geometrie, I, II. *Jahresbericht der Deutschen Mathematiker-Vereinigung* XII:319–324,368–375. Perhaps still available at http://gdz.sub.uni-goettingen.de/no_cache/dms/load/toc/?IDDOC=243972. Also in *Kleine Schriften*, I. Angelelli (ed.), Darmstadt, 1967.

Frege, G. 1906. Über die Grundlagen der Geometrie, I, II und Schluß. *Jahresbericht der Deutschen Mathematiker-Vereinigung* XIV:239–309,377–403,423–430. Perhaps still available at http://gdz.sub.uni-goettingen.de/no_cache/dms/load/toc/?IDDOC=243972. Also in *Kleine Schriften*, I. Angelelli (ed.), Darmstadt, 1967.

Frege, G. 1918. Der Gedanke. In G. Patzig, ed., *Logische Untersuchungen*, pages 30–53. Goettingen: Vandenhoeck & Ruprecht. Collection published in 1976.

Gabriel, G. et al. (eds). 1980. *Gottlob Freges Briefwechsel*. Hamburg: Felix Meiner Verlag.

Geach, P. 1976. Saying and Showing in Frege and Wittgenstein. In J. Hintikka, ed., *Essays in Honor of G.H. von Wright*, pages 54–70. Amsterdam: North-Holland.

Geach, P. 1981. Wittgenstein's Operator N. *Analysis* 41:168–71.

Geach, P. 1982. More on Wittgenstein's Operator N. *Analysis* 42:127–28.

Gödel, K. 1931. Über formal unentschiedbare Sätze der *Principia mathematica* und verwandter Systeme I. In Feferman, S. *et alia*, ed., *Kurt Gödel Collected Works Volume I Publications 1929-1936*, pages 162–195. New York and Oxford: Oxford University Press. Collected Works Volume I published in 1986.

Hacker, P.M.S. 1979. Semantic Holism. In C. Luckhardt, ed., *Wittgenstein Sources and Perspectives*, pages 213–242. Hassocks, Sussex: The Harvester Press.

Hacker, P.M.S. 1984. *Insight and Illusion*. Oxford: Oxford University Press. This is the second revised edition. The first edition is published in 1972.

Hacker, P.M.S. 1996. *Wittgenstein's Place in Twentieth-Century Analytic Philosophy*. Oxford: Blackwell Publishers.

Hacker, P.M.S. 2000. Was He Trying to Whistle It? In A. Crary and R. Read, eds., *The New Wittgenstein*, pages 353–388. London and New York: Routledge.

Heijenoort, J. van. (ed.). 1976. *From Frege to Gödel*. Cambridge, Mass.: Harvard University Press.

Hertz, H. 1894. *Die Prinzipien der Mechanik in neuem Zusammenhange dargestellt*, vol. 263 of *Ostwalds Klassiker der Exakten Wissenschaften*. Berlin: Verlag Harri Deutsch. This edition published: 1996.

Hintikka, J. 1956. Identity, Variables, and Impredicative Definitions. *Journal of Symbolic Logic* 21:225–245.

Hintikka, M.B. & J. Hintikka. 1986. *Investigating Wittgenstein*. Oxford: Basil Blackwell.

Husserl, E. 1925/26. Analyse der Wahrnehmung. In K. Held, ed., *Phänomenologie der Lebenswelt, Ausgewählte Texte II*, pages 55–79. Stuttgart: Reclam Verlag. Collection published: 1986.

Hylton, P. 2005. *Propositions, Functions, and Analysis. Selected Essays on Russell's Philosophy*. Oxford: Oxford University Press.

Ishiguro, H. 1969. Use and Reference of Names. In P. Winch, ed., *Studies in the Philosophy of Wittgenstein*, pages 20–50. London: Routledge & Kegan Paul.

Ishiguro, H. 1981. Wittgenstein and the Theory of Types. In I. Block, ed., *Perspectives on the Philosophy of Wittgenstein*, pages 43–59. London: The M.I.T. Press.

Janik, A. 1979. Wittgenstein, Ficker, and *Der Brenner*. In C. Luckhardt, ed., *Wittgenstein Sources and Perspectives*, pages 161–189. Hassocks, Sussex: The Harvester Press.

Janik, A.S. and S. Toulmin. 1973. *Wittgenstein's Vienna*. New York: Simon and Schuster.

Keisler, H.J. 1971. *Model Theory for Infinitary Logic*. Amsterdam: North-Holland.

Kripke, S. 1980. *Naming and Necessity*. Cambridge, Mass.: Harvard University Press. First edition 1972.

Laan, T.D.L. 1997. *The Evolution of Type Theory in Logic and Mathematics*. Etten-Leur: Private publication. PhD-thesis, Technical University of Eindhoven.

Lambalgen, M. van, and J. van der Does. 2000. A Logic of Vision. *Linguistics and Philosophy* 23(1):1–92.

Landini, G. 2007. *Wittgenstein's Apprenticeship with Russell*. Cambridge, UK: Cambridge University Press.

Linden, Th. A. 1968. *Tree Procedures for Infinitary Logic*. Ph.D. thesis, Yeshiva University.

Lokhorst, G.-J. C. 1988. Ontology, Semantics, and Philosophy of Mind in Wittgenstein's *Tractatus*: a Formal Reconstruction. *Erkenntnis* 29:35–75.

Luckhardt, C.G., ed. 1979. *Wittgenstein Sources and Perspectives*. Hassocks, Sussex: The Harvester Press.

Marion, M. 1998. *Wittgenstein, Finitism, and the Foundations of Mathematics*. Oxford Philosophical Monographs. Oxford: Clarendon Press.

McGuinness, B. 1988. *Wittgenstein: a Life. Young Ludwig (1989 - 1921)*. London: Duckworth.

McGuinness, B. 2001. *Approaches to Wittgenstein, Collected papers*. Oxford Philosophical Monographs. Oxford: Routledge.

Menger, K. 1994. *Reminiscences of the Vienna Circle and the Mathematical Colloquium*. Dordrecht: Kluwer Academic Publishers.

Pears, D. 1979. The Relation between Wittgenstein's Picture Theory of Propositions and Russell's Theories of Judgement. In C. Luckhardt, ed., *Wittgenstein Sources and Perspectives*, pages 190–212. Hassocks, Sussex: The Harvester Press.

Pears, D. 1987. *The False Prison, A Study of the Development of Wittgenstein's Philosophy, Volume One*. Cultural Memory in the Present. Oxford: Clarendon Press.

Post, E.P. 1967. Introduction to a General Theory of Elementary Porpositions. In J. v. Heijenoort, ed., *From Frege to Godel: A Source Book in Mathematical Logic, 1987-1931.*, pages 264–283. Cambridge, Mass. Harvard University Press.

Potter, M. 2000. *Reason's Nearest Kin*. Oxford: Oxford University Press.

Potter, M. 2009. *Wittgenstein's Notes on Logic*. Oxford: Oxford University Press.

Proops, Ian. 2001a. The *Tractatuss* on Inference and Entailment. In E. Reck, ed., *From Frege to Wittgenstein: Perspectives on Early Analytic Philosophy*, pages 283–307. Oxford: Oxford University Press. Perhaps still available at www-personal.umich.edu\~iproops.

Proops, Ian. 2001b. The New Wittgenstein: a Critique. *European Journal of Philosophy* 9(3):375–404.

Proops, Ian. 2004. Wittgenstein on the Substance of the World. *European Journal of Philosophy* 12(1):106–126.

Quine, W.V.O. 1962. *Methods of Logic*. London and Henley: Routledge and Kegan Paul. Second edition. 1978 reprint as paperback.

Russell, B. 1902. Letter to Frege. In J. v. e. Heijenoort, ed., *From Frege to Gödel*, pages 124–125. Cambridge, Mass.: Harvard University Press. Collection published in 1974.

Russell, B. 1903. *The Principles of Mathematics*. New York: W. W. Norton & Company. Re-published edition, February 1996.

Russell, B. 1905. On Denoting. In R. Marsh, ed., *Logic and Knowledge, Essays 1901-1950*, pages 39–56. London: Allen Hyman Ltd. Article first published in Mind. Collection published in 1987.

Russell, B. 1907. On Some Difficulties in the Theory of Transfinite Numbers and Order Types. *Proceedings of the London Mathematical Society* 2:29–53. Read December 14th, 1905.

Russell, B. 1908. Mathematical Logic as Based on the Theory of Types. *American Journal of Mathematics* 30:222–262. Reprinted in Russell (1956, 59–102).

Russell, B. 1912. *The Problems of Philosophy*. Oxford: Oxford University Press. Ninth impression 1980.

Russell, B. 1913. *Theory of Knowledge, The 1913 Manuscript*. London: Routledge. Edited by E.R. Eames in collaboration with K. Blackwell. Published 1984.

Russell, B. 1918. The Philosophy of Logical Atomism. In R. Marsh, ed., *Logic and Knowledge, Essays 1901-1950*, pages 177–282. London: Allen Hyman Ltd. Collection published in 1987.

Russell, B. 1956. *Logic and Knowledge, Essays 1901-1950*. London: Allen Hyman Ltd. Edited by R.C. Marsh. 1988 paperback edition.

Russell, B. 1987. *Autobiography*. London: Unwinn Hyman Ltd. One volume edition. First volume published: 1967, Second volume published: 1968, Third volume published: 1969.

Schulte, J. 2005. *Ludwig Wittgenstein. Leben, Werk, Wirkung.*. Frankfurt am Main: Suhrkamp Verlag.

Smullyan, R.M. 1968. *First-Order Logic*. New York: Springer Verlag. New edition available with Dover Publications, Inc., New York: 1995.

Soames, S. 1983. Generality, Truth Functions, and Expressive Capacity in the *Tractatus*. *The Philosophical Review* XCII(4):573–589.

Stegmüller, W. 1966. Eine Modelltheoritsche Präzisierung der Wittgensteinschen Bildtheorie. *Notre Dame Journal of Formal Logic* VII(2):181–195.

Stenius, E. 1960. *Wittgenstein's Tractatus. A critical exposition of its main lines of thought*. Ithaca, New York: Cornell University Press.

Stokhof, M. 2002. *World and Life as One: Ethics and Ontology in Wittgenstein's Early Thought*. Cultural Memory in the Present. Stanford: Stanford University Press.

Sundholm, G. 1992. The General Form of the Operation in Wittgenstein's *Tractatus*. *Grazer Philosophische Studien* 42:57–76.

Sundholm, G. 2009. A Century of Judgment and Inference: 1837-1936. In L. Haaparanta, ed., *The Development of Modern Logic*, pages 263–317. Oxford: Oxford University Press.

Themerson, S. 1974. *Logic, Labels, and Flesh*. London: Gaberbocchus. Second edition. First edition: 1973.

Tolstoy, L. 1894. *The Gospel in Brief*. Lincoln: University of Nebraska Press. The year mentioned is of its first publication in Russia. The translation is published in 1997; edited with an introduction by F. A. Flowers.

Ule, A. 2001. Operationen im *Tractatus*. In W. Vossenkuhl, ed., *Ludwig Wittgenstein's* Tractatus logico-philosophicus, pages 231–256. Berlin: Akademie Verlag.

Visser, A. 2011. A Tractarian Universe. *Journal of Philosophical Logic* pages 1–27. Perhaps still available at http://dx.doi.org/10.1007/s10992-011-9182-6.

Visser, H. 1999. Boltzmann and Wittgenstein *or How Pictures Became Linguistic*. *Synthese* 119:135–156.

Von Kibéd, M.V. 2001. Variablen im *Tractatus*. In W. Vossenkuhl, ed., *Ludwig Wittgenstein's* Tractatus logico-philosophicus, pages 209–230. Berlin: Akademie Verlag.

Wehmeier, Kai F. 2004. Wittgensteinian Predicate Logic. *Notre Dame Journal of Formal Logic* 45:1–11. Perhaps still available at http://sites.google.com/site/kfwehmeier/home.

Wehmeier, Kai F. 2008. Wittgensteinian Tableaux, Identity, and Co-Denotation. *Erkenntnis* 69(3):363–376. Perhaps still available at http://sites.google.com/site/kfwehmeier/home.

Wehmeier, Kai F. 2009. On Ramsey's 'Silly Delusion' Regarding Tractatus 5.53. In G. Primiero and S. Rahman, eds., *Acts of Knowledge - History, Philosophy and Logic*, pages 353–368. London: College Publications. Perhaps still available at http://sites.google.com/site/kfwehmeier/home.

Whitehead, A.N. and B.A.W. Russell. 1910. *Principia Mathematica*. Cambridge: Cambridge University Press. First edition of volumes I-III: 1910-13. Second edition: 1927.

Wittgenstein, L. 1913. Notes on Logic. In M. Biggs, ed., *Editing Wittgenstein's 'Notes on Logic' Vol. 2*, pages 5–32. Published: Bergen 1996: University of Bergen.

Wittgenstein, L. 1914. Notes dictated to G.E. Moore in Norway, April 1914. In G. Von Wright, G.H. & Anscombe, ed., *Notebooks 1914-1916*, pages 108–119. Oxford: Basil Blackwell. Published: 1979.

Wittgenstein, L. 1914-16. *Notebooks 1914-1916*. Oxford: Basil Blackwell. Second edition: 1979. First edition published: 1961. Edited by G.H. von Wright and G.E.M. Anscombe with an English Translation by G.E.M. Anscombe.

Wittgenstein, L. 1922a. *Logisch-philosophische Abhandlung / Tractatus logico-philosophicus*. Suhrkamp Taschenbuch Wissenschaft. Frankfurt am Main: Suhrkamp Verlag. Kritische Edition. Book first published: 1922 with Routledge & Kegan Paul. This edition: 1998. Edited by B. McGuinness and J. Schulte.

Wittgenstein, L. 1922b. *Tractatus logico-philosophicus*. London and Henley: Routledge & Kegan Paul. Published: 1961. Translated into English by D.F. Pears and B.F. McGuinness. With an Introduction by B. Russell.

Wittgenstein, L. 1929. Some Remarks on Logical Form. In Klagge, J. & A. Nordmann, ed., *Philosophical Occasions 1912–1951*, pages 28–35. Indianapolis & Cambridge: Hackett Publishing Company.

Wittgenstein, L. 1933. *'The Big Typescript', TS 213*. Frankfurt am Main: Zweitausendeins. Edited by M. Nedo, Wittgenstein Archive, Cambridge. Perhaps still available at http://www.wittgensteinsource.org.

Wittgenstein, L. 1967. *Bemerkungen über die Grundlagen der Mathematik*. Oxford: Basil Blackwell. Bi-lingual edition. English title: Remarks on the Foundations of Mathematics. Edited by G.H. von Wright, R. Rhees, G.E.M. Anscombe. Translated by G.E.M. Anscombe.

Wittgenstein, L. 1983. *Letters to C.K. Ogden*. Oxford, London: Basil Blackwell, Routledge & Kegan Paul. Paperback edition. First published in 1973.

Wittgenstein, L. 1984. *Ludwig Wittgenstein und der Wiener Kreis*. Frankfurt am Main: Suhrkamp Verlag. Conversation recorded by Friedrich Waismann. Edited by Brian McGuinness. Werkausgabe Band 3.

Wittgenstein, L. 1995. *Cambridge letters. Correspondence with Russell, Keynes, Moore, Ramsey and Sraffa*. Oxford: Basil Blackwell. Edited by Brian McGuinness & Georg Henrik von Wright.

Wittgenstein, L. 2000. *Wittgenstein's Nachlass: The Bergen Electronic Edition*. Oxford: Oxford University Press.

Wittgenstein, L. 2008. *Wittgenstein in Cambridge. Letters and Document 1911-1951*. Oxford: Basil Blackwell. Edited by Brian McGuinness. Renewed and extended publication of *Cambridge letters*, 1995.

General index

Index of names

Index of theses & early sources

Index of notation

Lightning Source UK Ltd.
Milton Keynes UK
UKOW01f0415080717
304903UK00001B/25/P